China's Peasant Agriculture and Rural Society

China's agriculture and rural society has undergone rapid changes in recent years. Many poorer farmers and younger people have moved to cities, and yet China has an immense challenge to feed a growing and more affluent population. This book provides a 'bottom-up view' of China's agriculture, showing how the many millions of Chinese peasants make a living.

China's Peasant Agriculture and Rural Society provides a vivid description of the mechanisms used by rural households to defend and sustain their livelihoods, increase their agricultural production and improve the quality of their lives. The authors examine the newly emerging trajectories of entrepreneurial and capitalist farming and assess whether such alternatives will be able to meet the enormous social, economic and environmental challenges that China faces.

The book also explores the paradigm that has underpinned the organisation and development of China's agriculture from ancient times to the present day. This shows the importance of *balancing* in the Chinese model as compared to the one-sided imposition of continual modernization in the Western model. It is argued that such balancing is at the core of the current *Sannong* policy, referring to the three ruralities of food sovereignty, well-being for peasant households and an attractive countryside.

Jan Douwe van der Ploeg is Professor of Transition Studies at Wageningen University, the Netherlands, and Adjunct Professor of Rural Sociology at the College of Humanities and Development Studies, China Agricultural University, Beijing. He is the author of the influential book *The New Peasantries* (Earthscan, 2008).

Jingzhong Ye is Professor of Development Sociology and Dean of the College of Humanities and Development Studies, China Agricultural University, Beijing.

Routledge Contemporary China Series

Nationalism, Democracy and National Integration in China
Leong Liew and Wang Shaoguang

Hong Kong's Tortuous Democratization
A comparative analysis
Ming Sing

China's Business Reforms
Institutional challenges in a globalised economy
Edited by Russell Smyth, On Kit Tam, Malcolm Warner and Cherrie Zhu

Challenges for China's Development
An enterprise perspective
Edited by David H. Brown and Alasdair MacBean

New Crime in China
Public order and human rights
Ron Keith and Zhiqiu Lin

Non-Governmental Organizations in Contemporary China
Paving the way to civil society?
Qiusha Ma

Globalization and the Chinese City
Fulong Wu

The Politics of China's Accession to the World Trade Organization
The dragon goes global
Hui Feng

Narrating China
Jia Pingwa and his fictional world
Yiyan Wang

Sex, Science and Morality in China
Joanne McMillan

Politics in China Since 1949
Legitimizing authoritarian rule
Robert Weatherley

International Human Resource Management in Chinese Multinationals
Jie Shen and Vincent Edwards

Unemployment in China
Economy, human resources and labour markets
Edited by Grace Lee and Malcolm Warner

China and Africa
Engagement and compromise
Ian Taylor

Gender and Education in China
Gender discourses and women's schooling in the early twentieth century
Paul J. Bailey

SARS
Reception and interpretation in three Chinese cities
Edited by Deborah Davis and Helen Siu

Human Security and the Chinese State
Historical transformations and the modern quest for sovereignty
Robert E. Bedeski

Gender and Work in Urban China
Women workers of the unlucky generation
Liu Jieyu

China's State Enterprise Reform
From Marx to the market
John Hassard, Jackie Sheehan, Meixiang Zhou, Jane Terpstra-Tong and Jonathan Morris

Cultural Heritage Management in China
Preserving the cities of the Pearl River Delta
Edited by Hilary du Cros and Yok-shiu F. Lee

China's Peasant Agriculture and Rural Society

Changing paradigms of farming

Edited by
Jan Douwe van der Ploeg
and Jingzhong Ye

Routledge
Taylor & Francis Group
LONDON AND NEW YORK

earthscan
from Routledge

First published 2016
by Routledge
2 Park Square, Milton Park, Abingdon, Oxon OX14 4RN

and by Routledge
711 Third Avenue, New York, NY 10017

Routledge is an imprint of the Taylor & Francis Group, an informa business

British Library Cataloguing-in-Publication Data
A catalogue record for this book is available from the British Library

Library of Congress Cataloging in Publication Data
Names: Ploeg, Jan Douwe van der, 1950– editor. | Ye, Jingzhong, editor.
Title: China's peasant agriculture and rural society: changing paradigms of farming / edited by Jan Douwe van der Ploeg and Jingzhong Ye.
Description: London; New York : Routledge, 2016. | Series: Routledge contemporary China series; 149 | Includes bibliographical references and index.
Identifiers: LCCN 2015047045| ISBN 9781138187177 (hbk) | ISBN 9781315643250 (ebk)
Subjects: LCSH: Peasants – China. | Farmers – China. | Rural population – China. | Agriculture – China. | Rural development – China. | China – Rural conditions.
Classification: LCC HD1537.C5 C4753 2016 | DDC 338.10951 – dc23
LC record available at http://lccn.loc.gov/2015047045

ISBN: 978-1-138-18717-7 (hbk)
ISBN: 978-1-315-64325-0 (ebk)

Typeset in Times
by Florence Production Ltd, Stoodleigh, Devon, UK

Contents

Notes on contributors

He Congzhi
Associate Professor of Development Studies at the College of Humanities and Development Studies, China Agricultural University, Beijing, China

Jan Douwe van der Ploeg
Professor of Transition Studies at Wageningen University, Wageningen, the Netherlands and Adjunct Professor of Rural Sociology at China Agricultural University, Beijing, China

Jingzhong Ye
Professor of Development Studies and Dean at College of Humanities and Development Studies, China Agricultural University, Beijing, China

Pan Lu
Associate Professor of Sociology at the College of Humanities and Development Studies, China Agricultural University, Beijing, China

Sabine de Rooij
Researcher, Centro di Sviluppo Agricolo e Rurale, Assisi, Italy

Wang Chunyu
Associate Professor of Development Studies at the College of Humanities and Development Studies, China Agricultural University, Beijing, China

Wu Huifang
Professor of Sociology at the College of Humanities and Development Studies, China Agricultural University, Beijing, China

Xiangdan Meng
Lecturer in Sociology at the School of Philosophy and Sociology, Jilin University, Changchun, China

Zhao Yong
Library Researcher, China Agricultural University, Beijing, China

Notes on contributors

[illegible]

Preface

This book is the outcome of a large and long-standing research programme of the College of Humanities and Development Studies at China Agricultural University in Beijing. Many different researchers have participated in the programme. They are mentioned at the beginning of each chapter. Nonetheless, we did the final editing and are responsible for any possible mistakes.

Our research has been driven by curiosity about how to explain the nature and dynamics of the immense but poorly understood phenomenon of Chinese peasant agriculture. We were also perplexed by the results of our comparative analyses: how on earth does China's peasant agriculture perform so well, when it transgresses all the given rules for 'successful agriculture'? This led us to embark on a search for the paradigm that is contained within, and simultaneously hidden by, the amazing and highly diverse reality of Chinese peasant farming.

We thank our colleagues and students from the College for their highly stimulating cooperation. We also thank Nick Parrott for the required 'healing' of the text. Special thanks go to Ao Song for allowing us to use his painting on the cover. We are also grateful to Tim Hardwick and Ashley Wright of Routledge for helping us through the process of getting this book published. We thank the anonymous reviewers who evaluated the original book proposal. Finally, we thank the publishers of the *Journal of Peasant Studies*, the *Journal of Rural Studies*, the *Wageningen Journal of Life Sciences* and the *International Journal of Sociology of Agriculture and Food* for giving permission to re-use fragments of articles published earlier in these journals.

Jingzhong Ye
Jan Douwe van der Ploeg
Beijing, 30 October 2015

1 The conundrum of Chinese agriculture

Jan Douwe van der Ploeg and Ye Jingzhong

There are several important reasons for examining and trying to understand Chinese agriculture. It is, to begin with, the largest 'agricultural system' in the world: it ranks number one globally in terms of farm output and embraces over 200 million smallholdings.[1] With just 10 per cent of all cultivated land in the world, these smallholders produce 20 per cent of the world's total food supply. This suggests that China's agriculture performs exceptionally well. Yet, we should be aware that if something were to go wrong in this agricultural system the consequences would be catastrophic. China's agriculture also stands out for its impressive development over recent decades. Production and productivity have grown continuously and this growth has gone hand in hand with an equally impressive alleviation of rural poverty. This suggests that the example of China's agriculture might also have a relevance beyond the country's boundaries, particularly in places where productivity has failed to keep up with population growth but also in other places where agriculture has reached an impasse. As implausible as this may first sound, it might also help to highlight some of the changes that are badly needed in agriculture in 'developed countries'. However, it is far from easy to understand the nature, dynamics and drivers of China's agriculture.

Coming to grips with Chinese agriculture requires a threefold inquiry. That is why it is so difficult and, from an intellectual point of view, so attractive. First comes the empirical inquiry into Chinese agriculture. *What* are its main structural components? What are the rates of growth in production and productivity? What are the main institutions? Questions like these are far from easy to answer, partly due to the immense size of rural China and to its impressive heterogeneity. It is especially difficult because it requires the understanding of something the academic world generally misrepresents or even ignores: the peasantry.

Examining the performance of China's agriculture almost automatically provokes a second line of inquiry. *How* has it been able to generate such unprecedented growth in production and productivity? How could it possibly perform in this way? When looked at from a Western perspective, China's agricultural performance appears almost impossible. From the Western paradigm of agricultural development[2] the accomplishments of Chinese agriculture appear to be structurally impossible. As such, rural China is an anomaly, a seeming deviation from the standard pattern – a deviation that conventional thinking would consider as doomed to failure, but which, in practice, is a remarkable success. There are several theoretical reasons why the Western paradigm views Chinese agriculture as destined to fail. First, the farms are far too small: the average holding is only 5 mu of land (one-third of a hectare). That is microscopic, not only from an American point of view, but even from a European or Dutch perspective.[3] The average Chinese farm is too small to provide an acceptable income, let alone allow for savings. And without savings there cannot be investments, thereby excluding the possibility of developing the farm. In addition, there is massive rural outmigration. It is estimated that currently at least 200 million people, mostly men, move every year from rural to urban areas. The result is that only women and the elderly remain to do the agricultural work. Many experts perceive such a pattern as intrinsically weakening agriculture (UNDP, 2003; Peterman *et al.*, 2010) – the outflow of so many men certainly does not help strengthen agriculture. As long ago as the late 1980s Lester Brown wrote an essay entitled 'Who will feed China?'. This title perfectly reflects the chasm separating the Western view and Chinese reality. We now know the answer to the question raised by Lester Brown. With a few exceptions, notably soy imports, *China is feeding itself.* And it does so in a proud and convincing way. China is now an important point of reference when it comes to debates about food and farming.[4] It is this that triggers the second inquiry: *How* is the seemingly impossible achieved? Put differently, *how* is the Western paradigm blocking (or blatantly distorting) our understanding of what is happening in China? In short, a serious inquiry into Chinese agriculture necessarily translates into a second inquiry that critically probes the dominant Western paradigm – an inquiry that needs to identify and challenge the seemingly self-evident 'lenses' that would, left uncorrected, strongly bias our understanding of what agriculture is and how it works.

Together, these two inquiries flow into a third one that focuses on the *why*. It centres on the reasons that *positively* explain the remarkable, but at first sight highly unlikely, performance of China's agriculture. Among the major ingredients that together explain the enigma are the *peasant* nature of China's agriculture, the *circularity* of town–countryside relations, the *horizon of relevance* used in agricultural research and development (which differs remarkably from the one used in Western agricultural sciences), the recent *turn* in the urban bias[5] and the omnipresence of *markets*. This last ingredient might surprise many readers. According to the Western view, Western agriculture is market-oriented and Western society is governed by markets (especially in this neo-liberal epoch). However, we should not forget that in reality the 'market' in the West is merely a front for large corporations and supermarkets that control the many transactions that bring goods and services from producers to consumers. 'Real' market-places, where many buyers meet many sellers and where an 'invisible hand' defines prices, volumes and flows, have been in decline for several decades and are now far from the main locus of transactions (Winders, 2009). By contrast, in China they are abundant – and this has an undeniable and positive impact. This will be discussed at some length in this book, alongside the effects of the other ingredients.

When properly combined and integrated the ingredients that emerge out of the third inquiry may carry an important promise. They can be used to delineate the contours of an alternative paradigm that may also be helpful in other parts of the world. They can offer new, unexpected solutions and help spotlight new empirical realities that have largely remained unnoticed. In short, this alternative (or 'Chinese') paradigm can be very helpful for those working in other situations.

Small-scale, but booming

When travelling through China one crosses, every now and then, huge plains that go from one horizon to the other, completely covered by grain crops. As extensive as these plains are, they are made up of tens of thousands of small fields, each tilled by a peasant family having on average 5 mu of land. Together they produce amazing amounts of grain. As already mentioned, from the Western perspective such units are viewed as too small to provide an income. As such, the peasant family should be living in poverty and trying everything it can to find job opportunities elsewhere and, having secured these, the family would probably dedicate less time and attention to the land or maybe even leave it lying barren. Even if all this were not the case, the unit would be too small to allow for profits, savings and subsequent investments. And with no investments being made, stagnation is the inevitable outcome. Nonetheless, while the average Chinese unit of agricultural production is very small, it also shows continuous growth and development. Elsewhere in this book we will try to unravel this conundrum, but here we wish to explore its broader paradigmatic aspects.

The exceptional nature and dynamics of China's agriculture clearly come to the fore when we compare it to agriculture in sub-Saharan Africa (SSA). As shown in Figure 1.1, China has achieved an impressive agricultural growth over the past 60 years. In constant prices (that is, after adjustment for inflation) the total value of agricultural production increased fivefold between the mid-1950s and the early 2000s (data derived from Li *et al.*, 2012). Since the early 1980s both production and total factor productivity have kept growing at a rate unparalleled elsewhere. Over the same period Africa showed a stagnation and decline. Production per capita has continually decreased over successive decades. Nearly three decades ago Mahmood Mamdani (1987) referred to the lack of an agricultural revolution as explaining this chronic stagnation. Whatever the reason for Africa falling behind, China has experienced remarkable agricultural growth. China and its agricultural trajectory diametrically differs from that of Africa.[6]

The ongoing growth in China's agriculture has occurred alongside impressive poverty alleviation, which again is in remarkable contrast with what has happened in SSA. As shown in Figure 1.2, poverty has been greatly reduced in China, in both absolute and relative terms. Both the rate and the amount of poverty alleviation are unique. Gulati and Fan (2007:17) refer to this it as 'an unprecedented achievement in the history of development of any country'. By contrast, in SSA poverty is 'a persistent headache' (Li *et al.*, 2012:1) or, as Henley and van Donge (2013:28) put it, 'it [i.e. poverty] is still the norm'. In relative terms poverty has

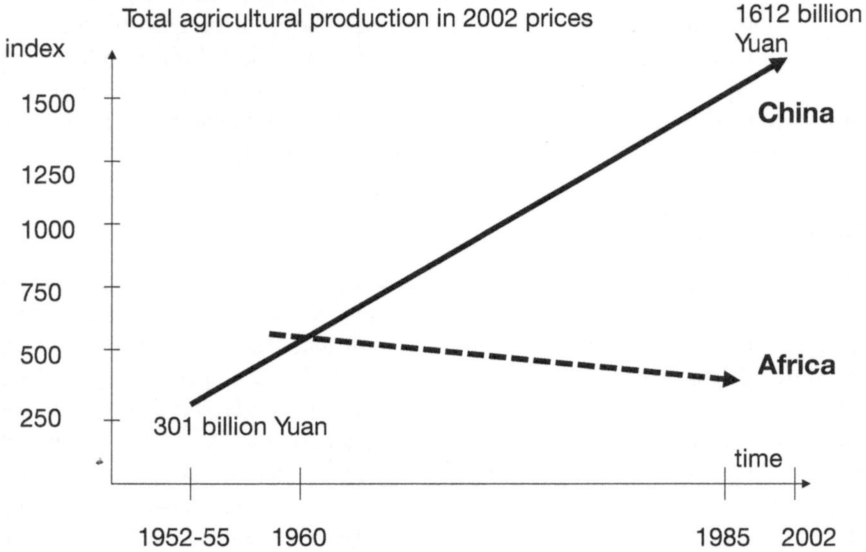

Figure 1.1 Agricultural growth patterns in China and Sub Saharan Africa (adapted from Li *et al.*, 2012)

remained at more or less the same level over recent decades. In absolute terms though, the number of those in poverty has nearly doubled: it went from 167 million in the early 1980s to 298 million in 2004. Today in China poverty is a phenomenon that is limited to specific groups such as abandoned, elderly or disabled people. In this respect China is again diametrically different from Africa.

There is, however, a clear downside as well. The rapid development of Chinese agriculture is, to a degree, built on decay. First, it is grounded on ecological decay. Extensive ecosystems have been negatively affected – sometimes in an irreversible way: sweet water aquifers are being exhausted and there is extensive contamination due to the widespread use of agro-chemicals (Rampini, 2005 and Zhang *et al.*, 2015). All this has allowed for quick gains in production and productivity, but at a very high price – one that may very well block future development. Massive attempts have been made to reverse these negative tendencies and to repair at least some of the damage. Whether this is possible remains to be seen. Second, there is socio-cultural decay: achieving these impressive growth rates has brought tremendous hardship for the many millions involved. For the 'left behind' women and elderly who had to take over most of the work in farming and for the children who do not see their fathers for long periods (Ye *et al.*, 2013). The same socio-cultural decay makes farming appear extremely unattractive in the eyes of many youngsters, which evidently threatens the long term continuity of farming itself. The hardship and suffering of men should also be mentioned. They temporarily migrate from their villages to the towns and industries in order to gain the money needed to finance the growth of agriculture and to maintain their families. In doing

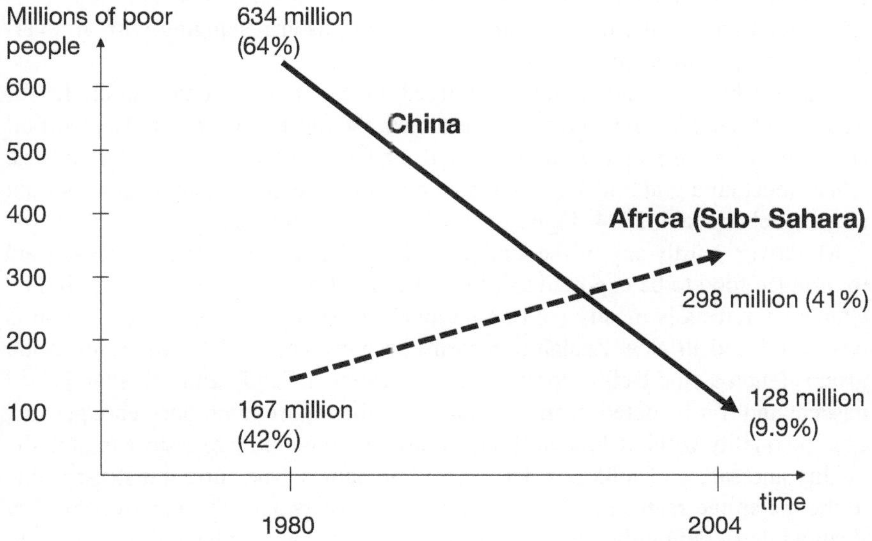

Figure 1.2 Contrasting patterns of poverty alleviation (adapted from Li *et al.*,2012)

so, they often have to face dangerous working conditions, long working days, insecurity about pay and, last but not least, solitude. Third, there is demographic decay. Although peasant families were allowed to have two children, demographic development has produced (in combination with the hardship of country life) an enormous risk for the near future: will there still be people to engage in farming? Finally, there is a fourth form of decay – a form that extends to and materializes within the large urban centres. This decay is the persistent degradation of food that translates into repeated food scares which sometimes result in many fatalities (and, sometimes, many dead and/or permanently damaged children and babies).

There is no doubt whatsoever that the success of China's agriculture is partly built on these different forms of decay. But there has been (and still is) considerable decay elsewhere in the world too. In SSA there is decay *without* agricultural growth and poverty alleviation. Hence, the need to explain the performance of China's agriculture sector and its success in alleviating poverty remains a central concern – for it cannot be explained solely as a consequence of multiple decay, of mortgaging ecological and social capital for rapid growth.

Land grabbing: a new token of failure?

Land grabbing is a hot topic these days, and while we do not intend to discuss the issue in detail, we need to make some reference to it, especially given that China's participation in this phenomenon is often taken as indicating the failure of its own agriculture. China's pursuit of land elsewhere, notably in Africa, is seen by some as a sign that the country itself cannot meet current or future consumption needs of its population – the more so since these patterns are changing with an ever-growing rate of meat consumption.

While China is undoubtedly involved in the international scramble for agricultural land it is not, as is often assumed, the most active actor. The Gulf oil states and Western capital groups are acquiring far more land than China. In Africa, where most land grabbing occurs, China only ranks as the eighth most important player (Schoneveld, 2014: Figure 4 and Table 4; Hofman and Ho, 2012:21).

Moreover, hardly any of the land now controlled by Chinese agencies is used to produce food to be exported to China. The food produced on land acquired by Chinese interests is mostly sold on national markets. The possible exception is milk produced in New Zealand on farms recently acquired by Chinese capital groups (such as the Beijing Agricultural Investment Fund). But this plan is still nascent and if it is rooted in any 'failure', it is the failure to produce enough milk of good quality within China itself and, above all, the failure to restore trust in the quality and safety of milk powder for feeding babies. The current distrust is due to the melamine scandal that swept China a few years ago. The takeover of New Zealand dairy definitely does not mean that China does not have the capacity to assure and expand milk production domestically.

Apart from soy, China is, on the whole, self-provisioning in food. It even exports considerable quantities of different agricultural products. Available forecasts indicate that it should maintain this position in the decades to come. Yet there is an urgent need for many profound changes in the way that Chinese agriculture is organised, including a radical shift to agroecological approaches, a sharp reduction in dependency on fossil fuels and in the use of sweet water, a drastic increase in product quality, a shift away from massive imports of soy and greatly improved town–countryside relations. China is not alone in facing such challenges: it is in good company. Such changes are urgently needed in most agricultural systems in the world – wherever they are located.

The phenomenon of land grabs is a worrying one (see also Borras, Franco and Ploeg, 2013), as is China's involvement in it. It cannot, however, be taken as indicating a failure of its own agriculture. It reflects, instead, the increased involvement of Chinese capital groups in the global scramble for scarce resources, including land and water, and the associated possibilities for the further accumulation of capital.

The difficulty of finding an explanation

As indicated above, at first sight all the odds are against China's agriculture, and not only because the average farm is 'far too small'. There are many other factors that seem to be highly adverse to any consistent and continued pattern of development (see Table 1.1 for an overview). Echoing Donaldson, we might refer to China's agriculture as a 'puzzling pattern [that still] remains a mystery' (2011:36). Apart from being small-scale (which is presumed to exclude investments and the application of new technologies), those involved in farming do not perceive it to be an 'economic activity'. Rather, they experience and organize it as a 'way of living'. In the Western view such a stance is associated with backwardness and inefficiency. Where is the search to maximize profits? Chinese farmers do not follow the 'logic of the market' when organizing and developing their farms. Beyond this, the men, i.e. the main part of the labour force, have migrated away and only women and elderly people remain. This is usually seen as yet another disadvantage. Thus, there is an extensive list of hindrances: some located at the level of single farms and farm households, others situated at the level of the agricultural sector as a whole and another set of obstacles is encountered at the level of the support structure, the institutional framework needed to make agriculture function well (see Table 1.1).

The variables set out in Table 1.1 cannot be understood as simple dichotomies. Rather, they represent ever-so-many scales which allow for extremes, but also for

Table 1.1 The many 'handicaps' facing Chinese agriculture

	Characteristics of China's agriculture	*Conditions considered to be necessary for ongoing growth*	*Level*
1	Small-scale	Large-scale	Farm
2	Farming as a way of life	Farming as an economic activity	Farm
3	Temporary migration	Full-time farming	Farm
4	Labour-driven intensification	Technology-driven intensification	Farm
5	Multifunctional	Specialized	Farm
6	Female	Male	Farm
7	Strong imprint of policy	No distortion of markets	Sectoral
8	Developmental tendencies seem to be chaotic	Developments are clearly outlined and planned	Sectoral
9	Agriculture as a whole is heterogeneous	Agriculture is standardized	Sectoral
10	Peasant population is 'uncaptured'	Farmers are regimented	Sectoral
11	Reciprocity is crucial	Availability of specialized and wage workers	Support structure
12	Informal credit	Strong and accessible banking sector	Support structure
13	Dependent on novelties that emerge in the countryside	Strongly developed research and development (R&D)	Support structure
14	Multiplicity of markets	Global market and strong value chains	Support structure
15	Peasant kitchen (including modest consumption of meat)	Convenience food (centred on high levels of meat consumption)	Society
16	Cultural disdain for peasants	Agricultural entrepreneurs have a high status	Society

many intermediate positions. Above all, they should not be understood as running from 'bad' to 'good' (or the other way around).

Nonetheless, the Western way of thinking about agriculture, farming and farmers typically frames these issues in a highly normative fashion. It is precisely this normative framework that distorts our understanding of China's agriculture, which by western standards should barely exist, let alone be thriving. In Western thinking the right-hand side of Table 1.1 represents 'superiority', while the left-hand side equates to 'inferiority'. Each single variable in the left hand column should trigger a sub-optimal performance. Taken together these features should make for a disaster. Nonetheless, all these features are highly evident in Chinese agriculture, yet its performance is far from disastrous. Indeed, it seems to function far better than many other agricultural systems in the world. Once again, this is the enigma that needs to be explained.

There are several issues underlying all this. The first is that these different variables are not independent from each other. They are not ever-so-many isolated causes that together constitute an efficient and prosperous agriculture (or the opposite). There are interactions, or more precisely *multiple causations*. Take an example from health. We all know that working in a dusty environment is bad for human health. It negatively affects the lungs. Working in a wet environment can also be damaging. However, if small drops of water are sprayed in a dusty environment this can have a positive effect: the drops catch and neutralize the dusty particles. This is an example of multiple causation: two 'negative factors' interact and neutralize each other.

There is the possibility of multiple causation when it comes to the variables synthesized in Table 1.1. We are not looking at a mono-causal model where single causes simply add up. There can, instead, be complex interactions that result in unexpected effects. A carefully constructed *combination* of, say, temporary male outmigration and the centrality of women in farming might turn out to be a powerful driver of agricultural development – especially when it translates into labour-driven intensification and builds on novelty production. These and several other forms of multiple causation will be discussed throughout this book, especially in Chapter 12 where we discuss the importance of multi-level balancing.

Second, it should be remembered that particular agricultural systems are always and everywhere structured in specific ways that reflect (at least partly) a specific way of *thinking* about agriculture. In turn this way of thinking (the core of which might be typified as a 'paradigm') helps to structure the future development of the specific agricultural system. Thus, the particular forms and dynamics of agriculture can be seen (again, at least partly) as a socio-material expression of the underlying paradigm – as much as this paradigm is informed by, and reproduced through, prevailing socio-material practices.

This applies to both Western and Chinese agriculture. The point is that the underlying paradigms differ enormously. The right hand side of Table 1.1 synthesizes the main features of the Western paradigm.[7] In this paradigm, agriculture is seen as being, first and foremost, an *economic* activity: governed by an economic calculus that centres on profit and income maximization (see row 2 in Table 1.1).

In consequence, the farm is, above all, an *enterprise* headed by an (agricultural) entrepreneur. Such a view is strongly internalized by today's farmers in the West: they perceive their farm primarily as an enterprise that is structured according to the logic of the market. And wherever, and whenever, this is not the case, farming cannot be but of 'inferior quality'.

By contrast, Chinese peasants do not perceive their farms as enterprises that primarily represent an economic activity. In the Chinese language there is not even an unequivocal term for 'farm enterprise'. *'Nongmin'* (农民) clearly refers to the peasant (or farmer), just as '*nongye*' (农业) refers to agriculture in general terms. However, there is no clear concept that outlines and specifies the farm *enterprise* as such. The term that comes closest is *'guorizi'* (过日子), in which the first character refers to movement through time and the last two to everyday life. This is close to (rural) *livelihood*, or to the *routine of the farmer*.[8] People sometimes say 'my business is my everyday life'. Work and life are tied together, each being identical to the other. Farming is 'what you do', it is, so to say, 'your life, your existence and your identity'. People expect that farming will provide them with enough food, an attractive price for the products they bring to the market (so that they can afford to buy other goods) and offspring who carry the promise of continuing the family line.

All this does not invalidate the *raison d'être* of Chinese farms. They are very real; each is a specific set of fields, each has its specific history and its boundaries, which are sometimes bitterly disputed. In addition, a farm also has a specific number of animals, a store of feed and fodder to get through the winter, available seed, the right to use water, the willingness of family members (and friends) to join the labour process, etc. Each farm also has a set of accumulated skills and is embedded in a complex network of relations, including the capacity to engage in market relations. Finally, there is the expectation that through hard work some material progress might be wrought. In all these respects each farm is very real. However, it is neither perceived nor ordered as an enterprise. Farms are perceived and defined *in sociological terms* rather than *in strict economic terms*. This reflects the way that the farm and farming are seen (and understood) *as integral parts of social life*. A farm is not perceived as located exclusively, or even mainly, in an economic sphere that is neatly separated from the sphere of social relations.[9]

Farming practices are understood as part and parcel of rural livelihoods. This, of course, does not imply that Chinese farmers aren't interested in economic gains (or, for that matter, are not worried about costs). Nor does it imply that they do not try to get the best out of their activities. The point is that these activities are not understood as constituting an enterprise, and they are not managed like one. Throughout this book many details of how Chinese peasants think about farming and how they actively organize it will be discussed. Here it suffices to reiterate that of course, there is production, transactions, incomes to be earned, investments to be made and productive forces to be developed. In these respects Chinese agriculture is an economic activity. The key point is that it is not explicitly *perceived* nor *organized* as an *economic* activity (i.e. according to the rules that we use in the West). Other concepts are used, other relations and balances govern the many practices and their unfolding over time. This, from a Western perspective, makes Chinese agriculture hopelessly out-of-date.

From the Western view of agriculture all the variables shown in Table 1.1 go from 'backwardness' (left-hand column) to 'modernity' (right-hand column) and this reflects, at least partly, the logic of Western agriculture. The Western view, and especially the agricultural paradigm entailed in it, is strongly based on linearity: that history only proceeds in one direction. This is also applied to the present and to the future. This approach excludes the possibility that the features listed in the left-hand column might also represent values (especially when taken together).

China's agriculture, taking into account its many problems, functions in a way that is unparalleled. Throughout this book we argue that it entails an implicit paradigm, the main dimensions and consistency of which we will explore in subsequent chapters. This paradigm allows us to come to grips with an agriculture whose main features are summarized in the left-hand side of table 1.1 and which, together, foster its impressive development. In brief, it helps us to see that what, in the Western paradigm, appears to be a structural weakness might very well be a major strength when viewed through the lens of the Chinese paradigm.

The question of whether or not one should view agriculture as being primarily an economic activity should also be applied to the other variables in our list. For instance, in the Western paradigm, markets and technology primarily define the way farms are organized and developed. The impact of policy is seen as only of secondary importance (and, in the more radical forms of Western neo-liberal thinking, as essentially a distortion). Farmers often share this perspective, which influences the way they think of developing their farms and how they relate to the state. Policies, farmers' strategies, social relations and cultural repertoires are at best of secondary importance – they may have some residual influence on the main trends, but these are primarily defined by markets and technology. In China, it is probably the other way around, as we will see in the empirical chapters of this book. China demonstrates that each of the features we discuss (policies, strategies, repertoires, etc.) has the potential to be a turning point. As Henley and van Donge rightly argue in relation to policy: 'State-led rural and agricultural development, leading to higher incomes for peasant farmers, has been crucial to south-east Asia's success, and we infer that its absence has been crucial to Sub-Saharan Africa's failure.' (Henley and van Donge, 2013:33).

In line with the previous axioms, the Western view holds that farm enterprises can only develop and prosper through ongoing enlargement. At the aggregate level this view implies that there will (and should) be a continuous and ongoing reduction in the number of farms and a constant shedding of labour. Fewer, but far larger, farms will remain. The rest will be outcompeted by the growing farms.

In the Western paradigm, agriculture is basically a male activity. It requires strength and courage to work in the fields in all weathers, face the markets and engage in the ongoing battle that will eliminate many and only reward a few. It requires entrepreneurs who are able to outcompete others. Here, once again, China represents a counterpoint: young males are absent, the work is done by women and old men. Nonetheless, despite (or thanks to) this, agriculture is blossoming.

Exploring and studying Chinese agriculture is exciting and intriguing. This is because it leads us to a search for another paradigm, one that probes into another way of conceptualizing, understanding and practising agriculture. In so doing the observer has to dissociate him/her self from the assumed universality of the Western way of thinking (this applies both to people from the West and to most Chinese observers). The observer has to accept that there are limitations to the Western way of conceptualizing agriculture and farming. S/he has to admit that there are other ways, and other paradigms, that might be as effective and efficient as the Western one. If travelling is, generally speaking, a rediscovery of one's own soul (Camus, 1998), then a Western specialist encountering Chinese agriculture cannot but be forced to critically reflect on the Western paradigm.

The self-definition of the Chinese peasant

Who are the men and women who fill the Chinese countryside? Who, and what, are the Chinese peasants? Who are these actors who play such a leading role in the remarkable performance of China's agriculture? No complete and satisfactory answer can be given to these questions (although see Little, 1989, for scientific debates; Day, 2013, for an overview of the socio-political debates in China during the last 60 years; and Schneider, 2014, for an overview of the current discourses). In this introductory chapter the discussion on the Chinese peasantry will be limited to their self-perceptions, i.e. the way they see themselves.[10]

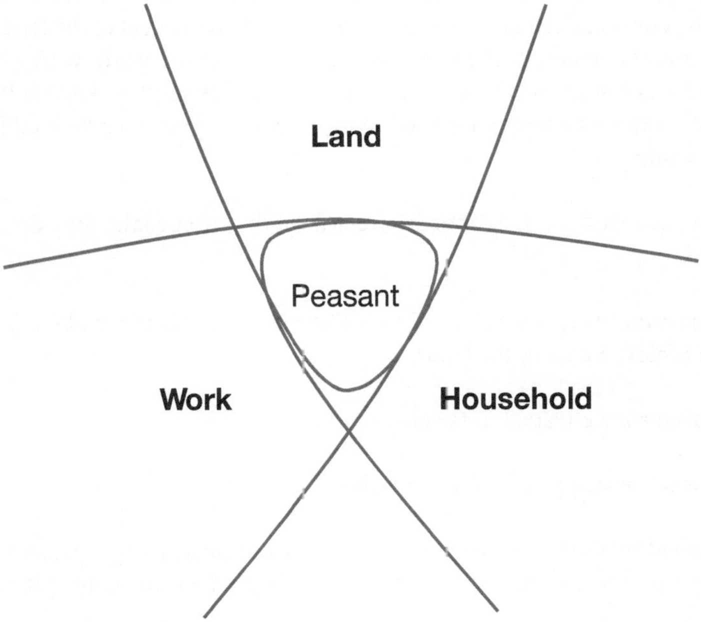

Figure 1.3 Land, work and the household as sets of social relations that specify the peasantry (own elaboration)

Following the basic principles of ethno-methodology, An Miao (2011) gathered rich material that helps us to understand how Chinese peasants define themselves. In the first place peasants perceive themselves as part of a wider collective. When asked to explain what a peasant is, the most likely answer is: 'a peasant is like me, myself', while the characteristics of one's own life are often described by making reference to other peasants.

The wider category of peasants is far from an unspecified whole. It is clearly defined by a comprehensive set of social relations, just as there is a clear delineation that separates and distinguishes the peasantry from other social groups. Nonetheless, the transparency of definition and delineation does not exclude ambiguity – which has been multiplied in recent years by the rapidly changing realities facing the peasantry.

An Miao's research findings highlight three sets of social relations that together summarize much of the world-view of Chinese peasants and which together help specify the notion of a Chinese peasant. As a folk-concept, the notion of a 'peasant' is located at the interface of land, household and work. This is illustrated in Figure 1.3. The notions of work, land and the household partially overlap – it is precisely at this point of 'overlap' that the peasant emerges as a socially-constituted actor.

Relations with the land

The first and highly important, albeit far from exhaustive, set of relations is that between people and the land. The peasant, in short, has access to the land (that is, under current conditions, the one with the *usufruct* of the land), works this land and produces his/her own food, alongside a marketable surplus. This multifaceted relation is expressed through the following interview fragments derived from An Miao's work:

> We have land [. . .]. That is different from the city people: they do not have land.

> A peasant [*nongmin*]? A peasant is like me, he works the land [. . .] and like the others, we work the land.

> Peasants are engaged in farming work.

> Peasants produce food for themselves.

The implied relation is a two-sided and reciprocal one. Being a peasant implies possession (or the usufruct) of land; but having the land, in turn, defines the peasant:

> I am a peasant so I must farm the land.

And vice versa:

> Whoever does farming work is a peasant.

As self-evident as this may appear to be (at first sight at least), it is important to note that the specified relations with the land are not strictly individual – they are mediated through the household. It is also true that these various components (having the usufruct of the land, working it and obtaining food from it) are not necessarily always all satisfied; their presence is mediated by both the social organization of time and state-peasantry relations. As will be shown later in this book, the fact that it is the peasant who does the farming work strongly affects the process of agricultural production, since this moulds the work practices in a specific way. To put it differently: peasants farm in a peasant-like way – one that materially differs from other ways of farming. It should also be noted that the notion of land is not restricted to the single plots being worked by the peasant – 'land' belongs to a wider socio-spatial notion: the countryside (or the rural [*nongcun*]). And finally it should be stressed that being a peasant implies the *right* to have access to the land. We will come back to each and every one of these points and their sometimes quite complicated implications later.

Relations within the household

The second crucially important set of social relations that defines and specifies the peasantry stems from the household. This is especially relevant when it comes to migration and/or generational succession. Those who migrate do not stop being peasants (see also Carillo and Goodman, 2012:6; Leung and So, 2012:63[11]). Through their household relations migrants remain peasants.[12] As a local leader of Sanggang village explains: 'They don't give up the land. Their women do the farming [. . .] and if they [i.e. the men] have time they come back and support farming'.

This is reaffirmed by other villagers:

> those who migrate, but whose family members are doing the agricultural work, are also farmers [. . .]; they [i.e. the migrants] let others plant the land but they do not give up the land, they remain as farmers [. . .]; over there [i.e. in the factories] they are also called peasants.

> Yes, other people over there [in the places of migration] call us peasants.

The food that is consumed within the household further strengthens the peasant identity. An old farmer says: 'those whose food is from agriculture, from the fields [of the household], are farmers; citizens, on the other hand, live from commercial grain.'

Thus, self-provisioning of food (or at least a part of it) is a key element in the many relations that together define the peasant identity.

Through their belonging to the household, peasants also relate to history. The same old farmer adds: 'Our roots are here, this is the place of our ancestors.'

The relations that characterize the labour process

The third set of social relations centres on work. But not on any type of work. Self-employment is central; wage-relations are (apart from some exceptions that will be discussed later on) critically excluded. A young peasant argues:

> We are peasants because our income is from our labour; the state does not give it [i.e. an income] to us. Citizens get their money from the state, they have a job [i.e. they are paid a wage or a salary], but farmers' money is from themselves. For instance the village teacher, is not a farmer, his money comes from the state.

And: 'The man who has a formal job is not a farmer, because even when it is raining [i.e. when the agricultural work cannot be done], they get their money from their job.'

Working the land is strongly perceived in terms of a moral obligation:

> I am farmer so I must do farming,
>
> I am peasant, I always work hard,
>
> Older people, such as myself, farm for as long as we can,
>
> As peasants we work as long as we can, as peasants we put off depending on our children for as long as we can. . . .

This is true even when there is no material need to keep on farming. A shop owner in Sanggang affirms: 'I keep farming, I do not need to, but nonetheless I keep at it as it helps me to support my house and family.'

This moral obligation (which often appears as a self-evidential feature) also links the different generations. A returned worker says: 'If I cannot work the land anymore, my son will come and do the work. How could it be otherwise? This is how things are.'

And a local leader speaking for women and the elderly said: 'They continue farming in order to reduce expenditure, the cost for food.'

Flexibility and persistence

Together, these three sets of relations define the peasant identity. In many places in the world (just as in many social science theories) peasants are defined in a residual or even in a negative way, in terms of what they are not. Such approaches can be found in China too. In the recent past (and especially in the cities) Chinese people were classified into five categories: 1) workers, 2) farmers, 3) business people, 4) students and 5) soldiers. Within this scheme 'peasants' (or 'farmers') easily become a residual category: those who are *not* workers, businessmen, students or soldiers are, by default, peasants. It is telling that the self-perception of Chinese peasants does not have residual or negative undertones. It is, instead, a *positive* definition: peasants are defined, albeit in a complex way, as what they are. This is a first important finding. Another equally important finding is that in a peasant's world-view being a peasant is a *relational* issue. It is a peasant's relations with the land, within the household and within the labour process that together define the peasant as social actor (and, consequently, as a member of a specific class).

Together these relations create a symbolic space that allows for many specific and highly diverse material expressions. The specific situation of peasants might differ greatly – and along many different dimensions. However, the resulting diversity is, as it were, embraced by and summarized in a widely shared common denominator: 'being a peasant'.

In this respect it is indeed crucial that it is not the individual characteristics of certain people (nor their level of income, the specificity of their culture, the particularity of their attitudes) that makes them peasants. It is the *relations* that tie together land, household, labour processes and people that *together* define the peasant identity. 'Peasants work for themselves' as we saw before. Being a peasant defines specific labour relations: self-employment instead of being engaged in wage relations. At the same time this applies the other way around: self-employment defines those involved in it as peasants. Especially when it is about working on the land. 'Working the land' means that you are a peasant and vice-versa: what peasants do is work the land (which assumes relations of e.g. usufruct). And of course, peasants are part of a household. When it is said that peasants work for themselves (implying that they themselves have the responsibility, that they themselves define what to do, etc.), this implies that they are working 'for the household': to feed the household, but also to contribute to the education, the marriage and the housing of their children and, finally, to take care of elder people. Being part of the household also implies that relations with the land are not easily interrupted: those who migrate remain as farmers (at least in their world vision).

Work, land and household together define the peasant identity. It is not one single aspect (or one single relation) that defines being a peasant; it is these three complex sets of relations that specify the peasant. It is precisely this complex nature of the social definition that allows for its flexibility (at the symbolic level) and persistence (in material terms). Being away (through migration) does not eliminate the peasant condition. It is continued through the sets of relations that define the household: their access to land and the work done by household members, neighbours and/or friends. As long as the migrant worker maintains access to the land and as long as s/he will return to the household (their own or parental), s/he continues to be considered a peasant. While labour migration might be a temporary break from 'working in the field' it does not imply a definitive adieu to the peasant condition. Equally, there may be periods when a peasant household has no access to land; this may occur when land is expropriated (for non-agricultural uses) and the peasant-family is awaiting – or claiming – new plots of land as compensation for the expropriated fields.

Some wider ramifications

Through these sets of relations that link work, land and the household, the peasant also relates to the wider world. Hence, the peasant condition is also specified by far wider contextual elements than the immediacies entailed in the direct interaction of work, land and the household (see also Figure 1.4). Through the 'land' one is part of wider spaces that embrace town and countryside and the many contradictory relations that link and characterize the two. Through the 'household' every single actor is part of complex social networks (village, clan, religious groups, schoolmates, production team, migration networks[13] etc.) and the particular organization of time that comes with these networks. These networks tie past, present and future together in particular ways and often imply a calendar of activities and meetings that occur with specific regularities. Through these 'work' structures and routines many peasants become part of complex movements that occur in both time and space.

Having the land means that one belongs to the rural: 'We have land, citizens do not have land.' However, there is a clear counter-image related to this: 'Belonging to the rural area and being engaged in farming work means that you do not have the right to share the welfare of the city [as represented by, e.g., medical insurance].'

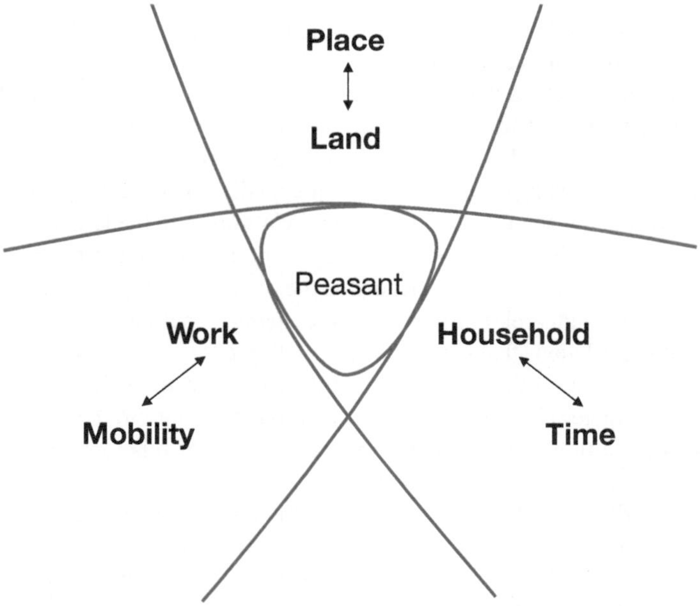

Figure 1.4 The wider linkages (own elaboration)

Thus, peasants consider themselves to be part of wider society – but a very specific part: they represent the underprivileged, the ones suffering deprivation, those who are still (at least partly) excluded from the riches and welfare that are concentrated in the cities.

Chinese peasants relate to both history and the future through the household. They preferably work the land of their ancestors (just as they take care of the tombs that are scattered throughout the fields). As regards the future, much of the current work is framed within a household logic that aims to improve the condition of the children. Peasants work hard to finance their children's education. Good schools which offer the promise of access to universities are expensive, the more so since they increasingly involve paying to stay at boarding school. Many peasants engage in migration specifically to meet the costs of educating their children. Beyond that they will invest in the construction of a house for their children as well as in their marriages: 'To engage in a good marriage, to get a good wife, implies that you have a good house . . . without a house, no household.'

And probably (in yet another axis that links the present with the future) they will develop new productive lands (for example on hills that are currently extensively grazed) for their children.

Many think that once their own children (now engaged in migrant work) grow older (and can no longer stand the hardship of industrial labour) they will return to the village (just as they did themselves) and start to work the land. Thus peasants are travelling through time: the younger ones move out while the older generation prepare the conditions for their return once the youngsters become older themselves. The moments for, and modalities of, departure and return might differ considerably – however, both the presence of a peasantry working in the fields and its continuity in Chinese society are secured through these movements through time. This is echoed, as already indicated, by many migrant workers who consider themselves as peasants.

A third ramification to be discussed here relates to work. Chinese peasants move through society and the economy for and through their work. They are able to do so because they are raised and socialized in rural villages (and can return there when there is a need to do so). They work in industries, construction sites, markets or whatever, in order to send remittances home; they will, sooner or later, have their children raised in the village and, finally, they will return more or less definitely in order to live and work the last 20 or 25 years of their productive life in the village. Migration is favoured because it allows them to earn higher incomes than in the villages and, especially at the beginning, for the escape it allows from strict social control. The return is favoured for the quality of life in the villages. This cycle of mobility is especially important in that it blurs sharp delineations between a hypothetical urban labour force and strictly rural peasants. The different conditions combine into one and the same person (albeit through a myriad of specific mixtures) – just as the working population as a whole has the characteristics of both peasants and workers.[14] At the same time, experiences from the industrial centres are brought to the villages, just as the sturdiness, the capacity to face hardship and (potentially) the explosiveness of the peasantry are brought to the large factories.

Tensions

The (self-) definitions described above relate in a complicated way to the changes that are currently reshaping considerable parts of rural realities in China. What is assumed to be an organic whole (of land, labour and households) is increasingly being torn apart into separate realities that are difficult to combine. Let us illustrate this with a series of questions:

a) What if the relation with the land is loosened and work becomes restructured (as is the case with new horticultural cooperatives and enterprises within which dairy farming activities are concentrated)?
b) What if out-migration tends to become a permanent phenomenon and the prospect of returning pales?
c) What if nobody can care for the land?
d) What if, through new housing schemes (as exemplified by the reconstruction after the big earthquake), farmers no longer have the storage facilities and stables needed for farming?
e) What if women do not find farming attractive anymore?
f) What if young children (increasingly going to boarding schools in townships) lose the opportunity to get acquainted with the joys, drudgery and pitfalls of farming?

We previously referred to the capacity of the social definition of the peasant to embrace many different situations. The point here evidently is about the limits. Where does the peasant self-definition stop being an adequate synthesis of one's own situation – in the sense that it offers promising pathways to the future and allows people to deal with contradictions in a satisfactory way?

The modernization of agriculture

For decades the Chinese authorities have been promoting a 'modernization of agriculture' (Ye, 2015). Although it is unclear how such modernization will unfold, it is evident that it will considerably heighten the frictions mentioned above. Modernization is generally assumed to entail a double process. A large part of the agricultural labour force moves from the countryside to the cities, and there is a simultaneous restructuring of agriculture – since the work is now done with fewer people:

> To achieve an orderly development process and ultimately to modernize, traditional society must be restructured so that the factors of production flow in fundamentally different ways. The labour force must shift from rural to urban occupations, capital must increasingly flow into modern rural enterprises.
>
> Ye and LeGates, 2013:239

These are the processes through which a 'modernized agriculture'[15] is assumed to emerge. This modernized agriculture is far more integrated with the wider processes of capital accumulation than peasant agriculture. This integration (which more frequently resembles submission) occurs through

1 increased indebtedness;
2 greater use of external inputs and new, more sophisticated technologies;
3 delivery to, and increased dependency upon, food industries and large retail organizations; and
4 taxing by the state.[16]

Thus banks, agribusiness groups and state apparatuses become more important – although not necessarily decisive – parts of the context in which farming takes place (Bernstein, 2010). They increasingly drain value from farming (this is evident in the 'squeeze on agriculture'), just as they increasingly prescribe and sanction the way farming is organized and developed (Benvenuti, 1982).

All this can lead to a profound repatterning of farming practices. It is widely assumed that a new model of 'entrepreneurial farming' (which is widespread in the north-west of Europe and Canada) or a model of 'capitalist' (or 'corporate') farming (as can be found in parts of the USA, Brazil and South Africa) are likely to become dominant.[17] There is a third possibility, which is far less discussed in the literature: that peasant agriculture is reproduced and renewed in a form that is able to face and resist this changing context (see Ploeg, 2008). An important part of this book centres on the question: Which models are emerging and might triumph in contemporary China (see especially Chapters 5, 6 and 7)?

The modernization of agriculture (at least as conceptualized in hegemonic theories and policies) is supposed to occur through, and to result in, the set of characteristics that are summarized in the right-hand column of Table 1.1. There is hardly any need to argue that the features of this model are capable of generating a wide range of frictions – and at many levels. The way in which China's peasantry, state and agriculture deal with these frictions (it is no exaggeration to compare them to a minefield) raises a number of questions that will pose many challenges to the peasantry, agricultural policies and farming in general in the coming years and decades.

The structure and nature of this book

There are several other elements and features that at first glance also seem to contribute to the enigmatic nature of China's agriculture, although after careful consideration they can help us to understand its exceptional performance. Some of these will be discussed throughout this book. Chapter 2 centres on the circularity of labour migration which implies a specific two-way flow of resources between the cities and the countryside.[18] Chapter 3 focuses on the symbolic relations between man and the land and how these relate to the social organization of time.

It presents, as it were, the cultural grammar (or 'logic') that governs rural livelihoods. Chapter 4 discusses labour-driven processes of agricultural intensification. These processes (that differ sharply from technology-driven intensification) underlie much of China's exceptional agricultural growth records. Together these three chapters explain, we think, much of the strength of China's peasant agriculture.

Chapters 5 and 6 approach the enigma from another side by asking, respectively, why entrepreneurial and capitalist ways of farming turn out to be only marginally feasible and successful in China. To put such ways of farming at the centre of agricultural policies (as is increasingly the case in China) seems to be a case of 'betting on the wrong horse'.

Chapter 7 discusses how contemporary peasant farming is increasingly being reinvented by young Chinese peasants. By doing so they seem to be laying strong foundations on which the future of China's agriculture might be grounded.

Chapters 8, 9 and 10 are of a more general nature. Chapter 8 discusses gender relations in Chinese agriculture while Chapter 9 looks at how young students in Beijing perceive the rural. Chapter 10 probes the issues of markets, marketing and the building of new markets.

Chapter 11 returns to the issue of the 'microscopic' nature of Chinese peasant farms by asking why they are so small in the first place. This discussion flows into another one: what is the current relevance of peasant farming – for peasant households as well as for Chinese society as a whole? Finally, Chapter 12 returns to the issue of the underlying paradigm and tries to tie together some of the threads developed in the previous chapters.

Over the last two decades much has been published about Chinese agriculture and many of these books and articles are of an exceptionally high quality. Overall these publications can generally be grouped into three categories.[19] First there are the studies that discuss China's agricultural sector in macro terms. These might be macro-economic, or politico-economic or sociological (focussing, for example, on new forms of resistance or changing gender patterns) but what they share is the focus on the macro level. Second, there are studies that focus on the micro level. These are mostly of the applied economics type. Third, there is a range of 'quick appraisals': articles and reports based on one or just a few cases. These are often rich in detail, but it is difficult to assess whether they describe the rule or the exception.

What is missing from this *tableau* are detailed studies that pay attention to the *grassroots realities and dynamics of farming*. To date there has been no serious representation of agricultural realities that reflects the strategies of the actors involved and the relations and encounters in which they are engaged. The same applies to the many interactions between the macro and micro levels. Such interactions are taken for granted (normally a simple addition or a one-sided determination is assumed). The ways in which heterogeneous micro-dynamics translate to the macro level and how the former are conditioned by the latter still require systematic inquiry. Such inquiries are not driven purely by curiosity. They are badly needed, we believe, to understand, and effectively intervene in, the different processes of transition that are occurring in rural China today. With our co-authors, we think that this book takes a few first steps towards filling these gaps.

A methodological note

It is not clear anymore whether there are, as Katie Melua sings, 9 billion bikes in Beijing. But it is a fact that there are 691,510 administrative villages in China. The number of 'natural' villages is far higher still. This book is based on extensive fieldwork, centred on seven peasant villages: *Sanggang*, *Yuangang*, *Nandugang* and *Baoshi* (in Pochang township, Yi county, Hebei province); *Huangjue* (in Ruifeng township, Qingshen county, Sichuan); *Sipai* (in Sipai township, Raohe county, Heilongjiang province) and finally *Qiaobei* (in Longji township, Shuyang County, Jiangsu Province). Of course these villages are not representative of rural China as a whole. Given the sheer size and impressive heterogeneity, no study can really claim to be representative of China's agriculture, countryside or peasants. However, we sincerely believe that the seven villages clearly show the main developmental *tendencies* that are occurring at the moment throughout China. We also think that they reflect at least some of the main developmental *issues* and *challenges* and also some of the *responses* to it. Alongside these seven villages several other locations have been visited to extend the comparative nature of the analysis.

The first author of this chapter, Jan Douwe van der Ploeg, has been visiting Sanggang twice a year (in autumn and spring) since the spring of 2007. He has carried out his own research there and accompanied MSc and PhD students, giving them in-field research training. He has also visited Huangjue on an annual basis since 2012. The second author of this chapter, Ye Jingzhong has been doing research in Sanggang since the 1990s (see among others Ye 2002) and, since 2010, has extended his research programme to include Huangjue. Zhao Yong (who contributes to Chapter 5) has done two extended periods of research in Sipai village, with each stint lasting for 6 months. Wu Huifang, Pan Lu and Wang Chunyu (who co-author different chapters in this book) have all undertaken long-term research programmes in these villages. Meng Xiandan did her PhD research in Qiaobei village and Sabine de Rooij, who co-wrote Chapter 8 with Meng and Chapter 9 with Wu Huifang, visits China every year. She combines her research among young students at China Agricultural University with a similar research programme in Italy. He Congzhi, who co-authored Chapter 10 on markets, is active in the construction of new, nested, markets.

Different methods have been used. These include extensive interviewing, construction of biographies, detailed case studies of single farms, surveys and participatory observation.

We tested our findings in seminars and lectures at the College of Humanities and Development Studies (COHD) at China Agricultural University (CAU) in Beijing. Here we could always count on the responses from a critical group of students and staff members who are highly knowledgeable about the Chinese countryside, agriculture, the peasantry and agricultural policies. These encounters with students and staff, in which we discussed the outcomes of our work, have been an important source of inspiration and reflection. Especially stimulating have been the comments of, and discussions with, Henry Bernstein, Jun Borras, Jennifer Franco, Ann and Norman Long, Harriet Friedmann and Joan Martinez-Alier.

We have also tested our findings by publishing scientific articles in both Chinese and international peer- reviewed journals. Several chapters of this book are (partly) based on such articles. Wherever this is the case this is indicated in the notes that accompany the text.

The chapters in this book have been developed over several years. As they have evolved the research team has used them as works 'under construction'. We used them in teaching, sometimes as the basis for articles and, always, as guidelines for further research. These chapters illustrate that research involves uncertainty: one is repeatedly trying to understand processes that are in constant motion and contain several possible, but mutually exclusive, outcomes. They are incomplete and do not carry the luxury of a definitive and solid end point that allows for telling the story in a straightforward and unambiguous way. Several chapters in this book illustrate the hesitations and doubts inherent in doing research into the real life situations that constitute Chinese agriculture.

Some chapters contain an update at the end. These updates describe changes that occurred just before completing the book and/or when new data was obtained. These updates also show (albeit indirectly) the contingent nature of China's agriculture. There are many on-going changes, which are often contradictory: what is happening in one place is often in complete contrast with the situation in another. The contemporary story of China's agriculture has not yet been told – and certainly cannot (yet) be carved in stone. It is a mix of alternating points of view and rapidly changing realities – and what seems to hold true today might well turn out to be wrong tomorrow. Precisely for this reason we use the updates as a stylistic device: they are meant to remind the reader that, as yet, the story is not complete.

Notes

1 According to the Agricultural Census 2006 there were just over 200 million rural households engaged in agriculture. This was an *increase* of 3.7 per cent on the previous census (1996). Of the rural households registered in 2006, 58.4 per cent derive the main part of their income from agriculture. For the remaining 41.6 per cent other economic activities provided the largest part of household income. However, the percentage of households deriving most of their income from agriculture is decreasing (by 7.2 per cent between 1996 and 2006). Source: *www.gov.cn/gzdt/2008–02/22/content_897216.htm*

2 We are well aware that this paradigm is no longer limited to the West in the geographical sense. This paradigm has been adopted by, and becomes dominant within, Chinese academia, think tanks, state apparatuses and in parts of the government. As we will spell out later, this might provoke dangerous frictions.

3 For most visitors from the USA, agriculture in the north-west of Europe is 'impossible' because its farms appear to be too small. This is repeated, as it were, in the way Western academics, especially those from agricultural universities and faculties, perceive south-east Asian agriculture. When the first author brought forward the Chinese experience in public debates in Wageningen most of his colleagues rejected such references

straight away saying: 'the consequences of doing that would be disastrous – there is nothing that you can do on such a miniscule scale'. Such views are also reflected in the far more sophisticated work of Hayami and Ruttan (1985) who recognize the specificity and rationality of the 'Asian trajectory' but maintain that there will necessarily be a convergence between the Chinese and American approaches in the future.

4 The first author witnessed this himself several times. In the early 1970s he worked in Peru where a radical land reform became increasingly contested. Peasants' unions argued that the land should be distributed to far more people than the state wanted to include within the new agrarian structure. The reply of the experts was telling: they argued that a wider distribution of land would create '*a Chinese situation*'. The land would be overcrowded with far too many people; the only possible outcome would be a sharing of poverty. Now, 40 years later, 'China' embodies the opposite. It is illustrative that, in a recent publication of the High Level Panel of Experts (HLPE) of the Committee on World Food Security (CFS) of the Food and Agricultural Organization (FAO) on 'investing in smallholder agriculture', China is referred to forty-three times (in positive terms). That exceeds the references to any other country, including the USA or France.

5 In general terms, China's agriculture was severely drained by the urban economy and the state between 1948 and 2004. The reforms from the late 1970s and early 1980s loosened the squeeze on agriculture a little, although the overall balance remained highly negative until around 2004. It is only through labour migration that an initial, and limited, flow of resources from the cities towards the villages emerged. The mounting social unrest, especially related to the growing imbalances between the urban and the rural, led to another set of reforms in 2004. These included the abolition of agricultural taxes and this helped drive a dramatic reversal in the flow of resources. The catchphrase for the 2004 reforms – 'Agriculture has given a lot, now it is time it gets some back' – was not completely mistaken. For an extensive description see Ye *et al.* (2010) and Ye (2015).

6 More generally speaking, the agricultural sectors of south-east Asia (not including China) also perform better than those of SSA (Berendsen *et al.*, 2013). However, China in its turn performs better than south-east Asia. It also performs far better than India (Gulati and Fan, 2007).

7 One has to keep in mind that the Western paradigm on agriculture has undergone significant changes. The 1930s were a major period of change in the agricultural sciences. The same applies to agricultural policies in the 1960s. Equally there was a 'geographical shift' in the main points of (empirical) reference: with the USA becoming increasingly hegemonic after WW2. Through Marshall Aid the American perspective was transmitted to universities right across the West. The Western agricultural paradigm currently and overwhelmingly reflects the dominant interests of agro-industries, banks and export groups. It paints a virtual image of how agriculture 'ought to be' (see Ploeg 2003 for an extended discussion).

8 This is reflected in another word that might be used: *'nongcun'*(农村). This word primarily refers to the countryside as a whole, or to 'the rural'.

9 The difference (and consequently, the dialectical relations) between 'farm' and 'enterprise' has also played an important role in agronomic and agro-economic traditions in many places in Europe. For a recent contribution, see Iacoponi, 2007.

10 The self-perception (or self-definition) as grasped through anthropological methods is not necessarily at odds with a more 'structuralist' categorization, constructed with politico-economic methods that focus on the social relations of production. The two approaches can strengthen each other.

11 Leung and So understand this self-identification of migrant workers as referring to a 'place-based status' (2012:63). It is thus linked to the circular nature of migration processes. Migrants leave the villages for the cities in order to return, in the end, to

their villages. Whether this implies a rejection of a class perspective or place and movement being constitutive elements of the Chinese peasants as a class is a subject of much debate (see e.g. Lee, 2007).

12 From here on the quotes are derived from our own research in Sanggang village in Hebei province.

13 See Rozelle *et al.*, (1999a) on the importance of migrant networks.

14 As in Latin America where peasants define themselves in generic terms (that include partial integration into labour markets) as 'we the poor people of the land' (*nosotros, los pobres del campo*).

15 This is also sometimes referred to as 'industrial agriculture' (see for instance Ye and LeGates 2013:278). This concept expresses the increasingly standardized character of the labour process and a shift in the ratio between capital goods and labour employed.

16 The possibility that capital accumulation might occur within, and through, capitalist agricultural enterprises is not mentioned here. Chapter 6 explores this possibility in detail.

17 Entrepreneurial agriculture might be understood as 'accumulation from below', while 'capitalist agriculture'(as it is manifesting itself in China) typically represents 'accumulation from above' (Bernstein, 2015).

18 See in this respect also Ma, 1999 and Taylor *et al.*, 2003. This specific flow might be part of more complex economic and financial flows that simultaneously run in different directions. In aggregate these flows may imply a draining of the countryside or, as is increasingly the case since 2004, a 'refuelling' of it.

19 There are many agronomic publications as well but these focus exclusively on the technicalities of agricultural production and/or the features of new technologies. The link with the socio-economic relations and dynamics is hardly ever touched upon in this type of publication.

2 The circularity of town–countryside relations and multiple job holding[1]

Jan Douwe van der Ploeg and Ye Jingzhong

There are many complex and sometimes even contradictory interrelations between rural China's peasant economy and the wider economy in which it is embedded. This chapter focuses on labour and how it circulates between town and countryside. Multiple job holding plays a key role in this respect. Drawing on research undertaken in a peasant village in Hebei Province, this chapter highlights some of the remarkable differences that exist between development processes in China and in other developing countries and traces these back to the strong linkages that exist between rural China and its urban 'global factory'. The chapter also shows that the commonly held assumption that the Chinese peasant farm is 'too small' is nonsensical. These farms are an intrinsic part of the multiple-job strategies of Chinese peasant households and therefore they cannot be disqualified from being viable units.

Multi-level circularity

The industrialization of China has not been based on, nor has it contributed to, a demise of the rural economy. This is in contrast to many other countries, particularly in Africa and Latin America, where industrialization has been accompanied, if not accelerated, by a far-reaching impoverishment, if not destruction, of the peasantry. By contrast China's industrialization is being accompanied by a revival of its peasantry. This has mostly been achieved through the introduction of the Household Responsibility System (HRS) in the late 1970s which led, in the following decades, to a notable strengthening of the peasant economy (McKinley and Griffin, 1993; Bramall and Jones, 1993; Lu, 1997; Howell, 1998; Gulati and Fan, 2007). This strengthening has also partly emerged from the peasantry itself. Eyferth, Ho and Vermeer argue that the reform 'was no policy imposed from above, but sprang from the grassroots of rural society' (2003:1).[2] Rural development policies (Li, 2003: 71–2; Brauw, Huang and Rozelle, 2004) and the reduction of the 'peasant burden' (Lu, 1997; Mons Walker, 2006; So, 2007) have also had a positive impact, especially in the past decade. However, it is primarily the *interactions* between industrialization and the nature and dynamics of the peasant economy that have led to both sectors developing in tandem. Compared to the experience of many other developing countries (and former socialist countries) this is a remarkable achievement.

China's rapid industrialization has led it to become known as 'the factory of the world'. It is largely based on relatively cheap labour provided by rural villages. However, the flow of labour from the countryside to this new global factory has not been accompanied by the creation of extended slums, as has occurred in many other parts of the world (Davis, 2006). The belts of misery that surround large cities and metropolises in the South are populated by people who have *permanently* left the countryside. While China also has sad areas where rural migrants gather, they only stay *temporarily* in these areas. They are part of an immense and often complex flow that goes from the countryside to the 'global factory' and then returns back to the countryside. A considerable part of rural migration (especially the part that supplies the industries that require cheap labour) is temporary; it is, at least partly, a *circular* process that moves from the rural to the urban and then back again to the rural.[3]

This circular pattern has a specific set of characteristics that positively feed into the process of industrialization and simultaneously help to avoid major social and political unrest and the high levels of deprivation typically associated with 'misery belts' elsewhere. First, *each* cohort of the industrial labour force (and not, as occurs elsewhere, just the *first* one) has grown up in and been educated, socialized and trained in rural villages. That is to say, the social and economic costs of creating a large labour force are paid for by the rural villages. They do not impose a heavy burden on the urban and industrial economy. Second, rural villages provide an informal, but extensive and stable, security net for the labour force engaged in the global factory. Workers who become sick, tired, unemployed and/or disabled can return home to the villages. The villages provide opportunities for alternative employment, and if these are lacking they can sometimes be created. People sometimes literally say: 'the land is our life insurance' (see Fei, 1939:181 for historical background). Third, the villages continue to take care of the wives and children; the negative side of this phenomenon is known as the 'left behind women and children' (Ye and Wu 2008; Ye and Pan, 2008). The circular nature of rural migration reduces the need to provide additional (social) infrastructure, such as school facilities, extra housing, medical care, etc. in the urban areas. In short, the 'costs' of urbanization are significantly lower.[4] Finally, the circular process (of rural migrants returning to the villages) implies avoiding much of the huge expenditure associated with building a pension system and care facilities for the aged and infirm.[5]

At the same time, the circular pattern also has potentially positive effects in the countryside. The (temporary) outflow of people helps to reduce the huge pressure on the land (Eyferth, Ho and Vermeer, 2003:5). The remittances that flow from the urban centres to the countryside greatly help to improve living conditions in the countryside and to sustain farming (Janvry *et al.*, 2005; Taylor *et al.*, 2003; Rozelle *et al.*, 1999b; Saith, 1995).[6] They also help to enlarge the internal Chinese market for 'wage goods'. Some of the returning migrants will invest their savings into agricultural activities, thus helping to further develop agricultural production (Wu and Meng, 1996a and 1996b).[7]

Evidently, the set of positive externalities linked to the circular flow of part of the labour force is very dependent on the maintenance of carefully tuned balances at different levels of Chinese society and its economy.

In contemporary China, circularity is a multi-level phenomenon. It exists at the macro level in the very large group of rural migrants who contribute to both industrial and agricultural development. It is estimated that China has a total of 225 million migrant workers (figures from the end of 2008). Of these, 85 million work in rural regions (other than their own region) and the remaining 140 million work in urban centres. Seventy-one per cent of this group is concentrated in coastal areas. At the micro level the circularity is manifested, at least partly, as multiple job holding. Circularity and multiple job holding do not completely overlap. Many people engaged in multiple job holding (the combination of agrarian and non-agrarian activities) carry out their non-agrarian activities (mining, industry, trading, transport, etc.) within their own village or its immediate vicinity. The possibility of doing so has been increased considerably through the decentralization policy for industry known as the TVE (Township and Village Enterprises; see Kung and Lin, 2007). When multiple job holding interacts with local industry, it evidently does not translate into rural migration. It is, though, an important expression of the circular flows that link agriculture and industry. Equally, many of the rural migrants, who leave for extended periods, remain linked with farming through a range of different arrangements.

Multiple job holding

The circular pattern of rural migration is closely related to the widespread phenomenon of multiple job holding in rural villages in China (Vermeer, 2003). This phenomenon has been a constant, albeit highly differentiated, feature of rural life in China (see e.g. Fei 1939:182, 236 and 285; and 1945:44–5). In recent years it has evolved into one of the *main axes* that links and partially explains both rural and industrial development (Ash, 1998:298–301). This axis allows rural people to obtain the funds they need to improve their living standards and to invest in the rural economy. At the same time, it provides the industrial sector with a relatively cheap and dedicated labour force. In this way it generates a multiple use of resources which induces an overall reduction in cost levels in the economy as a whole (Saccomandi, 1998).

The existence of the particular axis between the urban and the rural can be explained by the overall policy context in China, which has, at the moment, far less 'urban bias' (Lipton, 1977) than most policies elsewhere in the developing world. Since 2004 especially, China's rural policy has been oriented towards solving the '*sannong*' problems (the three rural problems): the need to develop agriculture (*nongye*), the need to simultaneously improve the living conditions of the peasantry (*nongmin*) and the need to augment the strength and quality of the rural economy and rural society (*nongcun*). The *Hukou* system (as unpopular as it is) has been instrumental in this respect as well. It has helped to create, frame and maintain the circular flow that is at the heart of the axis that links Chinese agriculture and the 'global factory'. *Hukou* and *sannong* have provided some of the balances that regulate the overall circularity.

Recent research (Ye and Pan, 2008; Ye and Wu, 2008; Ye and He, 2008) highlights some aspects of the circularity. First there is a daily circularity. This takes the form of farmers who, alongside their farm business, engage in another non-agrarian source of income within the same village (or its vicinity).[8] Second, there is a pattern that can be summarized as short-term circularity, with people returning to their home village each year for a certain period. Thirty-nine per cent of rural migrants working in urban industries and construction sites (and who have children in their home village) return at least twice a year to their village of origin to visit their children and parents and participate in agricultural activities. Fifty-two per cent return at least once every year. The length of the stay in the village varies considerably. For 78 per cent it is between one week and two months. The remaining 22 per cent of the return visits cover a period of between 3 to 5 months. Third, the same studies reveal what we can refer to as long-term circularity. Thirty-four per cent of urban migrants move away to work in the urban economy for up to 5 years; 36 per cent remain active in the urban economy for 5 to 10 years.

Long- and short-term circularity combine into a dazzling array of arrangements. In this respect it is interesting to note that, on the whole, women are away from their villages for fewer years than men. Data on men who leave a wife behind in the village reveal that 30 per cent of these men stay outside the village for less than 9 months per year. The *number* of visits that they make per year can vary considerably: 18 per cent go back at least four times a year and 25 per cent at least twice. Finally, there is data on the grown-up sons and daughters of 'left-behind elderly'. Twenty-three per cent of these people work between 6 and 8 months per year in another place. The rest of the year they are in the village with their parents. Fifty-seven per cent stay in the village for between 1 and 2 months and only 20 per cent stay away for more than 1 year at a time. It is probably needless to say that these frequent backward flows also imply considerable cultural change in the villages. As Leslie Chang observed, rural migrants find the cities offer them freedom 'from constant socializing and negotiation; [from] a conservatism that has been going on for a long time and [that] will continue after you are gone' (Chang, 2008:293). In the cities the factory girls become 'individuals' (ibid., 333): when they return to their villages they do so with a very changed cultural repertoire.

Daily, short-term and long-term circularity imply different time horizons and different spatial patterns. The longer the distances involved, the greater the probability of long-term circularity. However, there is no direct or fixed relationship between the organization of time and the organization of space. Empirically the patterns are highly differentiated and one type of circularity often flows into the other. Long-term circularity (implying rural migration) might be used to make the savings needed to create future daily circularity (characterized by permanently living back in one's own rural village). What is important is that all the different patterns relate, in one way or another, to multiple job holding and, especially, to the way that the household economy is socially constructed.

Circularity in everyday life

Our empirical research in Sanggang village showed that for the people involved in both farming and non-agrarian activities, 'multiple job holding' represents *a unity of two non-comparable, but supplementary domains*. Figure 2.1 shows how farming and what people call 'economic activity' (or 'business', or a 'job', or, if it is located elsewhere, an 'outside job') are two distinct domains that people combine together. The folk concept of 'economic activity' does not include farming. 'They are different and they are separated' according to one of our respondents. 'Farming' is, in the perception of villagers, primarily oriented toward securing family food consumption. Second, it is meant as a kind of life insurance. Only in third place is it meant to generate money. Most villagers recognize that one can earn 'some money from the land', but usually not enough to satisfy all the needs of the peasant household. Hence, some additional 'economic activity' is needed.

When discussing the notion of a 'good farmer' with villagers they stressed two qualities:

> A good farmer is to perform well in the work. A good farmer knows how to produce very well. But a good farmer must also take care to provide a good income for the family, so a good farmer also engages in other jobs.

This underlines the *unity* of 'farming' and 'economic activity' as perceived by the villagers.

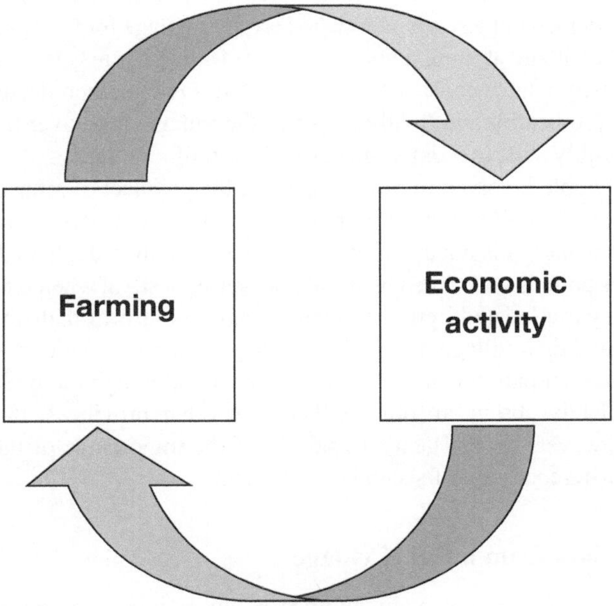

Figure 2.1 Folk concepts used in multiple job households

Just as farming in this Hebei village embraces a highly differentiated set of combinations (gardening, crop production, animal breeding, fruit and nut production, etc.), the notion of 'economic activity' also covers a wide variety of possibilities. These range from working in the big enterprises of the 'global factory', working in park maintenance in Beijing, trading in local and regional markets, iron ore mining, working in local factories refining iron ore, transport activities, road construction and calcium production to being the owner of a shop, restaurant or an iron ore refinery or working as a local craftsman. These activities imply different time horizons, different locations and they provide very different levels of income.

Although the two domains are unequal, Figure 2.1 illustrates that they are *linked* to each other through different feedback mechanisms. Involvement in 'farming' allows for engagement in 'economic activity' and involvement in 'economic activity' can also strengthen farming activities. As argued before, these circular patterns imply highly contrasting time horizons. People may shift between the two on a daily basis; for example, by working in their fields in the morning and then in the 'other economic activity' (whatever it is) in the afternoon. Sometimes the circular flow has a seasonal character (this coincides with short-term circularity). Some people will dedicate their time to farming during peak labour periods (e.g. land preparation and sowing in the spring and harvesting in the late summer) while being employed in other jobs during the low season. Clearly, the daily and seasonal flows imply different spatial patterns of organization. Rural industries, with flexible working schedules, are essential for the daily cycle. The 'seasonal' circularity allows a flow of migrants to larger regional centres. Finally, there is long-term or multi-year circularity as well. This pattern is mostly adopted by young people (mainly men but also women) who leave the village for long periods. Often they leave the village at quite a young age (i.e. 16 to 18 years) to work for some years in industry and/or construction, accumulate savings, then decide to marry, try to find a job nearby and finally return to the village, take over their parents' land and possibly shift to a daily or seasonal form of circularity.

For most people involved in rural migration, the prospect of returning remains an important option. For them, rural migration is not an *exodus* (as has occurred in other continents). It is not a definite adieu to the countryside, the village and/or farming. This point is underlined by the annual spring festival when most migrants, wherever they are located, return to their village to reaffirm their linkages with their family and their village. In the Hebei village, where we did this research, the daily and seasonal patterns of circularity dominate, although the multi-year type is not completely absent. In other villages, in other provinces, this might be different. However, for the general analysis of the importance of these circular flows such differences are of secondary relevance.

Empirical data from a Hebei village

Figure 2.2 summarizes the main findings of an enquiry into the structure of household economies in the village. The data for this research was collected in November 2008 and May 2009. The second period was especially interesting. It coincided with the sowing season. There is a notable contrast between the two periods. In May the fields are full of men, many of whom are absent from the village during the rest of the year. In spring they actively engage in the preparation of the land and in sowing and planting. Afterwards they return to their 'economic activities', sometimes located within the village, sometimes hundreds, if not thousands, of kilometres away. In these two periods we interviewed representatives of fifty-two households.

Figure 2.2 shows different income levels and the degree to which they are based on farming or non-agrarian activities. An observation located on the dotted line indicates that 50 per cent of the overall income (monetary and non-monetary) is derived from farming and the other 50 per cent from what villagers call 'economic activity'. Figure 2.2 shows that the majority of the households derive most of their income from non-agrarian activities, although farming almost always contributes to the household economy. This is not only from wage labour. Several households developed non-agrarian entrepreneurial activities in trading, mining and small industries. In this respect Saith (1995: 225) had already referred to 'a fresh opportunity created by the reforms'. He also signalled that, in China as a whole, millions of people followed this route.

Only seven out of the fifty-two households are not directly engaged in non-agrarian activities. These are mostly elderly people who receive at least some assistance, in the form of help and/or remittances from their children. On the other hand, there are just two cases in which there is no income from agriculture. These households voluntarily made this choice handing over their land to neighbours and friends.

The majority of households in this village depend on both agrarian and non-agrarian incomes: 81 per cent of households are engaged in multiple job holding. From a quantitative point of view the contribution of farming can range from minimal to considerable (in 24 per cent of all cases the contribution of farming to overall income is more than 50 per cent). However, from a qualitative point of view even a minor contribution should be considered to be important.

If there was no contribution from farming, and all the household's food requirements were bought in the market, the income available for other goods would be reduced considerably (due to the price difference between self-produced and bought food)[9] and the perceived quality would also be considerably lower. Several respondents who, from an economic point of view, do not have to grow their own food spoke of growing food for 'freshness, quality, taste and purity'. The gender dimension might also play a role here: vegetable growing (or more generally, gardening) is usually a woman's domain (although men quite often give a helping hand during land preparation and sowing).

Figure 2.2 also shows that there is considerable differentiation in overall income: 15 per cent of households have an overall income of at least 30,000 yuan/year, while 25 per cent earn 10,000 yuan/year or less. In most better-off households, the income derived from non-agrarian activities dominates over the income derived from agrarian sources. For poor households (< 10,000 yuan/year) the opposite tends

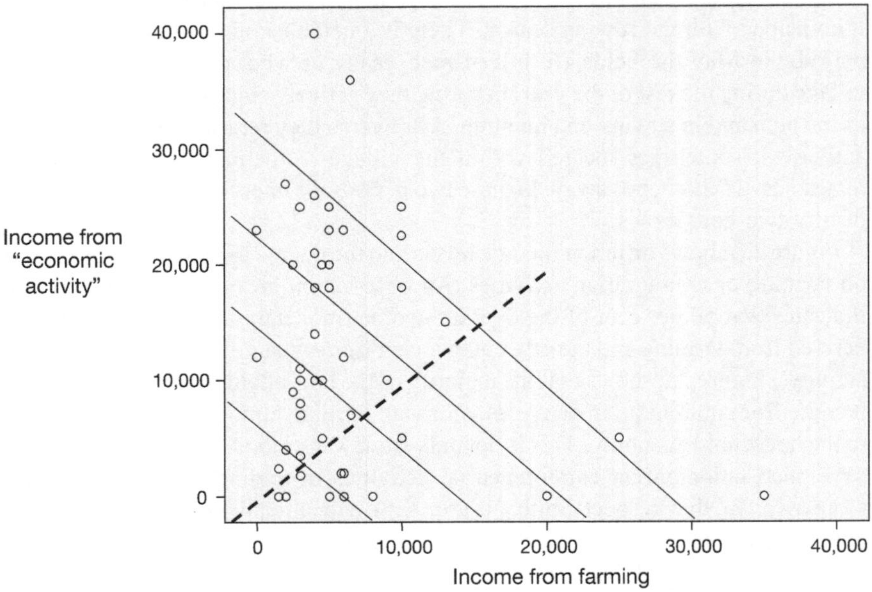

Figure 2.2 Income levels and sources of income in multiple job households[10]

to be the case. There are three households (6 per cent) that derive a relatively high income (> 20,000 yuan/year) mainly or exclusively from farming activities. These households have been able to go considerably beyond the average amount of land per family (3 to 5 mu: 5 mu = one-third of a hectare) either acquiring an additional piece of land, or through engaging in fruit production and/or animal husbandry. These agricultural activities allow a household to go beyond the strict limits implied by having a given amount of land.

Figure 2.2 describes the dynamics of multiple job holding in today's Chinese economy. If peasant families want to increase their incomes beyond the quite rigid boundaries of the limited land available to each family, they have to engage in non-agrarian activities.[11] Engagement in non-agrarian activities offers a form of emancipation. However, as many such non-agrarian activities (especially working in the 'global factory') contain high levels of risk, it is important for the participants to maintain a fall-back position. Thus, the multiple job holding family is repeatedly being reproduced, the more so since both family and gender relations contribute to such reproduction.

Meng (2014) and Hu (2014) did similar village studies in Jiangsu and Sichuan Provinces (n=100 and 225 respectively). Identical patterns emerged, although the specific distribution over the space defined with the axes used in Figure 2.2 differs slightly between the villages. Hu distinguishes four categories (2014:195). For the first category total agricultural income[12] constitutes more than 80 per cent of total household income. Twenty-one per cent of households fall into this category. In a second category the agricultural income fluctuates between 50 and 80 per cent.

This is the case for another 21 per cent. The largest category embraces 39 per cent of the village population: here agricultural income contributes 20 to 50 per cent of total household income. Finally, there are the non-farming households (<20 per cent). They make up 18 per cent of the village population. In the village studied by Meng only 7 per cent of the men were only farming, but even in this group everybody had previous experience of labour migration. Forty-five per cent of the women had migrant work experience (2014:54 and 51).

Table 2.1 (derived from Ye and Pan, 2015) gives data at a macro level. It shows that the relative contribution of agriculture as *part* of total per capita income is declining. At the same time though, the *absolute* contribution of farming is increasing substantially. Taken together this implies that without the many small peasant farms and without the multiple job holding strategies China would be in big trouble.

It is remarkable that all this very Chinese-looking data is, as a matter of fact, not at all unique to China. A wide range of empirical studies (that, however, have had hardly any influence on political discourses) show that on the majority of farms in the West, the man, the wife, or both, have a job outside the farm. Weersing *et al.*, for instance, report on United States Department of Agriculture (USDA) data for dairy farming which shows that 'almost 60% of [. . .] farm households had either the spouse, operator, or both, employed in off-farm work and approximately 80% had higher cash incomes from off-farming earnings than from their farm' (1998:128). Based on survey data they also note that 'maintaining the dairy farm business was the most frequently cited reason for off-farm employment of operators' (1998:138) and that multiple job holding is 'a flexible mechanism for coping with changes in the economic environment' (1998:127). In line with this finding, Kimhi and Bollman conclude, in an analysis of 24,288 farm records in Canada and another 20,122 farm records from Israel, that 'exit probability decreases with the extent of off-farm work' (1999:69).

Table 2.1 Composition of net income for rural residents in China (yuan per capita)

	Year			
Index	*1990*	*1995*	*2000*	*2009*
Net income	686.3	1577.7	2253.4	5153.2
1. Wage income	138.8	353.7	702.3	2061.3
2. Income from household operation	518.6	1125.8	1427.3	2526.8
2.1 Agriculture (cultivation, forestry, pastoral and fishery)	456.0	956.5	1090.7	1988.2
2.2 Secondary industry (manufacture, construction)	21.3	48.2	99.4	164.5
2.3 Tertiary industry	41.2	121.2	237.2	374.1
3. Property income and transfer income	29.0	98.3	123.9	565.1

At the empirical level multiple job holding (or pluriactivity) is the rule rather than the exception in both the East and the West. The difference resides elsewhere. In the paradigm that is dominant in the West, multiple job holding is considered to be an anomaly, a temporary disgrace. It represents, at best, a (gradual) shift towards the assumed end of farming. In the Chinese view, it is taken as a fact of life – it is how agriculture is structured. However, as we will see in Chapters 5 and 6, the Chinese view is increasingly being displaced by the Western paradigm.

Multiple job holding describes a very broad spectrum of differentiated livelihoods. Here we focus on the commonly shared feature: that the link with farming is maintained *over* time, or, as also occurs, is restored *after* a long period. Young people who work in nearby cities will return home to help their parents with land preparation and harvesting. Those who work far away (in the 'global factory') might return after they have accumulated enough savings and/or when the father dies and somebody has to care for the house and the land. Of course, not all rural migrants who enter the 'global factory' will return to their villages and re-engage with farming in association with another, suitable form of economic activity. However, those who do will:

- return with considerable resources (not only savings, but also new knowledge, experiences, crafts and skills) that might be of considerable help in further developing the local rural economy and the quality of rural society;
- help ensure that it is not only elderly people who remain in rural villages, which would considerably decrease agricultural production and productivity and the vitality of rural areas;
- contribute to the strengthening of the rural development process.

In synthesis, while rural migration initially shifts relatively cheap labour towards the industrial centres of the country, it also shifts wealth and development prospects back to agriculture and the countryside. Thus, multiple job holding links and combines industry and agriculture, both in macro-economic terms and also in the biographies of the actors involved.

Different cultural grammars

The two domains, which together constitute the economic basis of the peasant households of this village in Hebei, each follow their own distinctive logic and in a way reflect the opposed, but combined attitudes, of 'contentment and acquisitiveness' discussed by Fei 60 years ago (1945:84).

The farming domain is dominated by social values. These values include the already mentioned importance of having self-produced and fresh food. Equally they stress the importance of living in the village, of taking care of one's family, home and the land and having 'space to breathe'. The village is also valued for the security it provides as a fall-back position. This applies not only to social relations, but also to food availability. Older generations remember the years of terrible hunger and thus retain a particular attachment to farming and maintaining a modicum of autonomy and insurance against exogenous dangers and threats. Together these values reflect the centrality of the local; these values make 'place' an important issue in the minds of local people.

The farming domain is primarily a non-commodity domain. While there are many transactions in and around this domain most of these are non-commodity exchanges. They are socially regulated. Labour is exchanged on the basis of reciprocity, additional land is obtained 'for free', loans – if needed – are obtained from relatives and/or neighbours. Seeds, fertilizers and pesticides are mostly bought on the market (although self-selected seeds, manure, green fertilizers and traditional techniques for pest control are also used). Compared to the total value of production, the costs of these external inputs are relatively low (Zhao and van der Ploeg, 2009). In addition, they are mostly paid for by money obtained in the other domain, the one of 'economic activity' and thus they enter the domain of farming as non-commodities. They have already been paid for with the farmers' own money (*zi ji de qian)*. They are not '*she zhang*' (goods that are still to be paid for). This concept of 'own money' is frequently used in the discourses of villagers. It means that there is no need to recover (*huan zhang* – literally, to return) the costs through the production and selling of agricultural commodities.

Nonetheless, considerable parts of agricultural production might still be sold. However, this only occurs after the household's needs have been met (family consumption, animal feed, seeds for the next cycle and gifts for relatives and friends in order to sustain important social networks) and socially defined needs are met after exchange in kind (for instance exchanging certain amounts of corn for specific amounts of wheat or rice). Any remainder can then be sold. This can be a considerable quantity that provides relatively large amounts of 'own money'. However, the market is not the organizing principle that orients the main decisions about production. It is instead the household (and its needs and prospects) that functions as the organizing principle. A telling example of this is the explanation from one farmer who planted 3 mu of corn and only 1 mu of peanuts:

> Corn is far more useful. You can eat it, you can feed your animals, because if you have to buy the grain to feed them it is impossible to keep any animals whatsoever. You can sell corn, or exchange it for wheat. You can burn the stalks in the stove, or you can use them for fertilizing the land. Peanuts are only useful for eating and for oil. Peanuts are less useful.

In the domain of economic activity there are, to some extent, competing values. Looking for money and making an income are central, but obtaining one's own money is also very important. This often involves a step-by-step approach in which the aim is to avoid relationships of dependency and the associated risks. The step-by-step construction of a set of self-owned resources is also perceived as very important (i.e. buying a tractor or a lorry with one's own money).

Commodity transactions are central to this domain of economic activity. However, these commodity transactions remain embedded in a wider framework of values that stresses the importance of obtaining, maintaining and enlarging autonomy. The centrality of 'own money' and the step-by-step approach frame major decisions. Social networks are therefore highly important. They can help, for instance, in dealing with the authorities when seeking to obtain licenses and permits.

On synergy and competition: the perception of land

Although the two different domains operate under distinct logics and through different cultural repertoires this does not imply that they are independent from each other. On the contrary, the two influence each other through several interrelations, some of which are *synergistic* while others create *competitive* tensions between the two domains. The synergistic relations have a positive impact on agriculture, while the competitive relations have a negative impact. The balance between synergistic and competitive influences varies and this variability, as we will later argue, poses a major challenge for the future of China's agriculture.

The synergistic and positive relations, which provide benefits to both domains, are primarily grounded in *multiple resource use.* That is to say, particular resources are used not in one, but in both, domains. According to neo-institutional economics (Scherer, 1975) this implies a reduction in the costs of a particular resource *within each single domain* (precisely because they are divided over two domains) and creates synergy between the two.

Possibly the most typical example is the three- wheeled tractor with a heavy diesel engine. Such tractors are the workhorses of modern Chinese peasant agriculture and are almost the icon of contemporary rural China; they are often purchased with savings ('own money') obtained from work in the 'global factory'. Within the village they are used for transport to, from and within the farm (e.g. bringing manure to the fields, bringing the harvest home and then taking a part of it to the market). But alongside the main uses in the domain of farming a tractor will also be used in the domain of 'economic activity' (e.g. for transporting rocks with iron ore from the mines to the refinery, for road construction, trading, etc.).

The same principle of multiple resource use also applies to the labour force, which can be simultaneously engaged in farming and in local industries, or road construction, mining or whatever. The village inhabitants indicated that a key condition for this is that the working schedules in the local industries allow for flexibility (and are close to people's fields and homes). This allows people to do their farming jobs at the appropriate times. Social networks and knowledge (e.g. of rural markets) can also be considered as important resources that can be used in both domains, thus strengthening the economics of multiple resource use.

In this chapter we have already underlined the *macro level* synergy that exists between Chinese agriculture and the 'global factory', a synergy that is embodied, as it were, in rural migrants (and especially those migrants who return to the countryside). Alongside this macro level type of synergy we can also see a *micro level* synergy that is rooted in multiple resource use at the local level. Here again it is the *flows* that go from one domain to the other – and vice versa – that are decisive, just as it is at the macro level. Rural migration is, diachronically as well as synchronically, the key to both types of synergy.

Yet besides this mutual *strengthening* of both domains, which results from actively created synergy, there are also competitive tensions between the domains of farming and economic activity. These can *weaken* one of the two domains (most usually farming). These competitive tensions are most prone to emerge when labour time becomes limited. In our interviews with the villagers in Hebei province (all of them involved in multiple job holding) we asked them to comment on two contrasting situations that are summarized in Figure 2.3.

Most peasants from the village indicated that the first example is the most favourable one (66 per cent opted for example 1; 26 per cent for example 2; the remaining 8 per cent argued that it would depend on the specific conditions). This strong preference for the example that implies *intensive* farming (characterized by high yields) reflects the peasant nature of agriculture. This becomes more evident when the arguments underlying the preference for the second example are analyzed.[13] The respondents mostly said that:

> having 2 mu of land is of course far better. When production per mu is lower, then you have to improve the quality of the land and use more fertilizer so that you can get a similar yield and, increase production to 2,000 jin.

For many respondents this was a very self-evident position. It clearly reflects, from an analytical point of view, the limited availability of land in these areas (hence 2 mu is better than 1). It also reflects a typical peasant strategy: improving and strengthening the resource base of one's own farm unit. This is done, in this case, through improving the quality of land (through various strategies such as better land preparation, applying more manure, green fertilizer and chemical fertilizer and maybe improving irrigation or even digging a well, etc.).[14]

The analysis of the explanations for preferring the first example reveals that, alongside the typical peasant grammar (in which high yields figure as an important objective), a new one is emerging. Although it is, as yet, not clearly articulated (at least not in this Hebei village) it clearly reflects *competitive* relations between the domain of farming and other activities. It does not centre on the yield level but *on the comparatively small amount of land* (as expressed in the first example). Some people said that '1 mu is of course far easier to manage'. Others that 'working 2 mu requires more time, and it is better to use that time for economic activities, that can provide you with more money; the production from the land is not very

1 mu of cropland used for corn	**2 mu of cropland used for corn**
Yield is 1000 Jin/mu	Yield is 600 Jin/mu
Total production is 1000 Jin	Total production is 1200 Jin

Figure 2.3 Villagers' farming preferences

much'. Some also said that the advantage of the first example is that 'you have fewer costs'.[15] As a matter of fact some of these respondents (heavily engaged in marketing) had de facto reduced their farmland to 1 mu by renting out – 'for free' – the rest of their land to other farmers. They would have preferred to receive a monetary rent but 'since they are neighbours and relatives we cannot ask money for it'.

What we are witnessing here is the emergence of a new strategy that signals a considerable shift from the initial *balance* between farming and economic activities towards the *dominance* of the latter. This is associated with a *deactivation* of agricultural production.[16] In general, the villagers recognise that 'people nowadays focus less on agriculture; they focus less on the land'. This is giving rise to a new phenomenon, which the villagers call 'simplified farming'. It involves eliminating several steps in land preparation: double cropping disappears, external inputs displace the role of labour (e.g. herbicides are used instead of manual weeding and insecticides are replacing biological traps or protective paper sacks wrapped around the ripening fruits) and 'demanding' crops such as wheat disappear altogether. This deactivation is graphically illustrated in Figure 2.4. This figure (projected onto Figure 2.2) indicates the proportion of the total maize crop (the most important crop in the village) that is sold.

Poor and old farmers with no outside jobs sell an average of 18 per cent of their maize harvest. Farmers with non-agrarian sources of income have a larger harvest and sell a larger part (53 per cent). The higher the non-agrarian income, the more is sold (the percentage might increase to 67 per cent). However, there are other, contrasting, interrelations to be noted as well. In some instances a higher off-farm income combined with a relatively low agrarian production, translates in lower

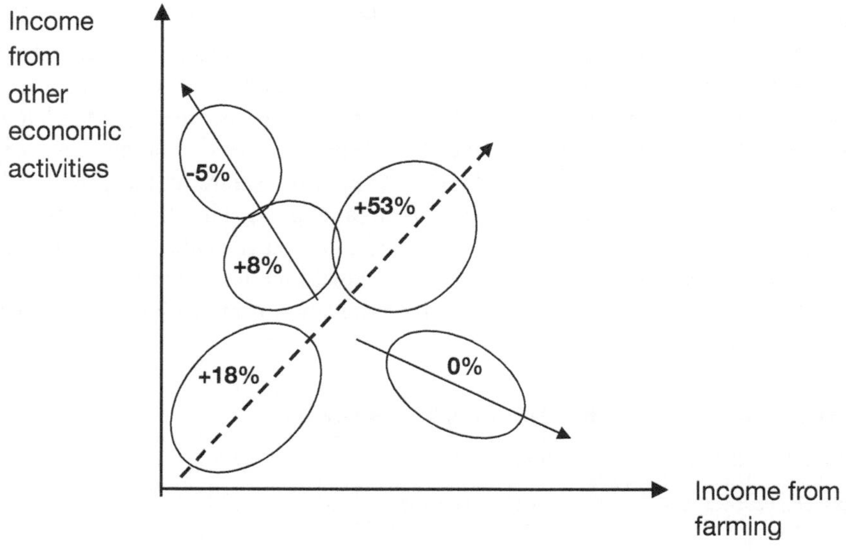

Figure 2.4 Part of the maize crop that is sold on the market

marketable surplus (just 8 per cent) and even into the need to acquire corn in the market (here the marketed surplus is minus 5 per cent). This pattern can be seen as a deactivation of agricultural production: some families following this pattern even deliberately reduce their farmland. A similar pattern can emerge as a result of increased specialization (cattle, goats, fruit trees). Farmers pursuing this path only grow maize for self-consumption (including feed for the animals); they don't have a marketed surplus. In theoretical terms, Figure 2.4 also shows that 'subsistence' is far too simple a concept to apply to the changing patterns that can be found among multiple job peasant households and those engaged in deactivation.

Differential development trajectories

In more general terms it might be argued that four contrasting development trajectories can currently be identified within this Hebei village. These are shown in Figure 2.5 which is also based on Figure 2.2 and involve the following trends:

- the deactivation of agriculture (*qunongyehua* type)
- the pauperization of rural households (*fanpin* type)
- the strengthening of peasant agriculture (x*iaonong* type)
- the creation of larger and more specialized farm enterprises (*dahu* type).

A deactivation of agriculture might occur when there is a generalized indifference towards agriculture and local food production and/or when prices of agricultural products become so low that it is hardly worthwhile continuing to farm as part of the multiple job holding portfolio. Pauperization occurs when the social networks that sustain the countryside erode. For instance, when access to another economic activity disappears, and/or children stop assisting their parents in agricultural work and/or stop sending remittances. Compared to the average situation (symbolized by the circle in Figure 2.5) pauperization implies a considerable reduction in the overall available income. It is not only the income derived from non-agrarian activities that disappears; agricultural production also decreases sharply, since it is no longer supported by the synergetic effects of multiple job holding.

There are three trajectories that might result in an increase in overall family income. They have the potential to result in comparable income levels (as shown in Figure 2.5). We will discuss these trajectories in the following two sections.

Until now farming in this village has developed along the lines of peasant agriculture. Intensification (based on the quantity and quality of labour and on the improvement of land, seeds and irrigation, as well as on the increased use of fertilizers, etc.) has been one of the important constituents of this development trajectory. As in Chinese agriculture as a whole (Ash, 1998:265–75; Gulati and Fan, 2007) yields have been continuously improving in this village. Maize yields, for instance, increased from 300 to 400 jin/mu in the period 1966–1974 and to 800–1000 jin/mu currently (2 jin = 1 kilogram). The average yield of sweet potatoes grew, in the same period, from 2,000 jin/mu to 3,000 jin/mu. Soybean yield doubled from 200 to 400 jin/mu. Multiple cropping (more harvests per unit of land during

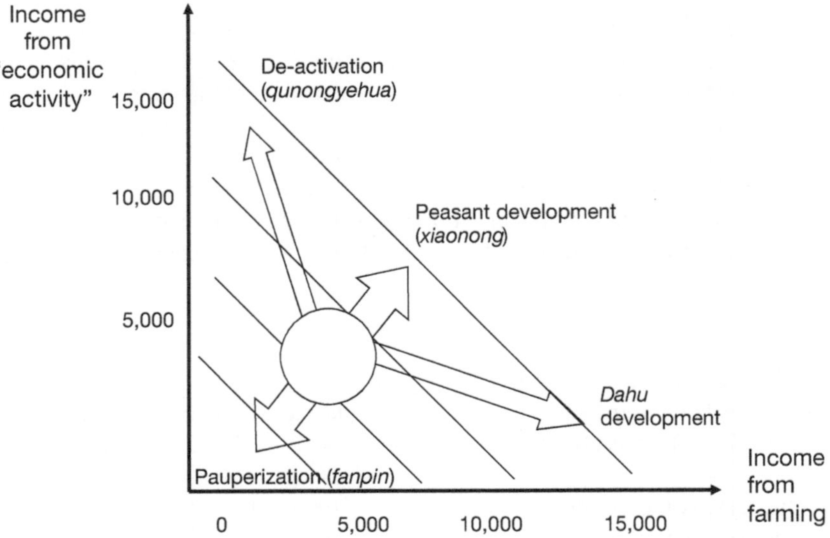

Figure 2.5 Potential farm development trajectories in Hebei village

one year) equally increased. Alongside this many-sided process of intensification (and the improvement of resources that is part of this process), multiple job holding and a strong orientation to local values and relations have been other important ingredients of this development pattern. This process was made possible through the Household Responsibility System reforms and was then strengthened and supported by an intelligent rural policy that represented the opposite of the urban bias which has created wreckage in so many other countries (Lu, 1997; Karshenas, 1996). The recently introduced support for grain cultivation and agro-forestry is just one expression of this policy. The same applies to the elimination of taxes, improved access to medical care in the countryside and recent changes to land law.

Is any further peasant development possible? We cannot give a definitive answer to this question here (we will return to this question in Chapter 4). However, it is telling that peasants in the village see a range of possibilities. These include:

- *The enlargement of herds* (notably goats). This possibility is interesting because it implies the exploitation of a thus far unexploited resource: the surrounding hills that are now covered with shrubs. This, therefore, represents an escape from the limits imposed by the availability of arable land in the river valley. The enlargement of goat herds may also increase the quantity of manure available for improving soil fertility in the crop lands. Promisingly, one recent state intervention in the village aims at introducing more goats.
- *Planting more fruit trees.* This approach involves adding new resources to the peasant economy. Some peasants already have suitable areas in the small

valleys among the hills for planting trees. These could be expanded considerably (a new strategy for planting trees in the *hills* will be discussed in Chapter 4).

- *New, higher yielding, varieties.* This prospect evidently depends on research and development and on the interactions between researchers and practitioners, which is decisive in developing varieties that are adapted to local conditions.
- *Introduction of glasshouses* (of the type encountered near Beijing, which are made out of an earthen wall and a plastic cover). This would imply a further intensification that strongly builds on the experience and know-how obtained in gardening.
- *More irrigation facilities:* at present only some of the crop land is irrigated.

The feasibility of such possibilities will depend substantially on market conditions and rural policy. Strengthening multiple job holding (which indirectly strengthens peasant agriculture) may also follow other, unorthodox, ways. As some peasant women in the Hebei village observed, most, if not all, 'economic activities' are only available to men. They often involve rough and dangerous jobs, such as road construction, transporting ore from the mountains to the iron refineries, working in the refinery, etc. However, many rural women would also be eager to engage in economic activities, provided that these match their needs and possibilities (see also Meng, 2014). Hence, *stimulating forms of rural industrialization that could provide light and flexible work* appropriate for rural women would also be a desirable policy.[17]

In the years that followed our first empirical inquiry (2008–2009), several of the possibilities indicated by peasants themselves have been realized – some individually, others collectively. The number and magnitude of herds has expanded. Several hills surrounding the village have been 'reconstructed' (see Chapter 4) to allow the planting of walnut trees. This is the outcome of a joint effort of the village committee and individual peasant families. New, higher yielding varieties have not (yet) been introduced but several peasants started experimenting with organic crops, free-range eggs (see Chapter 7) and the direct delivery of 'peasant food' to consumer groups in Beijing (this is discussed further in Chapter 10). Finally, the village committee organized and financed the construction of a new irrigation canal and an additional water reservoir (with electric pump) and pipes going to the hills. Bad weather (and/or design errors) led, however, to the collapse of the reservoir.

Back to the question of farm-size

The context in which Chinese peasant farming is embedded implies that it is virtually impossible to make judgements about farm size as such. The average Chinese peasant farm is part of a wider household economy, characterized by multiple job holding. Total household income might be quite satisfactory (especially since 2004 when the exchange relations between town and countryside started to improve) and agricultural activities contribute to this, just as labour migration does. Within the framework of multiple job holding a small sized farm might be the most appropriate. It delivers what the family needs (these needs range from fresh, home-grown food and additional income to social security and pension for old age) just as it allows the household members to engage in 'other economic activities'. The latter, in turn, help to maintain and improve the agricultural resource base (see also Chapter 4).

Alongside multiple job holding, other contextual elements also help to explain the performance of units that, in the eyes of the 'experts', are too small. As discussed before, reciprocity is an example here. Mutual help, the exchange of seeds, manure, tools and equipment strongly reduce monetary costs, just as informal credit (within the family, clan or other social networks) helps to avoid financial costs.

As Figure 2.2 showed there are also farms that do not have, or have hardly any, non-agricultural income. In Sanggang village this came down (in 2008 and 2009) from around 14 per cent to 8 per cent. This resembles the 7 per cent level reported in Meng (2014). Hu (2014) reported a level of 21 per cent. At first sight such farms seem to represent an anachronism if not an anomaly. While they are small the framework of multiple job holding seems to be absent.

However, it is here that the time dimension pops up and provides a key strategic context. These farms are highly intensified. Building on an improved resource base they are able to produce, *solely from the farm*, incomes that are identical or even superior to those obtained through migration work. And they are able to do so *because* they used their savings, obtained through *previous* labour migration, for investments in their farms (in Chapters 4 and 7 we will unravel the mechanisms underlying this).[18]

In short: when multiple job holding and the circularity of town–countryside relations are taken into consideration, it is clear that one cannot talk about Chinese peasant farms being 'too small'. They are integral and well-functioning parts of a more comprehensive household economy. Beyond that, multiple job holding generates savings which in turn are used to vitalize peasant agriculture. As Xu observed (in an econometric analysis of the 1979–1996 period): 'the most important factor [explaining the growth in total factor productivity in agriculture] may well be the expansion of non-farming activities.' (1999:119)

What applies to single rural households, their land and their agricultural activities also applies at higher levels of aggregation. Due to circularity labour migration does not function as a drain. Instead, it helps to generate an important flow of income (and other resources) to the villages just as it helps to avoid depopulation. For the urban economy, circularity provides a flow of cheap, and constantly renewed, labour.

A key factor in all this is, of course, the *balance* of the two sides involved: the urban and the rural. If income disparities and/or differences in living conditions and the quality of life become too large, circularity may turn into the one-sided rural exodus that has had such devastating effects elsewhere (Davis, 2006). This balance is the real issue at stake, not farm size taken in isolation.

Notes

1 This chapter is an updated and expanded version of an article previously published in the *Journal of Peasant Studies*, 37(3):513–30.
2 This characteristic is further discussed in Chapter 11 of this book.
3 This is different for those who have graduated from the universities, who most definitely do leave the countryside. They only return to visit their relatives and/or join the many newly-blossoming forms of rural tourism.
4 We refer here not only to public spending, but also to the massive social costs associated with the formation of slums.
5 Nguyen and Locke (2014) observe and discuss the same relations but present them, from a gender perspective, as negative phenomena.
6 Already in the late 1930s and early '40s Fei strongly argued in favour of a dense interaction between agriculture and decentralized industry: 'it is [. . .] essential to widen the industrial opportunities of the masses.' (1945:310)
7 This also occurred in the context of the cross-border migration process within Europe from the 1950s onwards. This strongly dynamized rural areas such as Tras-os-Montes in Portugal and the region of Krakow in the south of Poland. See also Fuller, 1975.
8 From an analytical point of view this closely resembles 'pluriactivity' as it is defined and practised in Europe (Bryden *et al.*, 1992).
9 The family labour used in the production of self-produced food is not considered as a cost. Beyond this, the margins and transaction costs associated with food trading are avoided.
10 The unit is yuan. When collecting this data, 1 US$ equalled 6.8 yuan. The income contribution of farming has been calculated by adding the economic value of all the agricultural produce as if it had been sold. However, in practice nearly all or at least a considerable part is consumed by the family or used for gifts (see e.g. Zhao and van der Ploeg, 2009).
11 On the whole, people in this village are convinced that a couple (man and wife) can work, with their own labour force, only a maximum of some 5 mu when most of the work is done manually. This could be raised to 10 mu if most of the work were mechanized. This limit on the land to be worked explains the skewedness of Figure 2.2: when total incomes tend towards higher levels, they are necessarily derived from economic activities other than farming.
12 Including self-provisioned food and gifts distributed through social networks.
13 Alongside the choice itself, the associated narrative (the explanation) provides useful insights. During the field research all the respondents were asked why they made their particular choice. This often provides highly informative explanations that reveal the 'grammar of farming' (see Ploeg, 1990 for an application of the same instrument in a different context).
14 From a more general point of view it is telling that none of the respondents replied that the second example *as such* (i.e. without any further improvements) would be superior to the first one. This would reflect a typical entrepreneurial view of farming (see e.g. Ploeg, 1990, 2008). Such a view is absent in this village where the peasant view and the associated mode of reasoning (or 'logic') dominate.
15 Some respondents related their choice to weather conditions. The second option would be better if there were plenty of rain, but with a drought the first scenario would be better.
16 Not only because agricultural production per family is lowered, but also because no (or few) improvements are made to the quality of land as this is too time-consuming. The transfer of the land to others will not compensate for this: those who rent land do not normally invest in improvements as the short-term time perspective involved in renting does not justify such long-term improvements.

17 In countries such as Portugal, Italy and Brazil such rural industries (e.g. shoe making, textiles, food processing) have played an important role in regional rural development.

18 Wang and Fan (2006) are more pessimistic. They argue that only 'failing' migrants return. According to our data this is not the case in Sanggang village. It is the 'successful' migrants, with sufficient savings, who return and take up farming again or start another local business. We discuss this further in Chapters 4 and 7.

3 Man and the land

The social organization of farming[1]

Jan Douwe van der Ploeg, Ye Jingzhong and Pan Lu

Land has multiple meanings in the peasant economy of modern China. This chapter[2] argues that the three-generation structure of the peasant family and the circularity of labour migration ensure that land remains a central and non-substitutable resource. This implies that the oft-articulated thesis that migrant work represents a definitive adieu to farming is a fallacy. Peasant workers remain attached to their land and consequently return to it despite labour migration. In the absence of their husbands, women care for the land together with their fathers-in-law, and in so doing they sustain both the productivity and continuity of agriculture. We also argue that a further softening of the Chinese *Hukou* regulation will not trigger a massive rural exodus. The strong relations between peasant households and the land suggest that this will not happen.

Paradigmatic issues

The paradigm that governs most agrarian sciences and policymaking strongly conceptualizes modern agriculture as men's work.[3] Men do the heavy work; they deal with complex technologies and are the main decision-makers. Women, at best, do 'work of the second kind' (Rooij, 1992; 1994). They assist the men, they give a helping hand. And when responsible for specific tasks (e.g. feeding the calves, accountancy), they follow guidelines formulated by the men.

This paradigmatic view of rural women's role is, of course, very much a virtual image – and it is only partially true. In many parts of Europe, women run the farm because the men are working elsewhere (Plas and Fonte, 1994). And, if the men are at home doing most or all of the farming work, it might very well be that the earnings obtained elsewhere by the wife make this 'centrality' of the man possible. Farmers' wives also have a strong, albeit somewhat hidden, say in decision-making (Ploeg, 2003). Women also often play a prominent role in the new fields of activity that are being created as part of new, multifunctional, farms (Rooij *et al.*, 1995).

When viewed from the perspective of this paradigm, Chinese agriculture seems doomed to being second-rate: men in the prime of their lives are absent from the farm. Farming appears to be mainly the activity of women and old, retired, men. The young and middle-aged men are away working in the cities. When combined with the very small acreage of the farms (on average one-third of a hectare),[4] this seems a highly fragile constellation.[5] As Huang (2011:459) summarizes: 'most social science theory and the currently powerful Chinese ideology of modernization assume that with modern development, family-based peasant farm production will disappear.' If it were not for legal restrictions (the *Hukou* which obliges migrants to return to the village to receive social services), there probably would be far fewer people in the countryside and agricultural production would decrease proportionally. In this view, a softening of the *Hukou* rules would trigger a massive rural exodus. If this were to happen China's best option would be to spur an accelerated process of scale enlargement. This, it would seem, would be the only way that China could feed itself with a greatly reduced agricultural labour force.

In this chapter we argue that this particular and quite widely held view (as exemplified in Quisumbing, 1994; Saito *et al.*, 1994; Cheng, 1998; L. Zhang *et al.*, 2006; Fan and Zhu, 2007; Li, 2009) is at odds with the way in which farming is perceived, practised and organized by the Chinese farming population itself (in this respect we follow Zhang and Donaldson, 2013; see also Meng, 2014). This view also ignores the very strong relations that tie peasants and the land together.

The chapter is based on twenty-four family biographies that we elaborated together with household members of different generations between 2009 and 2012. These villagers come from three villages located in a hilly area in Hebei province. They draw on extensive comments made by the villagers about their trajectories through life. Special attention was given to the questions of who worked where and for what reasons? The history of the family farm and the relations between the family, its farm and fields were also carefully documented. The interview fragments in this chapter are derived from these biographies.

The social calendar of peasant life

During life the relations between people and the land change several times – and each time they change significantly. This can be schematized in a three-stage periodization.[6] First, when a boy, the family's land[7] will be worked by the boy's mother and grandparents (who might be assisted by the boy's father when there are heavy tasks to be done, such as land preparation). The land will feed the family and generate a small surplus which can be used to pay for daily expenses. Most probably, the boy's father will be working far away. He does so in order to 'feed' the land (for instance paying for the considerable expense of fertilizers) and to save money for big expenditures, such as paying for the construction of a house (for when the young man has grown to the age of say, twenty-five) and his wedding. At the end of this period, the boy (now grown) will already be doing migrant work – unless he is studying.

This first period is followed by a second one in which the young man himself, now married and with one or more children, is almost permanently absent from the village and engaged in migrant work. His wife enters the social category of 'left-behind women' (Ye and Wu, 2008). During this second period which, roughly, extends from the ages of 25 to 50, the family farm's income (generated by his wife and parents) is complemented with monetary income from elsewhere.

'Feeding' the farm and saving for big expenditures are now the responsibilities of the man (previously the boy). Before it was his father who did so, now it is the grown man himself. He does this because, as is said in the countryside, 'the family is always the first priority'. In the second period this implies being engaged more or less permanently in migrant work, even though it is hard for the man to be separated from his wife and children. Equally his work can sometimes be dangerous, monotonous and risky (sometimes the boss fails to pay). As one villager explained:

> For a household with a migrant worker, it is easier to have a good balance of money and food ... Nongmin [peasants] can eat well but they have less monetary income, they have to exchange things to obtain money ... But usually that is not enough for the education fees, the weddings and other social fees and the farm inputs [. . .]. We pay for the farm inputs and the other big expenses with the money earned from migrant work.

When the strength of the man declines, the third period (from, say, age 50 to 75) starts. He comes back to the village and starts to work the land (probably assisted by his wife and daughter-in-law). Yuan Juanwen (2010) applied a cohort analysis to different groups of migrant workers who originated from rural Guizhou. She concluded that: 'none of the couples in the survey said they would not come back [. . .]. Most migrants never think of migrating to the city permanently.' (2010:140)[8] The same was observed by Lou *et al.* (2004), who studied the migration experiences of young women from Sichuan and Anhui. In this third period the returned men try, if possible, to develop the land further so that, as the countryside saying goes, a 'foundation' is created for his son (now involved, in his turn, in migrant work) when he returns in the future. For the elder man this is the period of his pension. But he does not pass it idly. He works and develops the land – that is how the first priority, responsibility for the family, materializes. When the third period finishes (schematically at the age of 75) his own son (now already aged 50) will come back. He and his wife will care for his parents who will be increasingly unable to continue with the hard physical work – and he will care for the land that his father developed for him. In this period possession of the land passes to the son: 'sons are both entitled to inheritance and required by both customary expectation and by law to provide maintenance for their aged parents.' (Huang, 2011:469) Formally, girls and women can have possession rights as well. However when a young woman marries, the rule of patrilocality implies that she moves to the village of her husband and thus her (formal) rights pass to her family, i.e. to her father and then her brother(s). In practice, the (allocated) land belongs to the family household and within the family it passes from one generation to another. This practice is also increasingly codified in China's steadily evolving legal system (Huang, 2011:475).

The reciprocal relations that govern the third period are explained by one of the villagers as follows:

> A good son must treat his parents well and take care of the land even if farming gives less money. This maybe a sacrifice for the son, but he does so out of respect for his parents. [If the son returns] the father will be very happy. If the son wants his father's share [i.e. the part of the land allocated to the father], the father will also be happy. Normally the son provides compensation: often 1 sack of rice and 2 sacks of wheat flour every year. This happens when there are several sons. If there is only one son the rule is that the son gives his parents whatever they need for food.[9]

But the third period is not only about food production per se. If there are possibilities, the returning man will develop the land (i.e. extend and improve the resource base):[10]

> All the men who come back develop the land; those who come back want to take good care of the land and develop it [. . .]. At the moment there are some twenty men who came back, who are now developing the land, making new ditches in the hills, planting walnut trees, developing fruit yards, breeding new herds [. . .]. I want to leave a legacy for the next generation; newly developed land is the economic foundation for our children.

Another villager told us:

> For sure, my sons will come back when they are older. They will have less strength and the companies will not want to employ them anymore. They have to come back then [. . .] and have to rely on crop planting, but crops only render a little money, so I am now developing this contracted[11] land, which I have planted with walnut trees . . . to earn money [zheng qian: literally 'money earn']. In this way I can sustain myself, I am not a burden for the children, and when they return they can sustain themselves with the walnut trees as well.

Evidently, there will be many variations in this sequence. The indicated ages (the boundary indicators of the different periods) might differ considerably in real life, while the notion of migrant work changes over time. Often, youngsters go to distant industries or construction sites; later on they will work in nearby county towns (if possible) and finally they may commute on a daily basis to a nearby mine or factory. Another shortcoming of this brief synthesis is the focus on men (and the absence of daughters). Equally, the story does not apply to youngsters who go to university and get a good job somewhere in the urban areas. There might also be adverse circumstances that block the unfolding of this idealized typical pattern. Nonetheless, describing the life cycle as having three successive periods (and the associated changes in the relations that occur between men and the land) captures well the realities of many rural people. It also provides a basis for some major issues that we will discuss later.

Each generation passes through these three successive periods. Each generation is first '*fed* by the land', then provides assistance to the family in order to work the land ('*feeding* the land' by providing the money needed) and, finally, it works and further *develops* the land. Thus each generation makes its own dance through time. However, no generation does this in an isolated way. As Figure 3.1 illustrates, there are interactions between the single trajectories. For example, while my generation is being nurtured by the land, my father's generation is nurturing the land (directly and indirectly) and my grandfather's generation is working the land (and further developing it). Thus, the three generations are tied together by a complex set of complementary rights, duties, tasks and expectations that together compose a moral code in which the land is central (see the ovals in Figure 3.1 in which each oval represents the unity of grandparents, parents and children).[12]

Figure 3.1 also illustrates that the land is constantly at the centre of this moral code. There is not one simple usufruct relation (between the land and those working it). There is, instead, a complex set of relations that embraces three generations (generally with several people per generation). Each individual will be cared for and receive good food in the first period. At this time these youngsters are supposed to dedicate themselves seriously to their schooling, give a helping hand in the fields if needed and obey and respect their parents (and grandparents). As an adult (i.e. in the second period) there is a duty to earn enough through economic activities other than farming the family plot. This is done in order to maintain the family, to make farming possible and to save the money needed for the education, house and wedding of the next generation (i.e. the children). These duties go together with the right to return when labour migration comes to an end (for whatever reason) and to use the family farm as a pension. This marks the beginning of the third period. At this time new duties and rights emerge. These include the

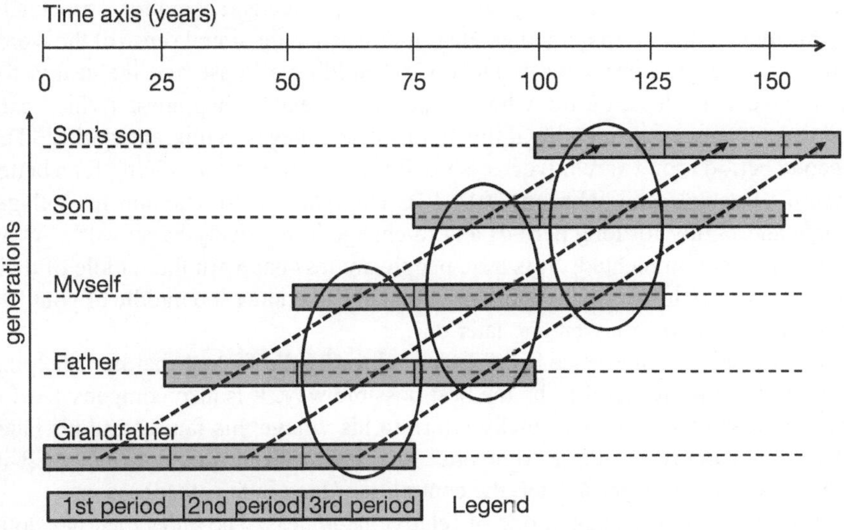

Figure 3.1 Land, the three-generation family and interacting life cycles

duty to further develop the land (if possible) and the right to be supported by the son(s)-and daughter(s)-in-law if or when the workload becomes too heavy.

Thus, the multiple set of duties and rights implies many relations between people and the land. These mutual relations change over time (according to a well-established pattern). The relations with the land also specify reciprocal relations between the people involved: relations that specify processes of giving and taking over extended periods of time. At the same time these relations (and the associated rules) govern the destiny of the land: its use, development, and fate. What is done with the land and the way land is understood and valued, is governed by the logic that reigns within the family – and definitely *not* governed by the logic of the market. The same set of duties and rights closely interlinks the life courses of father and son. In this respect Elder (1974) referred to 'interlocking trajectories'. For rural families, land is the connecting point that weaves together the different trajectories.

The three-generation family (me and my wife, my parents and our children) is not necessarily a unit where living, working and eating together is central (as was the case in the past). It is more a complex set of rules and conventions that function as a familial (as opposed to individualistic) land–labour institution, which systematically embraces three generations (and often relates symbolically to ancestors as well). Current world-views would see such a three-generation family as doomed to be replaced by a nuclear family (consisting of just me, my wife and our children) and expect the familial values to be replaced by more individualistic ones. However, both the three-generation family and the associated values 'retain a major position in modern Chinese society as well as the legal system, in sharp contrast to modern Western experiences, even as China's GDP has become the second largest of the world' (Huang, 2011:460).

The successive periods have different emotional loadings, just as they imply different mutual obligations. The first period is one in which the young boy (and later on the adolescent and the young man) is protected and cared for – especially by his mother and/or grandmother. He is nourished in the literal sense of the word, but also, later on, receives assistance with building a house and the search for a suitable wife. It is, on the whole, a period of relative happiness (which can, however, change considerably if the mother also engages in migrant work).[13] The second period is one of relative hardship. It is the period of 'sacrificing for a better future' (Eversole and Johnson, 2014:12). The man has to abandon his village, home and family for long periods and to engage in relatively harsh work. As an elder villager commented: 'It is true, people are less happy in the middle of their lives, they have to face many burdens. You have to endure the middle of your life in order to have a happy ending later on'.

Of course, this endurance is compensated for by the feeling that he is helping his family in what seems to be the best possible way. It is also compensated for by the prospect that he will finally return to his village, his family and his land. We are not implying here a 'romantic' love for the land; this does not exist in China. As we will discuss later, the centrality of land is for other reasons.

The third period is, again, one of relative happiness. The elder man has done his duty and now merits his pension. If additional money is needed, his son will help (that is his duty). But usually the man, now old and a grandfather, is proud not to have to ask for such assistance. Through farming he secures his economic independence (thus allowing the children to dedicate themselves to their own duties) and is even capable of constructing a 'foundation' for his son when he returns.

Figure 3.2 combines the temporal patterns in which both father and son are engaged and also illustrates the associated emotional loadings. The vertical arrows in this figure indicate how support flows from one actor to the other, thus contributing to emotional well-being. The figure illustrates how the life cycles of father and son are tied together to form a whole that repeatedly expresses a balance – but one based on alternating roles. Each individual experiences periods of relative well-being and others of relative deprivation. It is, however, only when the lives of father and son are analysed in their totality, i.e. as a *dual unity*, that the overall balance (shown in Figure 3.2) emerges. In her study on farming in Africa, Sarah Berry (1985) concluded that 'fathers work for their sons'. The same is true in China where fathers also work for their sons, but through a different and far more complex modus operandi: by 'feeding the land', so that the son himself might be nurtured, and then by 'working [and simultaneously developing] the land'. And then the son, having been nurtured by the land, starts to nurture the land itself (through migrant work) and finally develops it further (through land consolidation and improvement). In short, the different actors (and different generations) partly articulate to each other through the land. One cares for the land so that the land

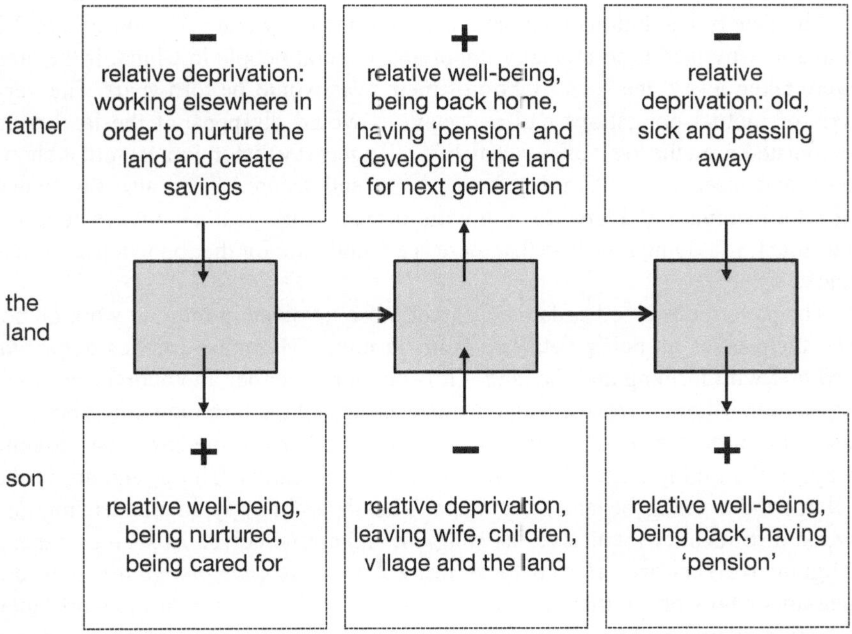

Figure 3.2 Balancing father–son relations through the land

can later benefit the next generation – just as one is able to enjoy the land because others (the previous generations) took care of it. Of course, things can be experienced differently. As a student at the College of Humanities and Development Studies at China Agricultural University commented[14] in an autobiographical context:

> Yes, the son often helps his mother and grandparents in farming. So the boy experiences the hardship of it and his parents and grandparents keep telling him that he should study hard and go to college or university so that he can move out of the fatigue and lack of future of the countryside [. . .]. He understands his father's decision to work far away but he misses him very much. And every time his father returns, he is told that he should study very hard to deserve his father's dedication. This strengthens his determination to study hard and move out of the village. Gradually the boy comes to hate the land and he never wants to work on the 'damned' land in the future.

Although this opinion is diametrically opposed to much of the story told so far, it *shares* the emotional loading of the man–land relations. Man–land relations are not neutral, based only on a utilitarian economic value. The associated emotions are central in the ordering of peasant life; just as strong emotions are present in the decision to create an escape from the land. These emotional loadings may very well change over a lifetime. It is possible that negative feelings developed during youth encourage labour migration, but that these change, as a consequence of hardship experienced during migrant work, into positive feelings later on.

The deeply institutionalized pattern synthesized in Figure 3.1 and Figure 3.2 explains why land is so crucially important for rural people in China. If the land were taken away, the social fabric of their lives would be torn apart. The very axis of mutual obligations and expectations would disappear if the land were eliminated from the logic of everyday life. Villagers say that 'migrant work is short-term and insecure, but farming is long-term and secure.' They also say 'when you lose strength and have to stop with migrant work you can always continue farming [. . .]. doing it well so that there is a foundation for the son when he comes back.'

The pattern discussed so far shows that those involved in migrant work do not see themselves as being detached from farming. Migration implies a *specific relation* with farming and the land – it is definitely neither a temporary nor final adieu to it. Migrant workers leave in order to support the farming family; but they leave in order to return. All this is expressed, perhaps in the strongest possible way, in the multiple use of the word *nongmin* (peasants). Those who are in the village and working the land are *nongmin* – but those far away, involved in migrant work, also identify themselves as *nongmin* and are perceived as such by others. Migrant workers are referred to as peasant workers (*nongmingong*): they are peasants who work in the city – but regardless of the location of the work, they are and remain peasants. Thus, farming and migrant work are expressions of one and the same livelihood – of being *nongmin*. But the two expressions are, of course, clearly distinguishable. A relatively young man,[15] previously engaged in migration work, but now investing in farming (and thus 'developing the land'), explains:

> A peasant is ongoing, there is continuity, it is long term, you have to endure . . . You have to invest, you need dedication, you need effort. By contrast, a migrant worker is about immediate profit . . . they get their money and that's it. You cannot harvest without effort and investment . . . farming is about long term securities. Farming is also, if not especially, about value adding [zeng zhi].

He added: 'That is why I want to have my farm instead of being a migrant worker. It allows you to progress and to reap the fruits of this progress yourself . . . .'

The pattern that ties the different generations and the land together also explains why villagers hardly ever abandon farming (in the worst case they ask others to temporarily care for the land). It is repeatedly argued (also by villagers themselves) that farming alone does not provide a decent living or cover 'big expenditures', especially during the second period in life. Migration is needed in order to enlarge the total cash flow. However, the potential extension of this argument (to abandon farming permanently and dedicate oneself completely to industrial and/or urban jobs) hardly ever occurs and certainly not to the extent that it does in many other parts of the world. Clearly, the *Hukou* regulation plays a role here but we do not think that *Hukou* in itself provides sufficient explanation for the non-abandonment of land. We will come back to this crucial issue later.

Land and food consumption

The dual urban–rural system is partly maintained by the *Hukou* system, which undoubtedly is a significant institutional factor that hinders the definitive migration of the rural population. However, returning to farming is also a voluntary choice made by *nongmin* themselves. It is not just related to the notion of further developing the land (as discussed above). It also clearly relates to the self-provisioning of family food. Peasants usually talk about this by comparing themselves with city dwellers:

> In the village we can survive and live well without spending even one cent. Grains for staple food are produced by the family and exchanged for flour with the trader. Vegetables, eggs and pork are all produced by ourselves and we don't need cash to buy them. In the city everything is different: cash is required everywhere and you have to buy everything with money. In the supermarket you can easily spend 100 yuan on only a few items even before you realize it. In the village we plant what we would like to eat and we also feel safe eating the food we have produced ourselves.

These sentences reflect two of the key advantages of farming – high autonomy and food quality. Peasants feel proud that they can satisfy the basic family demands for food without being dependent on the market. Although urban people earn more money than peasants, they rely heavily on the food market and need a cash income to meet these needs. This means that they have to work hard (with such incentives their attachment to their work will be quite different from someone farming his own land). Due to their low dependency on markets, peasants can maintain a decent living with a relatively low cash income and do not have the same worries about food quality as most city dwellers. In modern society people are bound to, and dependent on, a host of external factors. For a peasant (especially one who knows how to work the land well), the land is an essential instrument for pursuing autonomy and well-being. Under current conditions this translates into the possibility to produce one's own food that is safe and of high quality.

The capacity for self-provisioning and autonomy in food consumption is a fundamental prerequisite and also an important feature of being a peasant. This autonomy can even be expanded to the city through the non-commercialized circulation of food between rural and urban areas. In the villages where we are currently doing research this pattern is noticeable, particularly for glass noodles and free-range eggs, as well as fresh fruit and vegetables (glass noodles are a highly appreciated food product made out of starch that is obtained from sweet potatoes through a lengthy artisanal process). The existence and maintenance of a rural connection is fairly common among urban people who are rooted in the countryside, especially between rural parents and urban-based children. For rural parents, a considerable part of harvested food products is for their adult children who are officially registered as part of the urban population. This also allows their offspring to reduce their dependency on the food market. Food coming directly from rural areas is always appreciated and cherished by this group because of its high quality and the awareness of the dedication that has gone into producing it. It is also highly valued as it allows them to take a temporary break from the market. Thus even those urban dwellers who have passed the *Hukou* threshold, still much prefer the consumption of autonomous food which they intentionally embrace whenever they have the chance.[16] In synthesis, the preference for and pursuit of autonomy illuminates peasants' deep connection with their farmland.

The search for autonomy also explains why peasants refuse, or are reluctant, to completely commoditize their food production. First and foremost, the production of food is meant to satisfy the basic consumption demands of each family – the primary obligation and duty of a peasant.[17] For peasants, food is polyvalent: it has physical properties and is a corporal necessity, it has cultural property when bestowed as a gift, strengthening intergenerational bonds, a political property when given as a tribute to powerful individuals and many other properties. When put onto the market and converted into commodities, food is reduced to a monetary value (the main property of a commodity).

Over the last decade the value of self-produced food, consumed directly in the peasant household and/or distributed as a gift in wider circles, has increased further. This is due to a combination of high levels of contamination of food produced for the market (which have given rise to numerous food scares) and the fact that peasants hardly use agro-chemicals (of whatever type) for self-produced food.[18] Both these factors considerably increase the extra value of self-produced food.

Land and the struggle for autonomy

As is the case for peasants all over the world, land is a precious resource for Chinese peasants. It is a *sine qua non*, an indispensable requirement. This applies to those who actually work the land, but it also applies to the migrant worker. Maybe it applies especially to him. The land provides his parents, his wife and his children with food (just as he himself will take a considerable amount of food back with him to the city every time he leaves the village). But the land is also his fall-back position if, and when, major misfortunes (accidents, sickness, mass closures of factories due to recession or whatever) occur and it is his pension when he gets older. Beyond all this, the land often represents a concrete opportunity: it can be developed further, thus allowing for the possibility of making a better living and better supporting the family. In short, the land (together with other indispensable resources, such as the labour force, water, knowledge, seeds, networks, instruments, small tractors, etc.) is a major necessity. It is the central axis of an autonomous resource base that helps the peasant family face up and respond to a difficult, if not harsh, environment (see Figure 3.3). It is worth repeating that this not only applies to those currently working the land, but also, and maybe especially, to the *nongmingong*, the migrant workers.

The available resource base is used for different purposes. The labour force, for instance, is not only used in agricultural production, household activities and local non-agricultural activities; it also engages in labour migration. All these activities

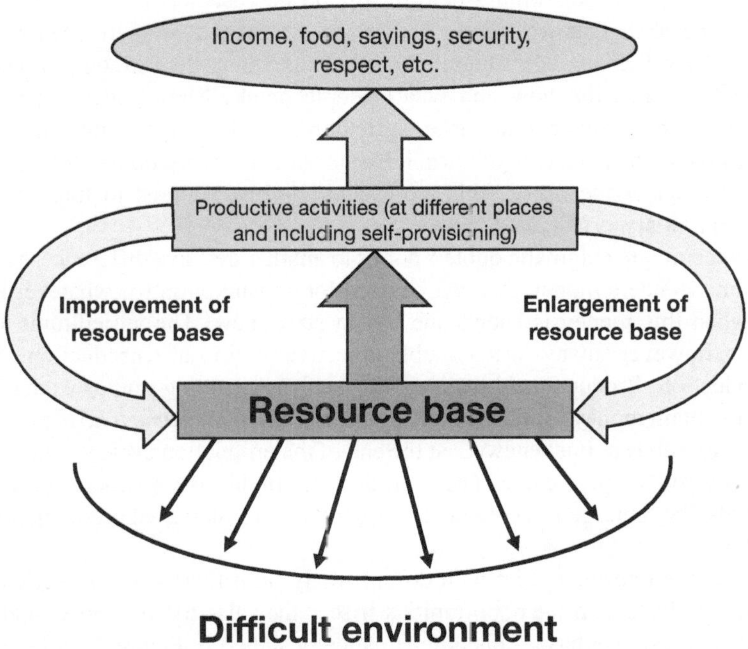

Figure 3.3 Having an autonomous resource as the basis to face a difficult environment[19]

are important and complementary. Together they secure food, income, long-term security and savings (to be used for weddings, houses, education, funerals) for the peasant family.

Land occupies the central place within this multi-faceted resource base. Its centrality and the way it is used is greatly influenced by the moral economy. Land is to be cared for. To have land and not to work it is seen as immoral. There are strong pressures, if not an outright moral obligation, to achieve a good production. At least, that is the way peasants perceive it:

> A peasant is supposed to care well for his land. There are some who go gambling and leave the land unattended. People say those guys are lazy men, and they do not want anything to do with them. They are *bu zhao diao*. That means that they are on the wrong road, they are deviating, they are lost [. . .]. Being a peasant is an ongoing struggle for a good harvest. When it is too dry or too wet you adjust your farming behaviour, trying to make the best out of it. Even the women, even when they are very tired, do what they can to get the best possible harvest [. . .]. But when you are talking about a good peasant you always have to take into account the weather and the family cycle. When the peasant has to dedicate a lot of effort to the problems of his children then he can dedicate less time to the land . . . .

There is a reciprocal relationship here as well. If a peasant works well, the yields will be high. That is, the land will be generous to him/her. It is about giving and receiving. Thus, a blossoming field or a good harvest says a lot about a hard-working peasant. This was nicely phrased by one of the villagers: '*Shi Zai Ren Wei, Di Zai Ren Zhong.*' This literally translates as 'things depend [on what] people do, land depends [on how and what] people plant'. More generally: success depends on the efforts people make, agricultural production depends on the care people give to the land. One of his neighbours added: 'Everything that I put in my farm will render me fruits; [this represents a strong contrast to migrant work because] a salary is just dead money.'

This seemingly enigmatic quote[20] points to another decisive difference. Migrant work only renders money that can be used for consumption (of whatever kind). And when this money is spent, one has to start anew. The agricultural labour process, however, always has a double objective: it aims at reproduction as well as production. Production is realized (for self-provisioning and/or exchange in order to obtain money for further consumption), *but at the same time the resource base is reproduced*. It is renewed: at the end of the production cycle it is not 'dead' but ready for the next cycle. The reproduction might take a simple form but it might also be 'enlarged', i.e. to lay the foundation for increased production in the near future.

Hence, the returning elder men do not solely limit themselves to agricultural production. Instead, if the opportunities arise,[21] they also try to improve and/or to enlarge the resource base. This is schematically shown in Figure 3.3 where agricultural production feeds back into improvements and/or enlargements. In the villages we have been studying this occurs through several mechanisms – e.g. terrace construction in the hills for walnut tree planting, the creation of herds, the development of orchards along the river and improvements in soil fertility. The technicalities of such improvements and enlargements are discussed in the following chapters – here we will limit ourselves to the underlying rationale.

'Developing the land' (an expression that is frequently used by the villagers and understood as an accolade for both making improvements and enlarging the resource base) is strongly related to the intergenerational pattern we discussed above (see especially Figure 3.1): 'Newly developed land is the economic foundation *for our children* [. . .]. I want to leave a legacy for *the next generation*' [italics added].

Monetary considerations and calculations are not absent from the strategic reasoning underlying the development of land. They are important considerations when comparing the alternatives. However, the justification for land development does not reside in the market as such. It centres, instead, upon continuity between the generations and the desire to hand over a 'stronger' foundation, i.e. an improved and enlarged resource base, to the children.

Working the land well and developing it further are, as discussed above, seen as moral obligations. These moral obligations have (and *sustain*) a counterpart: *the moral right to the land.* If a peasant family cares for the land, and develops it further, then it is morally entitled to the land. The land is theirs – not in a formal legal sense, but according to the basic rules of reciprocity. The strong socio-cultural bonds between man and the land that are actively wrought in this way also explain the tens of thousands of conflicts that occur every year in the Chinese countryside Walker, 2006; So, 2007). These conflicts occur as a result of repeated violations of what Chinese peasants consider as their moral right to the land. If the land is needed for purposes other than farming, peasants seek righteous compensation. It is telling that the peasant population more readily accepts expropriation by the state than land being allocated to private enterprises. Within the moral economy of the peasants, the state (still) has more moral authority than private enterprise.

At the micro level the attachment to the land reappears in daily life in many different forms. One telling expression is that the land is almost never left idle and that people do not sell the right to use it. If, for specific reasons, the family plots cannot be worked, other family members, friends or neighbours are asked to take up this responsibility. Normally there is no money involved in the transaction. It is just a temporary arrangement to take care of the land until the original users can do so again.

Land and the sense of belonging

Within the world-view of Chinese peasants, the relation between people and the land (i.e. the possession of an autonomous resource base that is worked and further developed with household labour – be it directly or indirectly) defines the contours of a social class (Webber, 2012:8–9). It is the axis that specifies a sense of belonging:

> All villagers are peasants, because they all have allocated land. You may have contracted land as well, but if you only have contracted land you are not a peasant. A big boss contracting 1000 mu of land is not a peasant. Such a boss, like other urban people, does not suffer what peasants suffer, nor enjoy what peasants enjoy. He does not belong . . . .

Somebody else observed: 'Urban people are not peasants . . . because they are not treated as peasants, they don't have heavy duties, but neither do they get the subsidies . . . .'[22] It is important, though, to note that this sense of belonging (or more generally, the class position) is not linked to specific fields (or to a specific property). The status of being a peasant is characterized by the possession of an autonomous resource base that one works and develops oneself. When discussing this issue with a village elder, he had this to say:

> I still can remember two pieces that formerly belonged to our family. Our family had quite some land in the past. The village was rich and my ancestors were among the better-off people. At that time people had a strong tie to their fields. There is also a family graveyard. It was destroyed during the Cultural Revolution but later on it was restored. It used to be on our family's land but this is no longer the case. It is not the case that we 'love our land' as you say European peasants sometimes do. Everything has changed here, there have been so many reallocations of land. That has been the great social transition of our country. Of course, we invest a lot in the land. But if we have to exchange it for another piece then that is alright as well, provided that the quality is equal or better. Then we won't feel disappointed. *We need land, not a particular field* [. . .]. When the commune was created we felt very depressed about losing our land, but now it is different.

Land is the cornerstone of the peasant's resource base. It is also a highly visible expression of autonomy. But we are talking about land *in general* as an indispensable part of the resource base and as a carrier of a degree of autonomy – and definitely not about a specific piece of land. Peasants wanting to improve their resource base might readily exchange one plot for another. Today's peasants do not define themselves by a specific plot, but by having direct access to the land, an essential means of production for them. That is what defines the peasant class and gives them a sense of belonging.

Land, the peasant family and continuity

The peasant family that works its own land is a land–labour institution (Pearse, 1975; Duncan *et al.*, 1977) that is remarkably stable over time (Mazoyer and Roudart, 2006). In China, just as in other societies, this institution has been 'impressively resistant to sudden change, though this does not mean that such structures were static' (Hobsbawn, 1994:320). In China it was even able to resist the two decades between 1958 and 1978, when production teams and communes dominated the countryside (Myrdal, 1965). This period only ended after the peasant revolution in Anhui in 1978 (see also Chapter 11). This revolution (and the subsequent formulation of the Household Responsibility System) restored the peasant family as the nodal point within the texture of village social life and the rural economy. The continuity between the pre-1958 peasant families and the ones that re-emerged from 1978 onwards is strong. Hobsbawn's (1994:335) observation seems to apply here more than anywhere else: 'Over most of the world the old textures and conventions, though undermined by a quarter of a century of unparalleled social and economic transformation, were strained, *but not yet in disintegration.*' (italics added)

Yang (1945:45), a Chinese anthropologist who did extensive fieldwork in the 1930s, described the peasant family before the 1950s as follows:

> A Chinese family, especially in rural China, is far more than a group of related individuals [. . .] it is a complex organization of family members, family property [including, evidently, the land], domestic animals, family reputations, family traditions, and family gods. It can be said that the family extends to the as-yet unborn generations and to the long-deceased ones.

And, with regards to the land, he noted that

> land is the most important form of property, for it belongs to all generations. It means much more than a piece of earth on which crops are cultivated; *it is the very foundation of the family.* Without land a family can never be settled and the family members will never have a sense of security. *People and land are the two pillars of the Chinese farm family* [. . .]. When we say a family is broken, we mean that the family's land is gone. Thus, land becomes a part of the life of the farmer and his family, and they have a very deep sense of attachment to it; their land is no less dear to them than are their children [. . .]. Land ownership also gives the farm family independent personality, spiritual inspiration. And a feeling of freedom.
>
> Yang, 1945:46, italics added

Although this was written over 70 years ago – years that saw unparalleled changes – these descriptions fit very well with the Chinese peasant family of today. As we argue in this chapter, people and land are, once again,[23] tied together through a dense and internally coherent set of rules that specify the relations between family members, the family and the outside world and, especially, between family members and the land. Land is once again a 'foundation'. While much has changed in the countryside Yang's observations remain valid: 'All three generations have a common interest in the family's economic security' (Yang, 1945:74) (note how Figure 3.1 shows the three generations being tied together through the land). And 'the continuity of a family line depends not only on having generation follow generation, but also upon the uninterrupted transmission of the family's common property. (Yang, 1945:81) This transmission assumes that

> if the property of the living generation is inherited from the previous one, then the present members are merely the stewards who keep it intact and hand it down to a new generation. Family ethics give a son the right to the property but also assign him an obligation.
>
> Yang, 1945:81

Labour migration and especially its circularity (leaving from, and in the end returning to, the village) can be understood as *re-emerging* (or *reinvented*) phenomena – rather than traditions that are disappearing. Yang (1959:71–2) observed that prior to the 1950s

> farming alone could not provide a livelihood for half of the villagers [. . .] life for the poorer half of the community was bleak and often impossible [. . .]. For these peasants, the solution lay in going beyond the economically restrictive confines of the village borders [. . .]. The emigrants were of two types: seasonal and long-term. Seasonal emigrants went to the city for a few months or shorter periods in a year, generally during slack periods at the farm. [. . .] Seasonal emigrants generally went to Canton [where] their predominant occupation was 'selling hard labour' (*mai k'oo li*, hence the pidgin English word 'coolie'). They worked as porters on docks, in warehouses [. . .] at pulling rickshaws and carts, at carrying and digging dirt [. . .]. Peddling vegetables into the city also engaged many villagers.

His namesake, M. C. Yang, drew attention to the circularity of these flows:

> In most countries, when men leave a rural community and find work in the city they settle there permanently with their wives and children and sever connections, for the most part, with their relatives back home. In China, however, most of the villagers who seek work in the city maintain close ties with their relatives and send their surplus earnings back to their homes to be used to buy land and build houses for the family. If they are married, the wives and children remain in the family home. [. . .] This custom makes it possible to have a society existing in a state of balance between city and country life. Such a population movement has advantages both for the village and the city. Returning villagers bring new methods, new ideas and new wealth to the village, and, since most of those who go forth to seek their fortunes eventually return, the countryside is not depleted by the lure of the city.
>
> Yang, 1945:229

Since 1978, the peasant family farm and the circular labour movements in which it is embedded, have once again become central features of the Chinese countryside. These features are not just contingent or provisional responses to temporary problems. They should be understood as the re-emergence of traits that are, as we have seen, deeply rooted in Chinese history (Fei, 1939; Huang, 2011). This re-emergence, and the underlying continuity, shows the strength of the peasant family farm as a land–labour institution that is able to face different problems, survive very contrasting periods and continually reinvent itself in order to cope with new circumstances. To expect such an institution to easily disappear would be ludicrous – to say the least.

Land, rules and flexibility

Although the historically persistent patterns (or 'social textures') that govern village life basically contain very clear sequences and well-outlined responsibilities, their application in real life is very flexible. Regarding the 'second period' (between say 25 and 50 years) a villager observed that:

> A good father has to construct a house for his son, be it in the village or in the county town. [But whether this is really done] depends very much on the family situation. If the father has limited capacities, then he cannot finance the construction of the house. Then the son might not complain. Beijing is another case. If they want a house in Beijing, that might be impossible. Prices are far too high there. Much also depends, on the wishes of the daughter-in-law.

The third period (from 50 to 75 years) is also characterized by the need to cope with specific circumstances:

> I have strong faith in my land, in my farm. If the economic conditions are good I can develop it further and will invest more. However, when the economic conditions are not good, providing no major disaster occurs, I develop the farm more slowly [. . .]. And the happy ending depends on many other things. If your children are married and have a house you do not have to face large expenditures anymore. But if the son is not yet married, then it is difficult to think about a happy ending.

We believe that the strength and endurance of these basic patterns lies in the very flexible way they are adapted in the face of highly variable conditions and situations. The moral guidelines of the Chinese peasantry allow considerable room for manoeuvre and help them avoid stark black-and-white choices of either sticking to rigid rules or rejecting them as impossible or unbearable. Each group of interlinked actors (see the ovals in Figure 3.1) assesses (often through lengthy processes of deliberation and negotiation) the best possible way forward. Thus, the many, heterogeneous micro episodes within which the man–land relations are assessed, lead to the reproduction of 'structural patterns' (Long and Ploeg, 1994). And, the capacity of the peasantry to flexibly adapt these patterns is a key mechanism in these overall patterns being reproduced.

Conclusions

Given the pivotal role of land in the tight web of relations, expectations and duties that ties rural families in China together, it is quite unlikely that Chinese peasants will abandon their land. It is equally improbable that migrant workers will cease returning to the countryside. Land is simply too important for them, and its importance is of a multi-faceted nature. It provides food, it promises security when there are problems elsewhere, it offers a pension, it allows for investing in the future of the next generation, it ties the extended family (three or more generations) together and, finally, it delivers the highly appreciated gifts (good, and often finely processed, food) that play a binding role in social relations. This tentative conclusion is supported by statistical data at national level. According to the Agricultural Census, China had some 200 million agricultural households in 2006. This was an *increase* of 3.7 per cent on the previous Agricultural Census (from 1996). Of the 200 million households with land, agriculture was the main source of income for 58.4 per cent, a decrease of 7.2 per cent on 1996.[24] In short, while other sources of income are increasing in importance, this does *not* translate in a diminishing number of households having land. Land remains important for households in China's countryside.

At the same time land is not a 'fixed' entity with narrow limits and its use-value does not automatically shrink to one of irrelevance during 'boom' periods. As we have shown above, during such times land is more likely to be developed as a resource for the future. There is also the phenomenon of a small, but increasing, number of adults returning from migrant work long before they reach the age of, say, fifty. They start to develop the land far sooner and create a living that provides them with an income equivalent to what they would otherwise obtain from full-time migrant work, while simultaneously building a legacy for their children. As we will discuss in Chapter 7, the emergence of new opportunities linked with activities such as agro-tourism, organic production and other types of multifunctional farming is strengthening this new tendency.

Within this context, *Hukou* (the obligation of rural residence and the associated benefits) evidently plays a role. However, we believe that its influence is largely over-estimated in the current debates about weakening the system. *Hukou* is, to an important degree, a formalization of underlying patterns that are deeply rooted in the Chinese economy and culture – as we have argued throughout this chapter.[25] Thus, even if *Hukou* is weakened or abolished, the underlying patterns will remain. It is even possible that an elimination of *Hukou* will be strongly opposed by parts of the rural population itself. For *Hukou* is a complex set of place-bound rights and obligations, part of which is that *every villager has the right to a fair share of the land.*[26] Abolishing the *Hukou* regulation could endanger this strategic right. Some students with rural backgrounds are already discussing the possible consequences of any change in *Hukou*. It is also telling that dwellers in towns and small cities are increasingly expressing a demand for land: even relatively small plots would be very useful in complementing their relatively low salaries and could help them avoid paying ever increasing prices for vegetables.

Since ancient times Chinese peasants have combined farming with other economic activities. They continued to do so during the period of collectivization. Afterwards this practice was continued through Town–Village Enterprises (see Kung and Lin, 2007) and, later on, through the massive migration to the large industries and the big cities with their informal economies and many construction sites. In the future for sure, this pattern will take new forms; probably through farms evolving into more multifunctional enterprises, more focused on the emerging needs of the urban population.

Be that as it may, the relations between peasants and the land are a cornerstone in the relatively harmonious town–countryside relations that characterize China (Long *et al.*, 2011). They help to create a balance[27] that cannot be eradicated without considerable social upheaval. The firm rooting of these patterns in the life cycle of rural people and the flexibility of those patterns suggest that such a balance will probably be maintained for many decades to come. This is supported by recent policy experiments in Chengdu (Sichuan province) where the *Hukou* regime was considerably 'softened' in order to better balance rural and urban development. In a critical review of this experience Ye and LeGates conclude that:

> It is noteworthy that this thorough reform did not have an immediate disruptive impact on Chengdu's economy and society. It did not lead to immediate mass migration into the city, the hollowing out of rural villages, or excessive demands on public services.
>
> Ye and LeGates, 2013:272

We want to finish this chapter with a methodological observation. In order to understand the dynamics and role of the Chinese peasant farm, and the three-generation family, research needs to depart radically from strictly Western world-views, and acknowledge that China is constructing a modernity that differs in several significant respects from the one that exists in the West (Long *et al.*, 2012; Long, 2014). It is not necessarily better or worse. The point is that it is and will remain different. This implies that a different (or adapted) paradigm is required to understand and represent the past, present and future of China: one that clarifies and highlights what remains invisible and/or irrational (and hence 'outdated') in Western views. This is especially true when it comes to farming and agriculture. Instead of continuing with ancient polemics about the demise or continuity of the peasantry, we would do far better to depart from Shanin's dictum that, above all else, the peasantry exists 'as a process' (1971). There are simultaneous processes of de- and re-peasantization, which articulate with each other in complex and often novel ways, giving rise to a plurality of forms and a wide spectrum of developmental tracks (Li, 2010). In this way the peasantry continuously reconstitutes itself. In the chapters that follow we will try to further uncover the peasantry as 'a process'.

Update

Travelling through rural China and talking with urban workers it is abundantly clear that alongside those who will sooner or later return to their villages and work the land, there are probably as many others who won't. Is that a problem? Of course not. As long as they have a good job, a good income and good housing that tie them to the city there is, on this side of the equation, no problem whatsoever.

It might be painful for their elderly parents, but in the end they will understand. And on the other side of the equation, in the village, it should not be a problem either, as long as the land passes to friends, neighbours or relatives and is cultivated, or returned to the village committee to be redistributed among other peasant households. It is indeed the case, as pointed out by Shanin, that the peasantry exists 'as a process'. Things change continuously. What is important is that the peasant way of farming is repeatedly reconstituted (largely because it is highly relevant to China – see Chapter 11). But there is also a fundamental flaw in this, a proverbial Achilles heel. The land being transferred increasingly doesn't go to other peasants willing to work more land, but ends up concentrated in the hands of others who are not making the best possible use of it (see Chapter 6). This is giving rise to a new agrarian question.

Notes

1 This title echoes an issue that was, and is, central to rural sociology. It is not intended to suggest a gender bias.

2 A previous version of this chapter was published as 'Peasants, time and the land: the social organization of farming in China' in the *Journal of Rural Studies* 36:172–81. The current version is updated and slightly expanded.

3 See Burt, 1987 for a classical explanation on how cohesion and structural equivalence contribute to paradigmatic framing.

4 This is also perceived as an indication of inferiority. Such a small acreage does not allow for investments. As a result, it is thought impossible to make such a unit highly productive (through the 'conventional' strategies of applying new technologies and high input levels).

5 Authors as diverse as Chen, 1987; Lu and Dai, 1987; Hinton, 1990; and Zhu and Jiang, 1993 have argued that the highly fragmented farm size blocks further increases in productivity and efficiency.

6 Yuan (2010) developed a four-stage periodization. The first and second stages she distinguishes are grouped here together in one single stage. Carter and McGoldrick (2005:384) arrive at a family life cycle consisting of six phases – each of which involve negotiations on 'the expansion, contraction, and realignment of the relationship system of family members in a functional way'.

7 Formally, women have the same rights to the land as men. Due to the rule of patrilocality, however, the land (i.e. the usufruct right) belongs, in practice, to the family of the man.

8 Before having a first child, man and wife sometimes leave the village together. This may also occur after the birth of a first child. In this case the grandparents take care of the child. There are strong regional variations here.

9 This comment was provided by an elder villager; younger people might very well be inclined to think differently.

10 Developing the land might mean two different things in China. It can refer to industrialization and urbanization but may also refer to investing in the land in order to increase its potential for agriculture. Here we take the second meaning.

11 Alongside the allocated land, peasant families may contract additional land from the village committee. This depends on the amount of land held in reserve.

12 It could be tempting to reduce the choreography illustrated in Figure 3.1 to mere 'tradition'. This would be partly true (Fei, 1939), but this pattern re-emerged from 1980 onwards, after the introduction of the Household Responsibility System (which reintroduced individual possession of land) and, especially, after the massive expansion

of migrant work in the 1990s. Consequently, this pattern is as much a product of modern times as it is tradition. It represents an adapted version of historical institutions.

13 This might be radically different when both parents migrate and the child is left to the grandparents. For an account of these 'left-behind children', see Ye (2011).

14 A draft of this chapter was distributed among and commented on by our students.

15 There is a clear tendency now for some young men, having accumulated sufficient savings, to return earlier to their village than others. They do so in order to start new businesses in their village. These often include 'new' agricultural activities: raising large numbers of pigs or poultry, developing fruit orchards, engaging in agro-tourism or in on-farm processing and direct selling etc.

16 For example in the apartment block where one of the authors live, people who live on the ground floor usually transform their little garden into a vegetable garden for domestic consumption and even some public flower beds have been reshaped for maize planting.

17 From the point of view of food security and nutrition, this is a strategic feature. In the case of China it is partly rooted in (or re-affirmed by) the large famines that struck the country in the 1950s and 1960s. It also relates to China's huge and decentralized food reserves. Every family has its own reserves.

18 Fertilizer is used for cash crops, not (or hardly) for producing food for domestic consumption.

19 Figure 3.3 is an adapted version of the 'choreography' described in Ploeg, 2008.

20 What we are dealing here with is, in more theoretical terms, the difference between capitalist commodity production (CCP) on the one hand and simple or even petty commodity production (SCP or PCP) on the other.

21 In the villages we are discussing the availability of additional land that can be contracted is often a decisive factor. The government's current policy to support the conversion of marginal agricultural lands (e.g. extensive and poor meadowland in the hills) into more productive land provides opportunities here. Yet the distribution of these newly emerging opportunities depends very much on local political relations. It may be a harmonious process, but it may also cause considerable conflict.

22 The subsidies related to e.g. grain production, or getting insurance paid by the state, are equally perceived as rights.

23 Texts that discuss features of peasant society, peasants' life and/or the peasant mode of production nearly always use the word 'still', when locating these features in time. It is refreshing that, when writing about Chinese agriculture, we can replace this expression with the word 'again'.

24 Data derived from www.gov.cn/gzdt/2008–02/22/content_897216.htm

25 We have to keep in mind also that arrangements such as the current Chinese *Hukou* have existed in many places. In the Netherlands, until the 1930s, it was forbidden to stay in the major cities without having a job and a proper place to live. Vagrants would be arrested and sent to workhouses in the countryside.

26 Another recently created right is that people working the land no longer have to pay taxes anymore (see Ye *et al.*, 2010).

27 From a theoretical point of view, we are witnessing here a kind of 'Chayanovian' balance – but one that operates at the level of society as a whole; it is not limited, as was Chayanov's original analysis, to the micro-cosmos of the peasant family itself.

4 Peasant-managed agricultural growth[1]

Jan Douwe van der Ploeg, Ye Jingzhong, Wu Huifang and Wang Chunyu

The impressive growth of Chinese agriculture is very evident at the macro level (see Chapter 1) but is also traceable at the level of the single village. As a matter of fact, without growth at village level, there would be no noticeable growth at the macro level. Take the villages in Hebei that have figured prominently in the previous chapters. Sanggang is one of these. In the years of the production teams and the commune (before 1978), 800 mu of land (15 mu = 1 hectare) was cultivated in this village – more than today. At that time there were 760 people living in the village and nearly everybody was working the land (the current population is below 700).[2] However, hunger was never far away, eggs and especially meat (sometimes only eaten once a year) were luxuries. The grain ration people received was 420 jin per person per year but only after a good harvest (2 jin = 1 kilogram). Beyond the food the village produced itself, it also produced a meagre surplus (in the form of peanut oil that was delivered to the state). This surplus amounted to around 5 per cent of total production.

Today, only 700 mu of agricultural land are cultivated,[3] and 170 households provide an agricultural labour force of 310 people (130 men and 180 women). Many households are also engaged in labour migration: 135 people, mostly men, are working elsewhere.[4]

Notwithstanding these reductions in the land cultivated and the labour force, agricultural production is now far higher. People eat far better, the menus are more varied and nutritious and include meat, eggs and fish: there definitely is no more hunger.[5] Beyond that, the village now produces much more surplus for the market than in the past. Estimates suggest that, between 1980 and 1995, total production doubled and in the 1995–2015 period it grew again by more than 50 per cent. These considerable increases have greatly contributed (together with labour migration) to a sharp increase in the well-being of the villagers. As villagers themselves say: 'there is now far more money in the village [compared to 1995]'. Most households have a television. One sees a satellite receiver on almost every roof – alongside the piles of drying maize. These two features symbolize a surprising combination of the old and the new, the modern and the traditional.

In this chapter we will discuss some of the mechanisms that explain the impressive and ongoing growth of agricultural production and productivity in smallholder villages like Sanggang.

Labour-driven intensification

Basically, Chinese agriculture is a peasant agriculture and agricultural growth is a peasant-managed process: it is rooted in the decisions taken within the many millions of peasant households about what to produce, by whom, how and for what reasons. Such decisions are taken in a contextual setting in which markets, policies, technological opportunities, ecosystems, gender relations, local and regional power relations and cultural repertoire all play important (and sometimes contradictory) roles. However, these contextual elements do not have a deterministic or unmediated influence in the places of production: they are always subject to peasants' interpretation, and subsequent translation, into specific courses of action.

Peasant-managed agricultural growth occurs mainly through labour-driven intensification. It is a process that results in steady increases in yields: i.e. the level of production realized per object of labour (land, animals, fruit trees, etc.) is augmented. The more value produced per object of labour, the more intensive production is. An intensive production shows an elevated Gross Value of Production (GVP) per unit of land. 'Labour-driven' means that both quantity and quality of (peasant) labour are crucial in driving forward this process of intensification. It is the opposite of technology-driven intensification, which often translates into a considerable reduction of the quantity of labour applied and into deskilling which reduces the quality of labour. Of course, both processes assume the presence of labour and technology. In labour-driven intensification, however, skill-oriented technologies characterize the process of agricultural production. Technology-driven intensification is characterized by mechanical technologies (Bray, 1986; Mazoyer and Roudart, 2006).

In this chapter we will discuss several mechanisms that play a role in, and explain, the remarkable growth of China's agriculture. The empirical data was gathered between 2008 and 2012 in Sanggang and Yuangang, two villages located in the south-western part of Hebei province.

Labour investments: improving the quality of available resources

The image shown in Figure 4.1 might be quite enigmatic to outsiders, yet in Sanggang village it is a self-evident part of daily life. It shows two peasant farmers digging mud out of a drainage canal that crosses the village. The canal accumulates a lot of detritus that is washed from the streets during the rainy season. This highly fertile mud is loaded into a small three-wheeled truck and taken to the fields in order to improve the structure and fertility of the soil. In more general terms, the picture shows the improvement of the resource base (i.e. the fields) through the investment of labour. This improvement is not achieved through the market – but is realized directly through the peasants' own labour.[6] An important consequence of this is that the improvement of the structure and fertility of the soil did not occur through a financial transaction. As such, it does not represent a financial burden that needs to be recouped later on. It has already been 'paid for': it is the result of the labour invested in it. It increases the chances of a better harvest which will directly benefit the peasant family without the need to pay a bill.

The history of peasant agriculture is largely the history of labour investments. Through such investments (in improved soil fertility, terraces, buildings, fences, irrigation canals, drainage systems, anti-erosion protection, levelling, mills, greenhouses, improved breeds, enlarged herds, etc.) peasants enlarge the value of their resource base. This allows for the introduction of more intensive cropping schemes (see Table 4.1) as well as for increases in yields (see Table 4.2). Several studies have documented that the value per unit of land of such labour investments (when expressed in monetary terms) often greatly exceeds the corresponding value of capital in large agricultural enterprises (see e.g. CIDA, 1973; Netting, 1993; Ploeg, 2008).

The increased value of the resource base is often discussed as an increase in capital. Improving the land is thus presented as part of capital formation. As such this interpretation is not wrong: the value of the available agroecological capital is increased, as is the value of the available economic capital. However, it is important to clarify that we are not talking about capital in the Marxist sense. There is no capital here that needs to produce surplus value to accumulate in order to be invested again as capital. Here, we are dealing with other values. The mud, once taken out of the canal and loaded onto the truck, acquires value as a fertilizer. Once applied, this fertilizer increases the fertility, and thus the value, of land. This allows for other values to emerge: a better harvest and more well-being for the peasant

Figure 4.1 Improvement of the resource base through the investment of labour

family. Mud is converted into fertilizer. This fertilizer is converted into more fertile land which, in its turn, is converted into higher yields (and/or more intensive cropping schemes). These conversions are non-monetary; they do not pass through the markets. They depend on labour. They are not capital investments – they are *labour investments* (HLPE, 2013; Ploeg, 2013). Consequently, the 'return on investment' is not an amount of money on top of the initial investment. The investment does not imply money, and the 'return' is not expressed in monetary terms. Both investment and return occur within and through non-commodity circuits.

Obtaining new resources through labour investments

Labour investments are not necessarily located within the farm. A decisive feature of current Chinese agriculture is that a considerable part of the labour investments take place in economic activities (industry, construction, mining, trade, services) located *elsewhere*. These activities contribute to capital formation *within* the farm and thus to further intensification. The three-wheeled tractor in Figure 4.1 is a good illustration of this: it is present on the farm because it has been bought with earnings obtained on a construction site far away. Thus migrant labour is converted into a labour investment (at least partly) because it contributes to capital formation *within* the farm (Mohapatra *et al.*, 2006; J. Zhang *et al.*, 2006; Ploeg and Ye, 2010).

The basic pattern is as follows. Young people leave the village and work for several years in industry or construction. This results in remittances and savings that are used in, e.g., the building of a house (in rural villages the house is not only a place for living, it is also an important and indispensable resource for farming) or the acquisition of a small three-wheeled vehicle. This vehicle is used not only for operations on the farm, but also for transporting stones from quarries (in neighbouring mountains) to the iron-ore factory in the village (Ye *et al.*, 2009). The savings thus obtained might be used subsequently to acquire a larger lorry that allows for heavier transportation and increased earnings. These increased earnings are then used to pay for the construction of 'ditches' (discussed below) that will allow for the planting of walnut trees. In the meantime, it is quite probable that the wife of the owner of the lorry is cooking for the people who are working in the iron-ore factory. The cooking may provide an income of 700 yuan per month. It is used to buy a cow and build, together with one of the neighbours, a small stable. Apart from agricultural work in the fields and cooking, the wife of the owner is now also alternating with her neighbour in herding their cows. The calves will be used to increase the small herd. This is the story behind Figure 4.2. The wife is herding the cows (along the small river), while the husband is passing on the road in his truck. The two are not engaged in meaningless routines: they are building a livelihood, in a knowledgeable and goal-oriented way. Resources are strategic in this livelihood (a house, the cows, the lorry, walnut trees) and *they are created through labour investments*.

The pattern clearly implies money, although it is not 'all-purpose money'. The major decisions are not driven by the search for the highest possible profit. The money is earned with a specific aim: it is to be converted into, e.g., a three-wheeled

Figure 4.2 Husband and wife involved in indirect labour investments

tractor and nothing else. We are dealing here with *socially defined* conversion processes. The role of money is subordinated to the socially defined strategy. Non-commodity considerations govern the use of this money. The operation of such strategies (and the resulting creation of 'capital goods') occurs, more often than not, through social networks (Zhao and van der Ploeg, 2009).

Intensifying cropping schemes

Even though they only have access to a very small amount of arable land, which is scattered across many small plots, the peasant families of Sanggang village cultivate an extensive range of crops. These include cotton, soybeans, peanuts, maize, sweet potatoes, millet, cabbages, many different vegetables and herbs, sesame, tree seedlings, etc. Alongside these crops there are several different fruit trees and animals including hogs, cows, chickens, goats and sheep. This 'package' does not remain fixed over time. Some crops, such as wheat, have disappeared, while others, such as chestnut trees and prunes, have been introduced.

Crops differ in yield and price. Consequently, the GVP per unit of land might differ considerably.[7] Maize, for instance, renders a GVP of 560–700 yuan per mu while cotton (including the oil) might result in 1,870 yuan per mu (2010 prices). When different crops are compared, the ones rendering a high GVP per mu are referred to as 'intensive crops'. Crops with a low GVP per mu are defined as 'extensive crops'. Intensive crops normally require more resources: more labour input per unit of land, more mechanization, more inputs, more irrigation water, etc. For the extensive crops the opposite holds true.

The notion of a cropping scheme describes the particular combination of crops within the farm. Cropping schemes might become more intensive, in which case the relative weight of intensive crops is high. When cropping schemes are moving towards more intensive crops we can talk about an intensification of the cropping scheme.

The intensification of cropping schemes can be attractive to farmers: it means they obtain more money and improve their incomes. It will also be attractive for a country as a whole: the intensification of cropping schemes implies an increase in the total wealth produced. However, the crucial question is always whether farmers have the means and the space to intensify and whether there is an acceptable balance between the required means and the obtained benefits.[8] The intensification of cropping schemes can occur at different levels: within the fields, at the level of a farm unit as a whole and at the level of a village as a whole.

Table 4.1 presents on overview of the main changes in the cropping patterns at village level. Relatively extensive crops (such as wheat) have been eliminated from the cropping scheme – in large part because they require considerable amounts of irrigation water which is becoming increasingly scarce in Sanggang. Villagers now obtain the wheat flour they need through barter: 72 jin of maize is exchanged for 50 jin of wheat flour (a jin is 0.5 kg). Another highly important element has been the elimination of animal traction, which has been replaced by mechanical traction. Thus fodder no longer needs to be produced for donkeys or oxen: this creates space for more intensive crops.

Yield increases: producing more per single crop

Once a cropping scheme is more or less defined, further intensification is possible through yield increases for each crop. This implies that, per crop, more production per unit of land is realized. Table 4.2 summarizes the results that have been achieved in Sanggang. Maximum yield levels (using maize here as an illustration) are often considerably beyond the average levels. This points to room for further improvements.

Intensification of this type is equally attractive to farmers; it means that the available resources are used in a more efficient way, thus generating more wealth. But achieving yield increases requires considerable effort. Soil fertility needs to be improved, more promising varieties must be selected, irrigation needs to be improved and cultivation fine-tuned. Development of the appropriate practical knowledge is an important prerequisite for achieving all this. Experimentation and the exchange of knowledge often play a strategic role in this respect.

Since the late 1970s Sanggang village has carried out a process of intensification based fundamentally on the mechanisms discussed above. We refer to this process as peasant-managed agricultural development. Evidently, new varieties (often developed through scientific research), the availability of new technologies and increased input use also played important, sometimes even decisive, roles.

Table 4.1 Main changes in cropping schemes in Sanggang village

a) Elimination of wheat cultivation
b) Introduction of fruit trees
c) Elimination of animal traction
d) Increase in vegetable production
e) Increase in animal production (goats, chicken, hogs)
f) Afforestation

Note: Behind these changes lie complex spatial reorganizations. Afforestation takes place partly in the valley but mostly in the hills. The introduction of fruit trees is currently taking place in the hills. The increase in vegetable production implies that the ratio between gardens and fields has changed.

Table 4.2 Yield increases in Sanggang village over the last 40 years

Crop	*1966–1974*	*2007–2009*	*Growth*	*Maximum*
Maize	300–400	800–1,000	+ 100%	1,200
Peanuts	200	200–300	+25%	
Wheat	200	n/a		
Sweet potatoes	2,000	3,000	+50%	
Soybeans	200	350–400	>75%	

However, it is the peasants themselves who have acquired and combined these new artefacts to produce effectively (Ye *et al.*, 2010).

The defining characteristic of peasant-driven intensification is its dependence on the quantity and quality of labour (as opposed to forms of intensification that depend on specific technologies and inputs). The labour input per unit of land is increased, while skills and knowledge (the 'quality of labour') are improved and increased. When skill-oriented technologies are central to the labour process (Bray, 1986), the development of skills and knowledge translates into higher yields. Fine-tuning is often a key word here: through the meticulous coordination of a wide range of growth factors, higher levels of production are reached and sustained. Another defining characteristic of peasant-driven intensification is that it strengthens the relative autonomy of the farm (again this is different from technology-driven intensification, which tends to introduce and strengthen dependency relations).

Alongside the mechanisms discussed thus far, new mechanisms are taking shape. These will, we believe, help to further strengthen the process of labour-driven intensification and thus contribute to food security in China, while simultaneously changing the *form*, but not the *nature*, of this process.

Embedded specialization[9]

The Wu family (see Figure 4.3) currently works 10 mu of land, which is all planted with prunes. In total there are just over 500 productive trees, which have

Figure 4.3 Specialization within a peasant farm

already been producing for 3 to 4 years. Good trees might yield up to 100 jin of prunes. Last year's harvest totalled 50,000 jin (i.e. 25,000 kg of prunes). Prices fluctuate between 0.80 and 1.00 yuan per jin. This makes for a GVP/mu that fluctuates between 4,000 and 5,000 yuan/mu/year, far higher than the average yield of some 600 to 1,800 yuan/mu realized by the average peasant holding.[10] Thus, the Wu farm is a clear exponent of ongoing processes of intensification.[11] If the enlarged acreage is taken into account it is a clear example of *Dahu* development (see Figure 2.5, page 40).

At first sight this farm appears to be completely different from the average peasant holding in the valley, which is highly diversified and oriented to household consumption – only surpluses are sold on the market. Instead, this prune producing farm is highly specialized and all its produce is marketed. A more careful look seems to affirm that the holding of the Wu family is indeed different – especially in as far as it goes beyond the limited scale of the average peasant holding. At the same time, though, it is still very much a peasant unit of production. It is managed according to the same strategic principles (and interrelations) that underlie the much smaller units:

- It has been constructed in a step-by-step way through labour investments, as discussed above.[12]
- It is combined with production for household consumption (located elsewhere).[13]
- The direct producers are not aiming at further expansion, at accumulation; their productive activities are embedded in, and limited by, a wider socio-cultural framework.[14]
- Reciprocity is strategic in many respects.[15]
- Risk avoidance is considerable.[16]

- Employed labour is family labour (the main exception being work done during the harvest).[17]

Taken together this implies that the plum orchard of the Wu family is based on a solid, autonomous and historically created resource base. It gives them dignity and pride: it is an expression of the conversion (step-by-step) of the hard work put into this orchard and its productive capacity. It does not represent dependency relations; it is, instead, a considerable asset that helps the family face and overcome potential dependency relations. It is, in short, a vehicle for independence and autonomy.

This autonomy is, of course, a *relative* autonomy. It is an autonomous resource base, but in order to produce the Wu's have to engage in more or less permanent commodity relations for the supply of some crucial inputs. Annually, they pay from 6,000 to 7,000 yuan for fertilizer, pesticides and electricity to operate the pump in their well. Compared to the overall sales of 40,000–50,000 yuan this is relatively modest (as discussed in Zhao and van der Ploeg, 2009).

It is important here to note that there are other, more general conditions that have also been helpful in creating this intensive plum orchard:

- The availability of good access roads (without these the marketing of the prunes would be almost impossible) and the availability of electricity. Both these conditions highlight the role of the state in the overall process of ongoing agrarian development.
- The availability and communication of farmers' knowledge as well as scientific support.
- The possibility of travel, to meet other people involved in promising activities.
- The role of the village committee in the (re)distribution of land.
- The existence of a strong social fabric in the village that allows for cooperation and reciprocity.

Another form of embedded specialization can be found with regard to livestock. Over the last 3 years we have witnessed a steady but ongoing growth of herds in Sanggang village. This applies to cows (for meat and calves) as well as to goats, sheep and to pigs and chickens that are held in stables located in the village. The development of these herds is also driven by labour investments. Herds are not bought (an initial animal or a sire being the exception); they are developed using the reproductive cycles of the animals, the leftovers of the harvest, and/or by increasing feed and fodder production within the farms. Herds have multiple functions. They represent a type of capital formation: they embody savings. They function as a security blanket, as a fund that might be used when unexpected events demand major expenditure. They produce manure. And they may sustain income flows (through the selling of new offspring or fattened animals) – they may enliven major events such as the spring festival.

Households without cattle can obtain their first animals through the Poverty Alleviation Programme. In Sanggang village this programme operates in a peculiar way (as defined by the village committee). It was decided that fifteen families could get one calf together, which, through breeding, should allow each family to eventually have their own animal. However, the villagers 'translated' this in their own way. One family in each beneficiary group 'bought' the calf by paying 100 yuan to each of the remaining fourteen families. Ironically, this interpretation of the system means that it is not the poorest families that obtain the calf, although they do receive an extra 100 yuan.

The reorganization of space

The geography of Sanggang village is typical of most mountain villages in rural China. A small river passes through a narrow valley lined with many small fields. The land is fertile and irrigation is nearly always possible. The valley is surrounded by hills. These have been deforested heavily in the past; until recently they were used exclusively for extensive pasturing (mainly goats and sheep). The village is located at the interface of valley and hills. Behind the hills lie the mountains where until recently much extractive mining took place. The mining has now extended into the valley: the soil contains layers with iron ore, and sand and stones are extracted for construction activities (see Chapter 5).

At present, the hills that surround the village are being transformed. The construction of 'ditches' (as they are locally called) is central to this process.[18] This involves creating strips of land along the contour lines. These strips of land are 2–4 metres wide and are separated from each other both in altitude and in distance (see Figure 4.4). The strips are constructed with mechanical diggers that not only level the land but also dig to a depth of 1–1.5 metres in order to loosen and mix the soil. Thus 'ditches' are created in very hard and impermeable soils and filled with churned and loosened soil. This technique has two enormous advantages. It greatly increases the water-retaining capacity of the soil (after the rains the ditches harvest the rain water) and it allows root systems to develop smoothly. An indirect longer-term effect is that this method also probably prevents soil erosion.

The construction of ditches, which is being organized and partially paid for by the village committee, allows for expansion and redefinition of the agricultural frontier.[19] Extensive cattle breeding (based on grazing in the hills) is being reduced while fruit tree cultivation is made possible without having to convert the fertile fields in the valley. This coincides with some major macro trends that are increasingly visible in China and strongly supported by the state. This creates a seemingly paradoxical situation: on the one hand, some marginal lands are being taken out of production (in order to allow for afforestation or widespread anti-erosion measures) while, on the other hand, other parts are substantially upgraded to allow for further intensification. This explains the apparent paradox of decreasing agricultural surfaces and ongoing production increases (see Qu *et al.*, 2009; Heerink *et al.*, 2010).

The ditches are highly suitable for growing persimmon, chestnut and walnut trees. In the first years, when the trees are still young, the land surface might be

Figure 4.4 Recently constructed 'ditches'

used simultaneously for peanut production, for example. This is a clear manifestation of the typical multi-floor farming practices that characterize peasant production all over the world: it allows for highly intensified, but sustainable use of the land.

In Sanggang village, walnut trees start to produce after about 5 years, with full production achieved after 10 years. At that point a tree will render a minimum of 2–2.5 kg of nuts, but harvests as high as 5 kg per tree are possible. Prices vary for different varieties but are generally more than 30 yuan/kg. A tree might produce between 60 and 150 yuan (or far more). There might be up to thirty-five trees per mu. Hence total GVP per mu could range between 2,100 and 5,250 yuan/mu. This represents an impressive intensification – especially compared to the marginal benefits of grazing these hills or the average GVP/mu in the valley.

Two hill areas are currently being reconstructed in Sanggang village. In the northern part some 5,000 metres of 'ditches' have been created. These benefit seventeen households. They already had the possession rights in this particular part (on average 3 mu per person). Inclusion in the construction scheme is voluntary. The participants pay 3.5 yuan per metre and the village committee adds another yuan. For some households the total costs are considerable. These are often paid with money earned in mining, transport and processing of iron ore. Beyond that, most villagers argue that 'two years of peanut production [on the newly opened strips] will pay for the digger'. The seedlings and the grafting that follows represent additional costs. The village committee obtains the seedlings and will sell them to the participants; if some help is obtained from the county, the seedlings might even be distributed for free. The southern part, which is currently being constructed, will benefit another nineteen households.

The reorganization of space is an important element in the ongoing intensification of local agriculture.[20] Total production of the village as a whole, as well as production per household, will increase considerably. Economic benefits are combined with ecological benefits. The erosion and degradation of the hills will be reversed, at least partly. Another effect might be, as villagers argue, that migrants (who will return anyway to the village when they get older) might now return earlier. A 40-year-old peasant explains:

> Working elsewhere might give you between 10,000 and 20,000 yuan per year. The fruit trees can give you the same or even more. Hence, you might come back earlier. Here you have fresh air and you can be self-sufficient. It is safe, pleasant and your vegetables will not be contaminated.

More generally speaking, the construction of the new 'ditches' is a source of *pride* for all villagers. It is a material expression of an enlarged resource base and, above all, of the capacity to enlarge the resource base through one's own efforts (including cooperation with the village committee). It is, as one of the villagers told us, about 'developing new *land*'.

The last element refers to an intriguing feature of local and regional patterns of governance. Village committees are locally elected (Howell, 1998). They have to prove, after being elected, that they can be useful to the village. The current committee invested first in paving all the roads in the village and then decided that 'something needed to be done for production'. Hence, the decision to construct ditches in the surrounding hills. This is important to the members of the village committee: it allows them 'to proudly present the project in Township and County meetings' (since the 1998 floods central government has strongly encouraged afforestation and anti-erosion projects). The position of the president of the village committee is also important. Being the owner of the local iron ore processing unit, he is one of the richest people in the village. This allows him to anticipate (i.e. pre-finance) many of the support measures that will come later. Thus, local development is not stifled by bureaucratic delays elsewhere.

On-farm processing: adding more value

At the end of October or the beginning of November (i.e. after the harvest) the fields of Sanggang are again flocked with people. They are working together in teams of twelve to fifteen people engaged in processing sweet potatoes into glass noodles. Glass noodles are a highly valued, high-quality product. They are used in many villages during the spring festival to cook special dishes and are also a highly appreciated gift. It is pleasant to observe the making of these glass noodles. It is like theatre: everyone knows his/ her role, and there is, especially in the beginning, a sense of mystery in the air. The observers do not know exactly what is happening; the players involved do not know, as yet, whether it will all unfold correctly. But you can feel the passion: the willingness to make the best out of it, to create a product that talks about the care and dedication with which it was made. Processing sweet potatoes into glass noodles involves more than ten steps (ever so many 'scenes'); it requires considerable skills and a lot of experience, and it extends over several weeks. Figure 4.5 provides an impression of one of the many scenes.

When sweet potatoes are sold directly, 1 mu of land might render 1,000–1,500 yuan. When processed into glass noodles (8 jin of very good sweet potatoes are needed to obtain 1 jin of glass noodles) the same mu of land with sweet potatoes might render 4,500–6,000 yuan. When conditions are very good, a plot of 2.6 mu may result in a gross income of 19,500 yuan.

Enlarging the value added per unit of end product (through the production of specialties, on-farm processing, direct marketing, etc.) is not a new phenomenon in China. It could be argued that China's agriculture has always been multifunctional. We noted, however, in the villages we are studying, that the range and magnitude of such activities is expanding rapidly. At present it embraces activities such as agri-tourism, the cultivation of ornamental trees, free-range chickens, the creation of new market segments in Beijing, the production of walnuts for carving, etc. All these activities are located at a particular interface. On the one hand, there are farmers involved in 'reinventing' agriculture and who are looking eagerly for opportunities to earn more with their multifunctional farm units. On the other hand, there is a rapidly growing middle class looking for distinctive food products and attractive rural services. At this interface new networks emerge, while other

Figure 4.5 Processing sweet potatoes

networks are acquiring new functions and values. The coming and going of migrant labourers, for instance, is used increasingly to transport the locally produced glass noodles to the urban centres and to commercialize them there, sending the earnings back to the village. If consumers are satisfied they may even use the same network to order high quality glass noodles again the following year.

Conclusions

In this chapter we have discussed seven mechanisms for agricultural intensification – some are classical (but still highly valid) approaches, others are relatively new. All these mechanisms contribute to, and strengthen, peasant-driven intensification. In both the classical and the newly created mechanisms, *labour* plays a central role and the consolidation or even the enlargement of *autonomy* is strategically important. Labour investments (especially when they imply migration) might indicate difficult and burdensome dependency relations – *but these dependency relations are mostly located in the alien workplace*. Peasants (especially the young ones) engage in such dependency relations *in order to create and enlarge autonomy in the rural workplace to which they will return*. They enter the factory in order to save for their own tractor, which allows them to operate on their own farm or in the wider rural economy as a peasant with an expanded resource base. They do not then have to rent somebody else's tractor to do farm work. They can participate in the wider rural economy (e.g. in transport or trading) using their own means of production.

The same goes for the construction of a beautiful plum orchard or for making ditches. They help materialize autonomy. Involvement in these activities means that the farmers do not have to engage in wage-labour relations elsewhere. They can make their own living and their efforts might well result in increasing their (and/or their children's) well-being. Yet they can only do so because they previously entered into dependency relations – they did so in order to *actively create* their current autonomy.[21]

The peasant condition has been defined as 'an ongoing struggle for autonomy' (Ploeg, 2008; see also Chapter 3 of this book). This struggle can take many forms and it might occur at different levels and in different places. The examples of the Sanggang and Yuangang villages reveal intriguing spatial and temporal dimensions: young peasants enter into relations of dependency and face considerable hardship in far-away places in order to construct enlarged autonomy in their own place in the future. This struggle is one of the main drivers of agricultural intensification. Through the *previous* struggles in far-away places (i.e. through indirect labour investments) and the *current* struggles to get the most out of their land (while simultaneously maintaining, if not enlarging, the resource base through embedded specialization and the construction of ditches) peasants are realizing an ongoing process of intensification – a process that follows the logic and rationale of the peasantry.

One theoretically important point here is that the two basic elements identified by Chayanov (1966) – drudgery and utility, which together structure much of the dynamics of farming – do not enter into a balance here in one single cycle of agricultural production and at one particular place (the farm). Instead, they are separated geographically and temporally. The drudgery is located in far-away industries and construction sites and experienced when relatively young. Utility follows later and is located in one's own village. But the two are neatly tied together. They are brought into equilibrium and a balance is achieved. The former makes the latter possible, while the latter gives meaning to and justifies the former. Thus, farming, or more generally being *nongming* (i.e. peasant), becomes an 'organized flow of activities through time' (Vincent, 1977), a flow that crosses time and space and bridges considerable distances. Farming cannot be defined, nor understood, as an activity that is just located at one particular place, unfolding solely through an endless repetition of the same routines specified by the agricultural calendar. Instead, farming is increasingly a *network activity* that extends in time and space and that implies complex balances.

The villages that we discussed in this chapter are far from 'hollow' (as suggested by Peng, 2007; Yang and Liu, 2009). Nor do they represent any 'demise of the rural economy' (Gudeman, 1978). One of the hidden strengths of the peasant economy of Sanggang and its neighbouring villages is precisely that nearly all peasant families[22] are involved in economic activities outside of the original unit of peasant production. Many work elsewhere in order to be able to make indirect labour investments in farming. Others participate in the opening of 'ditches' in the hills in order to extend the original unit. Some create space for specialized fruit growing that is embedded in the socio-cultural framework of the typical peasant unit of production, and another group is developing herds. Thus, peasant existence increasingly emerges as a patchwork, as an amalgamation of sometimes strikingly different, but cleverly interrelated and mutually reinforcing activities. As argued by Akram-Lodhi and Kay:

> It is now more common for rural livelihoods to be constructed from a plethora of fragmentary and insecure sources: petty commodity production in farming, to be sure; but also the sale of temporary and casualized waged labour, both on and off-farm; as well as petty commodity handicraft manufacture, petty merchant trading, the provision of petty services, and a reliance on remittances arising from migration.
>
> Akram-Lodhi and Kay 2010:179

This patchwork of activities and the associated creation of value added strongly contribute to the dynamic nature of the 'rural non-farm economy' – it contributes to it and simultaneously shapes the local market that further strengthens the rural non-farm economy (Mohapatra *et al.*, 2006; J. Zhang *et al.*, 2006; more generally, Haggblade *et al.*, 2007).

Involvement in a multiplicity of activities cannot be understood, as, for example, Kearney (1996) suggests, as the 'disappearance of the peasantry'. It is exactly the other way around (at least in the villages discussed here). Multiple involvement is used actively to *strengthen* the peasant unit. It helps to sustain the intensity of cropping schemes and the intensification of single crops. Multiple involvement equally helps to enlarge the resource base in which peasant farming is grounded. It contributes substantially to the creation of more autonomy (see Ploeg, 2008:23). Working elsewhere also generates the remittances and savings that are used for mechanization (thus allowing for more intensified cropping schemes) and for increased spending (on, e.g., inputs such as fertilizer, which allow for further intensification). It also translates into 'the availability of a lot of money in the village', which makes informal lending (based on reciprocity) far easier than in the past. In turn, the prospects of improved livelihood (related with, e.g., walnut production, plum growing, larger herds) implies that migrants will probably return earlier to the villages than in the past, thus enhancing the dynamism of rural and village life.

Update

Zhao Laisheng is a small man. He is often called Xiao Laisheng (xiao meaning small). But he is a very good farmer and a great craftsman. He is 67 years old but instead of enjoying his retirement he is busy and gives advice to many other peasants. He knows how to graft fruit trees and has designed very effective traps (using old tin cans and rotten sweet potatoes) to attract insects thus preventing them from damaging apple tree blossom. He works with grace and poise and it is always a pleasure to see how he works the fields and orchards.

In spring 2015 we meet again and Xiao Laisheng says it is now possible to realize yield levels in maize production of as much as 1,300 to 1,400 jin per mu (compared to the previous level of 1,200 – see Table 4.2). The average yield is still between 800 to 1,000 jin/mu 'but when the weather and soil are good and the work is well done, yields up to 1,400 jin are possible'. This is confirmed by others. In this respect, good farmers can make a significant difference, primarily by improving soil fertility. As Xiao Laisheng explains:

> Using manure, up to 10,000 jin per mu, will help very much although you need patience. In the first year there will be no effect or even a harmful one. It takes 3 years to get the full positive effect. And you can only do this if you have your own animals.

We also learn of another technique that Xiao Laisheng is using in his fruit yards. Stalks of corn are cut in long strips that are tied together and then buried in a triangular pattern around single fruit trees. This buried biomass absorbs and stores water from the rains in early spring and makes it available for the trees later on in the season. This can make the difference when there is a drought. Thus, even the weather can be mediated to a degree. Xiao Laisheng summarizes this by using a nice metaphor during a discussion about the difference between good farmers and those who do less well. The latter '*just catch the sesame but throw away the watermelon.*' Here the 'watermelon' clearly represents the potential. It is the promise of higher yields. Not being satisfied with 'sesame', peasants such as Xiao

Laisheng mediate the weather, improve soil fertility and develop novel biological traps (see also Chapter 7) in order to grow and harvest the 'watermelon'. In consequence the highest possible yield levels are constantly improving – in the case of maize from 1,200 to 1,400 Jin/mu.

Notes

1 This chapter was previously published in the *International Journal of Sustainable Agriculture and Food* 21(1):155–71. A slightly reworked Italian version was published as 'Terra, lavoro e produzione agricolo in Cina: meccanismi di intensificazione basati sul lavoro' in *Sociologia Urbana e Rurale*, in a special edition on the agrarian question (Anno XXXV, 102:33–50). Some minor changes have been introduced in the current version. A general discussion of labour-driven intensification is given in Ploeg, 2014.

2 Due to changes to urban *Hukou*, deaths and patrilocality (families only having daughters 'disappear' as it were), the population has decreased. The number of people active in agriculture has decreased far more (due to labour migration).

3 Actively cultivated agricultural land has been reduced over the years, as the combined consequence of mining in the valley (see Chapter 5), house construction, highway construction and the abandonment of infertile plots. However, in the hills, new terraces have been constructed.

4 The total population of 702 people also includes 135 children, 122 elderly (older than age 60) and 42 disabled persons.

5 Sixty-six people have a subsistence allowance (mostly two per household). These are mostly elderly with no children or children who cannot care for them.

6 Phrased differently, the improvement of soil fertility is, in this case, not created through the acquisition of fertilizer and, therefore, dependent on a monetary transaction.

7 Furthermore, many crops also render non-monetary benefits, normally not calculated as part of GVP (e.g., stalks of maize production used as cattle feed or for heating houses).

8 Here, several other considerations play a contributory role: the maintenance or improvement of soil fertility (some crops are highly demanding, while others restore soil fertility), pest and disease suppression, the availability of water, labour, etc. and finally the consumption needs and preferences of the family.

9 We use the adjective 'embedded' here to stress that specialization does not imply an adieu to peasant farming. It is, instead, embedded in it.

10 With maize a GVP of 560–700 yuan/mu is possible (in this village and with the price levels of 2009 and 2010). The GVP might fluctuate between 600 and 900 yuan per mu in the case of peanuts. If cotton is sown some 1870 yuan/mu (including oil) might be realized (using the 2009 price level).

11 Farmers are often explicit about this: 'Fruit trees provide more benefits than crop production.'

12 Mr Wu, who is now 56 years old, has travelled through large parts of China. He has had many different jobs, including trading. His wife (aged 53) always stayed behind in the village taking care of the scattered plots with vegetable production, maize, etc. The land they are currently cultivating with prunes consists of two parts, each with its own history. The first part was rented some 8 years ago for 830 yuan/year. The village committee required an advanced payment for 5 years. This totalled 4,000 yuan. Another large investment was payment of 16,000 yuan to a specialist who did the grafting. Hence, the total investment was 20,000 yuan. This investment could be paid for, according to Wu's calculations, by the first harvest. The second part could be obtained through an initial payment of 30,000 yuan, most of which could come from savings obtained through work elsewhere. The remainder of the money was lent by friends and relatives. Mr Wu works 2 months a year in tile making, which renders him 70 yuan/day.

This money is earmarked for paying back the loans. All this implies that currently there are no financial costs on the land.

13 Elsewhere, the Wu family has a range of scattered plots of 5–6 mu in total. Here they plant vegetables, soybeans, sweet potatoes, maize and millet. Part of the maize is sold to the 'hog factory' in the village (a plant for intensive pig breeding). The rest is for self-consumption. When asked why they do not plant prunes here as well, they explain that fruit trees bring more (monetary) benefit, and that wheat, rice, flour, or whatever you eat can indeed be bought. However, 'the products you produce yourself are green, they do not contain pesticides, and they are tastier.'

14 When asked about the possibility of further increasing the number of fruit trees, the Wus give a straightforward answer: 'This is already sufficient, our daughter is already married and our son already acquired a house . . . so there is no need for more trees, this is enough.' More generally speaking, production is oriented at: (1) self-provisioning; (2) guaranteeing the required main investments in social life, e.g., weddings and the construction of houses for the next generation; and (3) the creation of reserves that might help to bridge difficult times (diseases, funerals, etc.). Other people, in similar circumstances, might argue: 'We do not prefer a larger land area [than the one available at the moment], because we have limited strength, we can work the land we have now but not more; and hiring people is not convenient because the wages are too high.'

15 A lot of labour is organized through *hu zhu* (mutual help), especially for products (such as sweet potatoes) that need to be planted in a short time period and/or demand heavy work. 'Without mutual help a farm can hardly function', say the people in the villages. Mutual help is arranged through family bonds, friendship relations, between neighbours and between owners working on adjacent plots. Sometimes the work brigade (a remnant of earlier times) remains as the framework for labour exchange. Mutual help is, according to the villagers, especially important for the women. Equally there is considerable interchange of seed material. And finally, lending and borrowing between relatives and friends also occurs very frequently. Generally, no interest is paid on these loans even though the amounts involved are often considerable. Loans up to 20,000 yuan are no exception. If larger amounts are needed they can be obtained by asking several households to lend money. This practice is explained as follows: 'People are getting rich nowadays, so they can easily lend to others.' Mutual trust and having a good reputation are, of course, essential in this respect.

16 One expression of this is that more than forty-five different varieties are planted in this prune orchard alone. There is an ongoing search for the best and most resistant varieties. The family paid 700 yuan to obtain one variety from Japan.

17 Nowadays, all work throughout the year is done by the Wu couple themselves. This also includes difficult tasks such as pruning and grafting. They obtained these skills from the specialized worker who did the first grafting. In general terms, the Wus explain that they 'prefer to be independent' (no loans, own savings; no salaried workers, mainly family labour). Being independent gives them 'a sense of dignity and pride'. Although borrowing is not losing face, they prefer to have their own money. The same applies to labour: 'We prefer a somewhat smaller piece of land, then we can do all the work ourselves. I myself know what I do; workers maybe don't know, then they do bad pruning, or bad pollination, which can cause considerable damage.' The only exception is the harvest period when they contract twenty-five workers for 2 days.

18 These 'ditches' differ from terraces, which are locally referred to as 'ladder fields'. The latter are meant for crops (notably rice), the former for trees.

19 There are elements of rebelliousness and endurance in this. It is the second time that fruit trees have been planted here. Several years ago persimmon trees were planted (without preparing the land by digging ditches). At this time the peasants used a programme for afforestation that was not intended for fruit trees (the main objective was giving the land back to nature in order to prevent erosion and flash flooding) to

plant fruit trees anyway. However, quite a few of the persimmon trees died due to drought or because the land could not hold sufficient water.

20 Investment in ditches triggered another collective investment, consisting of the construction of a large basin in the riverbed (always containing a lot of water), the installation of an electrical pump and the construction of a pipeline that goes from the basin to the ditches higher up in the hills. It is to safeguard the newly planted trees in the occasional years when there is no (or far too little) rain and, consequently, no water harvesting in the ditches. This could lead to the destruction of all the trees in the new ditches. Again, there is no direct financial return on this investment – however, its value and significance are evident.

21 The theoretical complexity here is that being a worker or being a peasant are not separate or antagonistic categories. One role flows into the other in order to strengthen it.

22 The telling exceptions are peasant families that have lost in one way or another the support of their children (or do not have children) or whose members are physically or mentally disabled.

5 Entrepreneurial trajectories?[1]

Jan Douwe van der Ploeg, Ye Jingzhong, Zhao Yong and Pan Lu

This chapter asks whether new, entrepreneurial forms of agriculture, as they are known in the West, are also being developed in China. It first discusses, in detail, three examples that – at first sight – seem very close to the entrepreneurial model. As discussed in the previous chapters, peasant agriculture is generally grounded on a *self-controlled and autonomous resource base*. This resource base is delivered through socially-regulated exchanges that take place within the family farm (see Chapter 3) and/or through reciprocal networks that operate at village level. The resource base is usually improved and expanded through labour investments (see Chapter 4). Entrepreneurial farming, by contrast, is based on *commodity flows*. It mobilizes the resources required (or at least a significant part of most resources) through different markets. This means that land, capital, animals, seeds, fodder, knowledge, machine services, etc., enter the process of production as *commodities* that need to be repaid for during, and through, the production and subsequent marketing of the obtained produce. Both peasants and entrepreneurial farmers produce for the market. But in peasant agriculture the market is only an *outlet*. In entrepreneurial farming the market is the *ordering principle*. The two approaches also imply very different relations with markets, especially the upstream ones, which have another consequence: peasant farming is basically grounded on ecological capital, which is converted, through peasant labour, into food and other products and services. Entrepreneurial farming, on the other hand, is based on financial capital that is converted into commodities through a labour process in which technological scripts, developed elsewhere, play a central role. The greater integration with, and high level of dependency upon, the markets (especially the upstream ones) also implies that the margin per unit of end product is, under *ceteris paribus* conditions, lower in entrepreneurial agriculture than in peasant agriculture (simply because the monetary costs are higher). This explains why entrepreneurial farms are generally larger scale and why market dynamics drive such farms to seek ongoing scale increases (for an extended discussion, see Ploeg 2008). Wage-labour relations, though, are only of secondary importance in these farms. Consequently, they cannot be understood as capitalist farm enterprises (the latter are discussed in Chapter 6).

The three Chinese cases to be discussed here are, indeed, all relatively large and rapidly expanding farms. One is located in the village of Sanggang, in Hebei

Province; a second is located 'on the other side of the Great Wall' as Chinese people say. For historical reasons the population pressure on the land in that part of China is relatively low and farms are relatively large. Extended areas of as yet undeveloped land (notably wetlands) allow for further expansion. This specific case is based on an accountancy booklet, which was discovered by one of the authors of this chapter during his field research. The farmer, Sun, who currently works 726 mu of land (which is, for China, an incredible acreage: 15 mu = 1 hectare) seems, at first sight, to be an archetypical expression of the agricultural entrepreneur. The same is true of Zhao, a farmer from Sanggang. He has contracted a lot of land in the hills surrounding the villages where he planted poplars, chestnut trees and walnut trees. He currently controls some 350 mu. This is modest when compared to Sun's farm; but on this side of the Great Wall, it is a lot. Thus both farms come close to what a recent Number One document (the most important policy document issued each year by the Communist Party) has proclaimed as 'commercial family farms'. Although this Party document does not quantify this notion, it is generally assumed that such a commercial farm has at least 50 mu, some even refer to 100 mu. This is ten or twenty times as much as the average smallholding.

These first two cases represent one-off acquisitions of land. Here the *contracting* of land was key. The third case is based on ongoing land transfers: on *renting* land. This third case is located in Huangjue, a prosperous village in Sichuan Province.

The ability to transfer land (i.e. contracting land and/or renting the use right linked to it) generally plays an important role in entrepreneurial farming. It is an important feature in all three examples. Without the transfer of land (in whatever form) it would not have been possible to create these large farms. The discussion of these three cases is followed by that of another, highly important form of land transfer: the transfer of use rights in order to facilitate mining. The chapter concludes with a more general discussion of the possibilities for, and obstacles to, an entrepreneurial trajectory playing a significant role in China's agriculture.

Reclaiming land: the case of Sun

Sun is a 47-year-old farmer who moved to Sipai village where he has lived since 1985. Due to his poor family situation, he had to leave home after graduating from high school and go to north-east China, where his relatives live, to earn a living. In 1986 he married a local woman from a Hezhe ethnic household. Sun reclaimed 150 mu (i.e. 10 ha) of wasteland in 1989. This reclamation, i.e. the creation of new farming land out of swamps, was basically done using his labour and that of his wife and her family. This is an important detail: the land entered the farm as a 'non-commodity': it was not bought, but was created with the family's own labour.[2]

Since 'creating' this new land, farming has become the main source of livelihood for the family. In 1999, the farmer became the village clerk, with some support from his wife's brother, who is an official leader in Sipai Township. In August

2005, Sun was appointed as a village leader (second in command). His accountancy records describe the period between June 2004 and July 2008. His farm enterprise consists of 326 mu of land[3] that he owns and, since 2008, an additional 400 mu that is rented. Altogether this makes a total of 49 hectares (ha), although for the first 3 years on which this analysis is based (2005–2007) Sun had only some 22 ha of land. The work is currently done by Sun, his wife, her brother and some hired workers. He and his wife have two sons who attend high school in their father's home village. Figure 5.1 shows two pages of Sun's accountancy book.

Sun is highly respected in the village. He is a source of information for many other peasants. He is often consulted about the best seeds to use, about fertilizers, pesticides, etc. People have a lot of trust in him. He says he wants to help develop the village. In Western terms we could refer to him as an agent of change and/or an entrepreneur. There are no literal equivalents in the Chinese language. In the village the word *tou* (character 头), which refers to the head of the human body, would probably be used. *Tou* thus means 'he who guides the others' or 'he who makes the other elements move'. In 2007 a Chinese journal (*Kai Fa Yan Jiu Yu Lao Qu Jian She*, 2007, no. 2) published an extensive article about Sipai village in which this farmer is mentioned and described as 'head of becoming rich' (i.e. the leading example for those who want to make economic progress).

Sun's farm is large and it is quickly expanding. It is highly mechanized and, above all, it shows a remarkable *multifunctionality*, combining a wide range of different economic activities. In Table 5.1 (derived from the farm accountancy

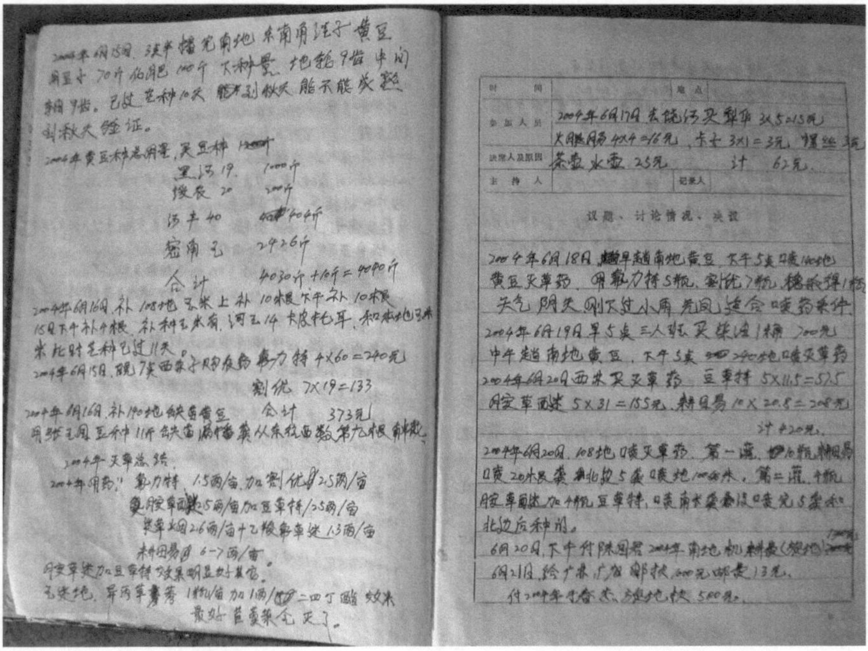

Figure 5.1 Two pages of the farmer's notebook

record) we distinguish these different activities, each of which provides a flow of monetary receipts into the farm. It is important to note that these different branches or aspects that we have distinguished are not organized as single and separate 'enterprises'. Rather, they compose one single unit. This might seem an irrelevant difference. But this is far from the case, since it implies that single transactions (or sets of transactions) do not necessarily have to be profitable. Their value might well reside in them strengthening another part of the farm, or an aspect of the networks in which the farm is engaged, or that they might provide some advantage that can be realized in the future. This also explains why the different categories are not separated from each other in the accountancy record. The record contains just payments and expenses that relate to one and the same set of activities: they are different *aspects* of one and the same rural livelihood. Their meaning or value cannot be understood in isolation, or derived from monetary calculations alone.

Benefit/cost relations

Table 5.2 gives an overview of yearly income (the data for 2004 only covers the last 6 months and the data for 2008 only covers the first 6 months, which implies that the harvest is not included in the data), of farming costs and of living expenses (typically, the portion of production used within the household is not registered: it is part of the non-monetary flows).

This data (which we have synthesized) reflect, in the first place, the relatively favourable position of Chinese farmers and peasants. In general, the gap between monetary income and monetary expenses provides a positive and, relatively high,

Table 5.1 The multifunctionality of Sun's farm (in yuan/year/activity)

Year	*2004*	*2005*	*2006*	*2007*	*2008*	*Total*
Agricultural production	–	107,725[4]	119,796	106,040	60,225	393,788
Animal breeding	2,566	1,400	–	–	–	3,966
Machine work	–	–	–	14,881	10,808	25,689
Salary for leadership	2,000	3,250	5,469	4,680	–	20,649
Reselling inputs	–	–	–	–	8,914	8,914
Food subsidies	816	3,269	3,268	5,440	–	12,793
Others	–	1,200	65	6,900	1,257	9,422

Table 5.2 Overview of farming income, farming costs and living expenses

Year	*Farming income*	*Farming costs*	*Living expenses*
2004	107,725	10,261	8,823
2005	119,796	39,073	26,335
2006	106,040	38,740	21,185
2007	60,225	26,753	17,834
2008		76,295	12,840

surplus (for a discussion on different notions of surplus see Martin, 1977). The 'clean part'[5] in 2005 was some 80,700 yuan. In most years the positive surplus derived from farming not only covers the living expenses of the (extended) household, but also allows for significant savings and accumulation. Thus, 'the heads that guide to richness' are emerging.

The relationship between the monetary expenses incurred in farming and the monetary income obtained is especially impressive when compared to the squeeze that farmers in Europe are currently suffering. For example, in the Netherlands farming costs (for the agricultural sector as a whole) are as high as 64 per cent of the Gross Value of Production (GVP) (LEI, 2008). According to the farm record, for this particular Chinese farm the figure is only 37 per cent (the average for the 2005–2007 period). This relatively favourable position of China's agriculture[6] is due to two factors. First, it reflects the current Chinese policy of favouring the countryside and the agrarian economy (recently through the exemption of taxes and extra payments for grain production). Second, it reflects the strong peasant-like nature of Chinese agriculture and the Chinese peasantry's search to minimize monetary expenses and dependency on the markets of factors of production (land, capital, labour) and non-factor inputs (feed, seeds, fertilizers, etc.). In Chapter 11 we will come back to this strategic feature.

Peasant or agricultural entrepreneur?

Is this farmer an 'entrepreneur'? He surely is very enterprising and is engaged in many different transactions. But are we witnessing the emergence of a new, entrepreneurial approach? Or do his activities follow the logic of a peasant?

Table 5.3 focuses on agricultural activities alone. It gives the average monetary receipts (calculated over 2005, 2006 and 2007), the average monetary costs and the resulting gross margin (i.e. the difference between monetary receipts and monetary expenses associated with farming). This gross margin or 'clean part' is, on average, 76,300 yuan/year (equivalent to around €9,500). Table 5.3 also provides information on *calculated* costs. These are the (hypothetical) costs associated with the flow of self-controlled resources.

The first item is the *calculated* salary.[7] As with the other calculated costs there are arbitrary aspects within such a calculation. Here we use the salary of a school teacher in a rural village: 1,800 yuan/month. Given the relative wealth and status

Table 5.3 Farming sales, monetary expenses and calculated costs (yearly average based on 2005–2007)

Monetary receipts (sales of products)	111,200 yuan
Monetary costs	34,900 yuan
Gross margin	+76,300 yuan
Calculated salary	21,600 yuan X 2 = 43.200[8]
Calculated interest for land property	65,200 yuan
'Net margin'	–32,100 yuan

of the farmer, this seems to be a reasonable point of reference.[9] The next item is land. In the village's land market, people normally pay around 200 yuan annual rent for 1 mu of land, although this will vary according to the quality of the land and the relationship between the parties engaged in the transaction. This farmer owns some 326 mu, and if we take a shadow price for the land of 200 yuan/year/mu (identical to the average local rent), the *calculated* financial cost of this land would be some 65,200 yuan/year.

Even without calculating all the other elements that probably exist within the flow of resources that are produced and reproduced in the farm unit (seeds, improvements to the land, animals, manure, etc.),[10] Table 5.3 clearly shows that the calculated total value of these internal resources is far higher than the value of resources mobilized from the market. The proportion of the resources that enter the farm as commodities (34,900 yuan) is only some 24 per cent of the value of the total resources used (34,900 ÷ (108,400 + 34,900)). The relatively high reliance of the farm on non-commodity resources highlights that this farm has a peasant nature.

If the farm did not have these non-commodity resources at its disposal (i.e. if market prices were paid for land and labour) then it would make a significant loss (calculated, as shown in Table 5.3, to be some 32,100 yuan). However, the farm is a feasible entity, precisely because a large part of the resources used are internal to the farm unit (and no money is paid for them) and another portion of the resources is mobilized through social networks (at no cost, or one that is below market level).

It is precisely this specific position between producing for the downstream markets and yet being autonomous (or distantiated from the upstream markets) that makes this farm, from an analytical point of view, a peasant-like unit.

Using Chinese characters and the underlying notions, we could refer to this farmer as '*zhongdi*' (种地). This means cultivating the land. It specifies a *relation*, i.e. the relation between man and the land. You have land (the most visible and most important of a wider set of resources) and it is used ('cultivated') in order to obtain a range of products, some of which will be sold on the markets. As Figure 5.2 shows, there are two flows of resources entering the production process and these are converted into two output flows: reproduced resources and marketable output. The first flow consists of non-commodity resources. These are produced and reproduced in the farm itself. Here the concept of '*zhongdi'* applies. The second, very different flow, is that of commodity resources. These are obtained from the markets. Here another notion (and an associated set of relations) applies: '*maimai*' (买卖), which literally means 'buy and sell' and, in more general terms, 'business' (business is indeed the combination of buying and selling). In such a case the flow of bought resources would dominate.

If we locate our farmer in this conceptual space, we might conclude that he is located somewhere between '*zhongdi*' (种地) and '*maimai*' (买卖), although more closely aligned towards the '*zhongdi*' (种地) side of the equation. If the farm were completely aligned with '*maimai*' (买卖) it would be impossible for it to exist.[11]

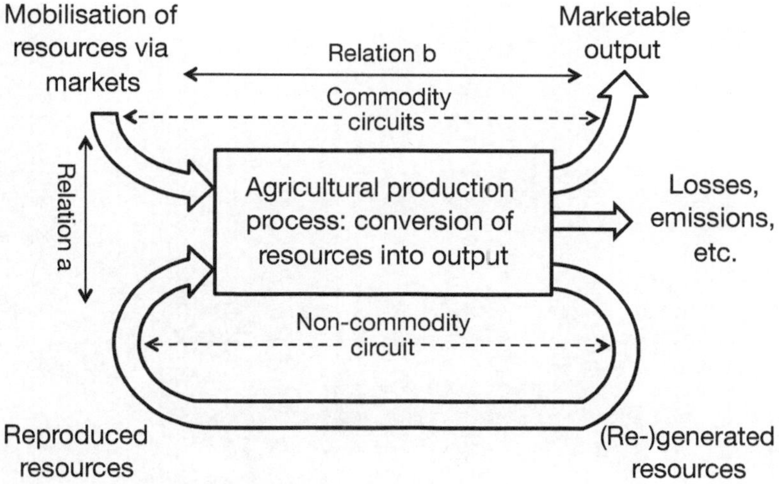

Figure 5.2 The flows entailed in farming

Contracting land

Whereas Sun reclaimed land, Zhao (see Figure 5.3), the farmer who is central to our second case, is a clear and in a way convincing example of the strategy of 'buying'[12] as much land as possible. Zhao farms in Sanggang in Hebei province (that is, on '*this* side of the Great Wall'). He contracts pieces of land, just as others buy tomatoes or whatever. He is able to do so because he developed an extended network as a driver of horse carts (in the epoch of the production brigades) and later as a trader: he knows many places, observes interesting differences, identifies opportunities and enjoys the trust that many people have in him.[13] Zhao worked also in industry in the south of the country, where he earned, at that time, 4.50 yuan per day. 'What I did was contract here [in his home village] two workers to whom I paid each 1.50 yuan/day and who worked my land here. Thus I saved 1.50 yuan every day and the work was all done.' Transactions like these are characteristic of the way in which Zhao operates.

In the early 1980s, like many others in the village, Zhao obtained some 5 mu of land in the valley. He exchanged a part of this land for another, far larger piece in the hills in order to convert it to a walnut producing area. He invested 40,000 yuan to cover the costs of paying the workers, buying the trees and the grafting. He is sure that once the trees enter into production he can quickly recover this investment. The calculations that lead him to this conclusion are not very precise. But he relies on what he saw while travelling: 'I have seen other farmers who earned some 23,000 yuan per harvest with less trees than I have here. Therefore, I think that I can repay my investment with three harvests.' Zhao acquired several other large plots. One is a 100 mu plot located far away from the village. Due to the distance other people were not interested in it. Zhao rented it (for 50 years)

Figure 5.3 Zhao

from the village committee and planted more than 10,000 poplar and robinia trees on it. Another plot (200 mu) is planted with 20,000 robinia trees. These trees already render a yearly income that currently reaches the 5,000 yuan level but which will grow considerably in the years to come. Another plot (20 mu) has been rented by Zhao to one of the local iron ore processing units to store its waste products. This renders him 6,000 yuan per year.

In Chapter 7 we show that not all farmers are involved in ongoing expansion per se – the current amount of land matches their physical capacity and their production matches their social needs and obligations. 'Enough is enough'. For Zhao it seems that such limitations are non-existent. Wherever and whenever needed, he contracts labourers (although he himself also joins in the work) and whenever an opportunity arises for obtaining and/or renting land from others Zhao will use it. While others were basically *working* the land (or working elsewhere in order to improve their land), Zhao was *buying* land (i.e. contracting it for very low prices) and looking for its most remunerative use. 'Normal things like grain production and raising pigs will never make big money for you. That will not work. It does not make you rich,' according to Zhao. He adds: 'However, in my way of doing things, I will be rich by the time I am 70 years old.' And, with a big smile: 'I am a *Nongmin Qi Ye Jia* [a peasant entrepreneur]. Peasants will never get rich, but I will be rich in the end. That makes me an entrepreneur.' He adds: 'And I will never sell it . . . How could I do that? That would be like throwing away the foundation of my richness.'

From an analytical point of view, Zhao occupies a nodal point that is at the intersection of different markets. He has a nearly permanent herd of goats and loves to go to the mountains with his herd. He raises the young animals and sells them just before the spring festival when prices go up. But the main function of having this herd is that it allows him to engage in large-scale goat trading. His herd is just a temporary station for many of the animals. Zhao gets orders from different places (as far away as the county town Baoding) and then starts looking for animals to fill these orders. He is able to find them because of his extended network. He takes the goats, sells them and only after having been paid himself does he pay his suppliers. This goat herding results in a permanent cash flow that allows him to continually invest in planting and maintaining his trees. He and his (extended) family provide much of the necessary labour. Anything else is mobilized in the local labour market. Thus the trade in goats is converted in constantly expanding woods with valuable trees. Over the last 5 years or so he started including walnut trees in this scheme as well. Currently, Zhao has more than 1,000 walnut trees. The extended hillsides needed for tree growing were obtained through contracting land for low prices at a time when hardly anybody recognized their value.

The goats also produce much needed manure: one basket every night, which suffices for two (walnut) trees. The rest of the manure he needs is obtained from relatives on the basis of reciprocity.

Zhao is a keen entrepreneur: he acts wherever he sees opportunities not yet perceived by others. He is the early bird catching the worm. Other people see him as a 'shrewd guy'. People do not like him very much since he is always trying to get the most out of his deals. People often feel cheated. There is envy as well. The jokes are telling: Zhao is often referred to, with a laugh, as being 'a new large landowner'.

Zhao controls a considerable amount of land. But these acquisitions did not follow the pattern that is typical for entrepreneurial agriculture. The lands were not acquired through a land market. Instead they were granted by naïve village committees that contracted the land to him for too low a price and for too long a period. And it was his herding and trading of goats (plus some other activities) that financed these acquisitions and the subsequent planting of trees. When asked, Zhao will say that he absolutely won't sell (the use rights of) this land. He will further develop it. The whole operation with the trees is framed within a family strategy:

> I will only sell my trees and my land when my grandson [now 9 years old] marries. Then I can build him a nice house and pay for his wedding. And give him the money to invest in something new . . . . Or maybe I will give it to my own son.[14]

Land transfers: the farm of Mrs Xiao and Mr Wang

While the previous two cases are relatively easy to locate on the axis that goes from peasant to entrepreneurial agricultural, the third case is more difficult to understand. The first and second cases (of Sun and Zhao respectively) show forms of expansion that go into the 'land of nobody', i.e. wetlands and harsh hillsides. They share one feature in common: at the start these spaces were not yet recognized by most people as representing an economic opportunity. They still needed to be converted, through hard labour, into resources that provide the foundation for a farm. Sun and Zhao are exceptions – and were thus able to create large farms that are now of considerable value. A shared distinctive feature here is that these resource bases[15] were not primarily constructed through transactions on the land market (nor through associated transactions on the capital market). They have been constructed through labour investments (and by using social networks and the political support of local authorities). This has allowed our two protagonists to construct large peasant farms that are, as we have seen, very much operated in a peasant way. What is more, if they were run in a fully fledged entrepreneurial way (that aims to valorize all costs that have been made), it would be hard to continue with them.

The case of Mrs Xiao and Mr Wang is more enigmatic. They actively operate in the land market and have constructed (and continue to reproduce) their farm through land leases. They 'buy-up' the use right for various family plots and pay, in compensation, an average of 1,050 yuan per year per mu. The lease contracts are renewed annually and, as a rule, the rent increases from year to year. So they have an ongoing involvement in the land market. The same applies to the labour market. During part of the year Ms Xiao and Mr Wang can do all the work by themselves but there are several peak periods that require them to hire up to six additional workers a day, sometimes even up to fifteen. But even then Xiao and Wang work alongside these workers.

The farm is located in Huangjue village in Sichuan province. It consists of 80 mu of land (and was expanded to 90 in 2014)[16] that lies alongside the river. Here a range of vegetables is grown. There are at least twelve different crops including watermelon, spring onions, tomatoes, pumpkins, courgettes and different types of cabbages. Alongside these marketed crops there are others that are being tried out: if they can be grown here successfully and if the prospects for marketing look good, these might soon be grown on a larger scale and for the market. Vegetable prices (most are marketed at the wholesale market in the neighbouring county town of Leshan) are highly volatile. One year spring onion prices were very high (2 yuan per jin – 1 jin = 0.5 kilogram) and thus: 'people with just a few mu planted with spring onions could earn enough to afford build a new house.'

Mostly, however, prices are not so good. Xiao and Wang (Wang is the one who does the marketing and he is very well informed about prices, trends and expectations)[17] are very clear about 'not following the market'. They see that when prices for a specific crop are very low in, say, year n, this crop will produce significant profits in year n+1. They definitely follow an anti-cyclical strategy.

Mrs Xiao and Mr Wang both have entrepreneurial experience elsewhere. Alongside their own small farms (now run by their partners and other family members) Mrs Xiao produced insects used in the pharmaceutical industry for making specific medicines, while Wang ran a labour team that worked in the

Figure 5.4 Mrs Xiao and Mr Wang

construction of derivation canals for water schemes. Both also worked as migrant labourers. In addition, Wang learned tomato growing in a large enterprise and thus became an experienced technician. The savings made by Mrs Xiao in the production and selling of insects have been the main source of capital (together with loans from relatives) to rent the land, finance the inputs and pay the workers. After the second year of running this farm, the costs could be paid with earnings from the previous year. Currently early crops from one year finance the planting of later ones. The cycles are intertwined.

The waste produced by the farm goes to a neighbouring peasant lady who feeds her five cows with it. The manure produced by the cows goes on the vegetables. Thus, there is no monetary exchange – instead it is based on reciprocity.

The farm of Xiao and Wang is not only bigger than a typical peasant farm in this area, it also has a different cropping scheme and different production techniques. On the whole it is more extensive, and this is evidently done to minimize the need for external labour. First, they produce the 'cheaper vegetables'. The surrounding peasant farms, with small plots, typically grow high value vegetables that require more labour. Second, they use plastic on the land to suppress weed growth. On neighbouring peasant farms, frequent manual weeding is the rule. Third, peasants more often use greenhouses (plastic tunnels). This is not done on the Xian/Wang farm because building and then opening and closing these tunnels each day would require too much labour. Thus this peasant farm has adopted several mechanisms to adapt to the vastly expanded acreage. But is it still a peasant farm? Or an entrepreneurial one? Or, for that matter, a capitalist farm?

We believe it can be argued that the Xiao/Wang farm is definitely not a capitalist enterprise. Although workers are hired, this farm is not run as a capitalist enterprise. The hired workers replace, as it were, the family members who are not available (partly because of the demographic profile). It is an 'additional and substitutive utilization of hired labour.'[18] It substitutes the missing family members, just as it substitutes sophisticated mechanization and automation (which is not available here and/or is far more expensive than the use of hired labour). And it is additional to the farmers' own labour employed in the process of production. This 'additional and substitutive utilization of hired labour' is not oriented towards (or restricted to) the production of surplus value. The farm promises to generate (in those years with adequate market prices) a very good income.[19] But it is a labour income (Chayanov, 1966; Ploeg, 2013) grounded on the conversion of ecological capital (land, water, seeds, manure, etc.), by means of peasant labour, into marketable products.

The Xiao/Wang farm is also not a typical entrepreneurial farm. It is not built on financial capital (loans from banks) but mainly on savings resulting from their own labour and used to obtain land. Nor is it oriented towards ongoing expansion. It is not built on the takeover of other farms.[20] It leases land, and whenever the owners want it back it will to be returned. Xiao and Wang are extremely 'entrepreneurial' in the sense of looking for new and daring opportunities. They are ready to take considerable risks, just as they are willing to invest their own savings and their own labour. As a matter of fact they work very hard. But all this does not make for a farm that is organized in an entrepreneurial way. We are talking here about entrepreneurship as a *personal* feature, an *individual* attribute, of both Xiao and Wang – not about a *structural characteristic of the farm*.[21] As entrepreneurial as Xiao and Wang might be (in identifying and exploiting economic opportunities), their way of farming very much remains a *peasant way*. It differs from the neighbouring ways of peasant farming developed by the smallholders in this village, but is nonetheless peasant farming, albeit in a relatively large enterprise. This is because the Xiao/Wang farm, as much as those of Sun and Zhao, rely on self-provisioning, reciprocity, step-by-step development, a self-constructed resource base and the centrality of labour (be it own or additional hired labour) as its foundation stones. But, just as Marcel Jollivet argued when talking of French agricultural entrepreneurs – 'within each entrepreneur there hides a peasant that might come out one day again' – one may wonder whether or not these large Chinese peasants hide an entrepreneur that might come to the fore, probably at an unexpected moment (Jollivet, 2001).

Mining the land

Peasant farming is built on a self-constructed and self-governed resource base. Entrepreneurial farming, by contrast, is built on commodity flows and the main resources are mobilized through the relevant markets, entering the production

Figure 5.5 The 'mining' of agricultural land

process as commodities. Land is a particularly important case here. When land (or the usufruct right related to it) is systematically sold, bought and resold (or, for that matter, leased) and its use is increasingly governed by the relations that reign in the market, land becomes a commodity[22] whose allocation and productive use become oriented towards the most profitable possibilities.

Turning resources such as land into commodities is reflected in another activity that is drastically reshaping the landscape in several parts of China. Sanggang village is one of these places (it is the village where Zhao, the second case discussed here, operates). All along the valley in which Sanggang is located there are mining activities. These have a strong tendency to commoditize the land. While land transactions between peasants are not (as yet) leading to the emergence of a land market,[23] mining represents a powerful intervention that de facto converts much of the land into a commodity.

There are several iron ore processing factories in the valley in which the village of Sanggang is located. Some are old, others are no longer working (these are former state enterprises). A few are brand new and awaiting the necessary licenses. Most factories, however, are small to medium size units that employ some twenty workers (three shifts of seven men), which may daily process 60 tons of rock containing iron ore that possibly will render 15 tons of iron ore.

All these different factories are located in the valley – simply because they need a lot of water for their production processes. The rocks are excavated in the mountains and transported by lorries or three-wheeled tractors from the mines to the factories. The mining, transportation and processing offer a range of employment opportunities to the villagers.

In the sandy soils of the valley there are also layers (often located at a depth of 3 to 5 meters) that contain iron ore. This has brought about a new practice. Businessmen offer 6,000 yuan per mu for the right to 'mine' the land. All the topsoil is taken away. The iron ore containing sand is taken apart and loaded onto trucks to be transported to a factory where it is sold. The deal mostly includes the return of the washed sands[24] from the factory to the 'mined' field. However, this part of the informal contract is not always respected. The businessman might go broke or simply disappear. There are also informal contracts that do not include delivery of the washed sands. Then the farmer himself needs to arrange for transport of the washed sands to his field in order to restore it (to fill the excavated holes). In addition to these kinds of arrangements considerable amounts of sand (that do not contain iron ore) are accumulated in large man-made hills that will be sold later for highway construction. This latter practice is certainly illegal, although there is also much uncertainty about the legality of mining the valley lands. The businessmen involved are clearly operating in a grey zone.[25] There is also considerable insecurity and ignorance about the final results. It is evident that with washing all the organic material in the soil will be lost. Thus it will probably take considerable time to restore soil fertility (and there will be complete dependency on chemical fertilizers). However, several farmers claim that the washed sands are very clean and that all the rocks have been taken out of it. They argue that the soils are especially good for peanut production.

Overall there are considerable ambiguities in all this. Mining might be favourable (bringing more resources into the village), it might equally be a disaster (destroying the land). One cannot know beforehand. Experiences in other parts of the world seem to confirm this delicate balance. Take for instance Friesland, the prosperous agricultural province in the north of the Netherlands. In several places depressions in the landscape testify to the harsh years during and after the agricultural crisis of the 1930s when impoverished farming families had to sell the fertile top soil of their lands that they had enriched during decades of hard work. To save their families they literally had to convert soil fertility into a commodity and sell it – after which it was excavated (i.e. 'mined') and transported elsewhere. Equally, in the same province there are villages (such as Eastermar) where in the nineteenth-century farmers sold parts of their commons converting their turf into peat for burning. Now, some 150 years later, peasants in this village are still better off than those in the neighbouring villages. This is reflected in a statue in the village of a young man holding up his own trousers – a proud symbol of what we now call autonomy.

One never knows whether things will turn out for the best. The only thing we know for sure is that it all depends on balances: the balance between peasants and miners (see also Long and Roberts, 1984), between confusion and determination, between long-term development and short-term exploitation and profits.

These balances notwithstanding, some villages (such as neighbouring Dabeizhuang Cun) have decided not to develop mining in order to avoid the destruction of the landscape and the degradation of water, air and soil qualities, while preserving the prospects of farming, agro-tourism and rural well-being.

Is entrepreneurial farming developing?

Just as there are degrees of 'peasantness' (Toledo, 1995; Ploeg, 2008), there are also degrees of entrepreneurship. A style of farming is more entrepreneurial when it is highly reliant on commodity flows, while peasant-like farming is built upon a self-controlled resource base that is primarily non-commoditized. Entrepreneurial farming builds (and is dependent) on financial and industrial capital (for credit, inputs, technologies): land, animals, seeds, nutrients, machines, etc. enter the process of production as commodities that need to be valorized through the process of production. By contrast, peasant farming is more reliant on ecological capital which is part of one's own, self-controlled resource base. Being built on industrial and financial capital means that entrepreneurial farming is strongly embedded in a complex web of commodity relations. Another distinguishing element is the strong orientation towards the market as the organizing device (and ordering principle), as opposed to farming being governed by cultural repertoire, collective memory and Chayanovian balances. The market generates a need for ongoing expansion (especially when there is a high degree of indebtedness).

The three examples discussed in this chapter cannot, we believe, be understood, as representing agricultural entrepreneurship. Yes, the land has been contracted. Zhao, Sun and Xiao/Wang all operate in the 'land market'.[26] All three do so in order to enlarge their own resource base. Sun and Zhao obtained the land for a very low price – and they were successful in obtaining it because they had access to extended networks and good *Guanxi* relations. However, once the use-right was obtained, the land *stopped* being a commodity. Zhao in particular is adamant that he will never sell his land to others. The mobilization and use of land, labour, capital and various inputs are governed by social networks and the rules that apply in these networks. In both cases the market is an indispensable *outlet*, but it is not the ordering principle: it does not govern allocation of the factors of production.

Sun and Zhao have expanded their farms considerably. They were, as the saying goes, the early birds that caught the worms. Now there is hardly any land left. Hence, any repetition of this particular strategy is hardly feasible.[27] More importantly though, is that peasant farming cannot be directly equated with smallness. A larger farm is not necessarily an entrepreneurial one. It is not magnitude that defines the mode of farming.[28] Entrepreneurial and peasant farming differ in the way that the relations between man and the land are constructed; in the way the land is obtained and in the land use strategies employed. They also differ in the way the unit of agricultural production is linked to the outside world. It is precisely here that the relevance of the Xiao/Wang experience emerges.

In general terms it might be argued that the politico-economic conditions required for entrepreneurial farming are lacking. First, there is no fully developed land market within which land is a 'free floating commodity' that may pass from seller to buyer without restriction. Even if the appropriate regulatory schemes were introduced, we doubt whether Chinese peasants would sell their land en mass (see Chapter 3). Second, it is hard for small and medium entrepreneurs to access the capital market; they prefer to lean on informal credit from friends, neighbours and relatives. The presence of, and active involvement in, the markets for land and capital are decisive for entrepreneurial farming precisely because this style of farming can only proceed, and develop through, ongoing expansion.[29] Time and again the available amount of land needs to be enlarged and this requires continued involvement in the capital market in order to finance land acquisitions. Entrepreneurial farming implies takeovers: the takeover of the productive and income-generating activities of other farms.[30]

A third politico-economic condition concerns the labour market. In rural China, the price of labour is very low. This means that when scale enlargement is considered, building a capitalist farm based on salaried labour is a self-evident alternative. This is more plausible than opting for an entrepreneurial farm. Fourth, there is no rural culture that supports entrepreneurial agriculture. These four conditions all apply in the case of Xiao/Wang: they could become agricultural entrepreneurs, but the prevailing conditions basically exclude or prevent such a change.

For agricultural entrepreneurs running a farm is primarily a *financial* operation. It is about using money in order to make more money. The agrarian process of production is structured so as to allow, in the most profitable way, for such a financial operation. The typical entrepreneurs perceive all their resources (whatever their particular biography) as representing money. Land embodies money, just as cattle, crops, buildings, nutrients, soil biology, etc. represent money. Everything is perceived as a commodity. It is tradable. It has an exchange value that predominates over its use value. Everything is known by its monetary value – and is subsequently used to generate more money. The return on the money embodied in the different resources needs to be at least equal to the average profit level. Thus, calculations are needed that embrace and include all the details and all the elements that play a role in the process of production.

Such views are, however, alien to the cultural repertoire that governs large parts of social life in rural China. Although particular parts of the entrepreneurial model might be present, such a view is, as a whole, absent. Not everything is for sale, not everything is known primarily through its price in the first place. And the market is far from the only mechanism that governs the allocation, combination, use and exchange of resources. Even in those instances where the market does play an important role, there are wider social relations within which these markets are embedded and that make them function.

As Zhao, who comes close to at least some elements of an entrepreneurial position, argues:

> The difference between me and a businessman? I think a businessman calculates very carefully. A businessman considers not only the price, he also considers the costs. Peasants do not consider the costs, neither do I. How could I calculate the price of my own labour? How should I count the worms and insects that the hens eat? How should I count the residuals the pigs eat? How should I count the grasses when I go with [the herd of] goats into the hills? All this cannot be accounted for. We cannot consider this [in the same way]

> that businessmen consider buying and selling. Prices are very important for businessmen. Their business is all about buying and selling and thus about prices. For peasants it is different. Peasants only consider the price when selling.

In short, the world according to peasants basically differs from that of entrepreneurs. Where peasants see resources as being useful in order to produce, entrepreneurs see a commodity that might be converted into another commodity that represents greater monetary value.

The irony is that those who are closest to perceiving the world around them primarily in terms of commodities do not engage in entrepreneurial farming but, instead, opt to sell their land for mining. That is far more profitable. We should, therefore, perhaps rephrase our tentative conclusion about the lack of 'external' conditions allowing for entrepreneurial farming: *maybe Chinese peasants are too shrewd to become agricultural entrepreneurs.*

What will the Chinese state do?

Although we think a new entrepreneurial style of farming is unlikely to emerge spontaneously from the Chinese peasantry itself, it could possibly be imposed by the Chinese state. Ironically, the emblem of the 'family farm' is being used as carrier, and as camouflage, for entrepreneurial farming. The term 'family farming' is being used increasingly in official discourses and it is a concept that is rooted in a virtual reality. In reality peasant farming is identical to family farming, but this virtual notion refers to something that barely exists. A 'family farm' (in official state discourses) refers to a farm with at least 50 mu (that is 3.33 ha), but preferably 100 mu or more (>6.66 ha). That aside, it is a vague notion. Definitions, standards and criteria vary among different provinces, municipalities and even counties.

In March 2013, the Ministry of Agriculture in China carried out the first national survey on the development of family farms. An operational definition was used which is summarized in Box 5.1.

In this national survey, 877,000 family farms were counted (this is in stark contrast to the 200 million peasant units of production mentioned in the background paper for *State of Food and Agriculture* 2014). These 877,000 family farms[31] (less than 0.5 per cent of the total number of farms) are working 13.4 per cent of the available farmland in China. They are the larger farms, notably located in Heilongjiang and Inner Mongolia. The average labour input amounts to six persons, of which 1.7 are long-term hired workers. Most of these 'family farms' specialize their production and operation in either crop cultivation or husbandry; only 6 per cent have taken up diversified operations. The average acreage of all 'family farms' is 13.3 ha. This is nearly twenty-seven times the average national farm size. Table 5.4 shows the differences within this category of 'family farm'.

To put the categories distinguished in Table 5.4 into context it would be interesting to locate the cases discussed in this chapter within the different size categories. The vegetable producing farm of Mrs Xiao and Mr Wang falls in the

Box 5.1 The concept of family farm according to the Ministry of Agriculture

Criteria for a 'family farm' in a survey by the Ministry of Agriculture:

- Operators of a family farm should be registered as rural residents (*hu kou*).
- Family labour dominates and there is no long-term hired labour – or the amounts of hired labour are less than family labour.
- Income from agriculture provides a major source of income for a family farm and net income from agriculture accounts for over 80 per cent of the total income of the farm.
- The scale of production reaches certain standards and is stable. The size of farmland (with a contract period of more than 5 years) in grain production should be more than 3.33 ha (double cropping) or more than 6.66 ha (single cropping). For farms with cash crops, husbandry or both forms of production, the operational size should reach the standards defined by local departments of agriculture.
- Operators of a family farm should have received technical training in agriculture.
- A family farm should have complete financial records.
- A family farm should act as a role model for other farmers and agricultural households.

Table 5.4 The scale of officially identified 'family farms' in China[32]

Scale (unit: ha)	*Number of farms*	*Percentage (%)*
3.33	484,200	55.2
3.33–6.66	189,800	21.6
6.66–33.3	170,700	19.5
33.3–66.6	15,800	1.8
66.6	16,500	1.9
Total	877,000	100

second category (3.33 to 6.66 has). The tree producing farm of Zhao, who controls several extended hillsides, falls into the third category that ranges from 6.66 to 33.3 has. Finally, the maize producing farm of Sun (the one located on the other side of the Great Wall) falls in the fourth category: it has some 600 mu (which equals some 40 ha).

It is, of course, impossible to generalize the characteristics of the three cases discussed in this chapter in relation to the characteristics of 'family farms' as defined by the state. Nonetheless, the cases discussed here at least suggest the possibility that not all of these 'family farms' are entrepreneurial ones. It is quite possible that many of them are simply large peasant farms. But then within the current discourse of the Chinese state it is not necessary that these larger farms are, de facto, entrepreneurial farms. The point is that they are supposed to be transformed into entrepreneurial farms. This will occur by making them the focal point of agricultural and rural policies and the main recipients of state programmes and subsidies. Farms with less than 50 or 100 mu are supposed to disappear. Their land is to be reallocated to the (expanding) 'family farms' while their labour force should move permanently to the cities. There are already preludes to this in the current state discourse. In recent Chinese history the peasant has been represented in different, often contradictory, ways (Day, 2013). Currently peasants are depicted as a major obstacle to development (Schneider, 2014). It appears that the classification of peasant farms as passive and backward looking and the large 'family farms' as entrepreneurial, is needed to construct some legitimacy for this policy of selection and exclusion.

The formal definition of 'family farm' in China should be understood in the broader context of agrarian change in China. Two forms of capitalist transformation of China's agriculture can be identified over the past two decades. The first development path concerns the commoditization of both agricultural production and the reproduction of farming households. The second path centres on the submission of peasant units to enterprises that operate beyond the household level (variously known as vertical integration, industrialization and scaling-up) (Zhang and Donaldson, 2010; Schneider, 2014). Recently a third path has come to the fore. That is the creation of capitalist farm enterprises, often financed by urban entrepreneurs who increasingly discover agriculture to be a profitable field for investment. We discuss this newly emerging pattern in the next chapter. Here we limit ourselves to the observation that the still immature (and much contested) policy of 'family farming' carries the very real threat that it becomes a vehicle for rich urban people to construct a new 'elite agriculture' (as is happening at the moment in many other countries in the Global South – see Sitko and Jayne, 2014). 'Capitalization' and 'capitalized farming' will become key words that refer to a massive use of new technologies, high investments and elevated debts to assets ratios. This may lead to the emergence of an entrepreneurial agriculture. But its emergence as a trajectory will not be wanted, tested, adopted or further developed by the large majority of peasants; rather it will lead to an expropriation of their resources and their livelihoods. In the Songjiang district near Shanghai there is a demonstration area for 'family farming'. It is based on the premise of land transfers. By 2011, 99.4 per cent (!) of the land here had been transferred to 'family farms'.[33] One must have serious reservations as to whether this makes for a solid agriculture or will contribute to food security. During the current period of prosperity we might well expect to see such a strong upsurge of 'elite agriculture' – but when prices go down and price volatility emerges, we can also expect that this 'elite' will be the first to stop farming and that it will do so very quickly.

Update

The type of mining described in this chapter has already been declared illegal for several years and the state currently uses satellites to spot illegal mining. But it is difficult in practice to stop it, partly due to the entanglement of mining interests with local and regional authorities. The valley of Sanggang village has been badly damaged by both large artificial hills of washed sands from the iron factories and other hills of sand and gravel waiting to be used in highway construction. Around these hills an intriguing drama is now developing that sometimes seems to be derived directly from *One Hundred Years of Solitude*, Gabriel Garcia Marquez's famous novel about the many transgressions that occurred in Macondo – a fictitious village in Colombia. The washed sands (from the iron factories) are now sold to newly emerging 'sand factories'. The latter use the sands to fill the many holes in the landscape. This is done under the cover of 'river bed reconstruction'. Thus, in a way, they restore the fields – but not without taking another part of the land and selling it as sand to be used for brick making, or to obtain yet more sand, rocks and gravel for highway construction. Misery (the washed sand) is converted into a commodity, sold, and used to obtain more land that is converted into another commodity. Thus misery is reproduced and made into a permanent phenomenon. Within this logic, the possibility that farming in these fields may be impossible in the future is considered to be unproblematic. Some people are already thinking about developing them into parks for luxury recreation and second houses for Beijing residents. The rhetoric of Deng Xiao Ping about 'getting rich' unleashed an incredible energy that helped to change China. But it seems also to have unleashed some seemingly untameable and voracious animals. Mining is one of these. It might be registered through satellite surveillance, but it cannot be stopped by it. This is partly because these extractive activities are given new names that obscure what is really happening – as was the case in Macondo.

Notes

1 The first part of this chapter is based on an article published in Chinese and English. See Yong and Ploeg 2009 and 2014. The second part owes much to Ye, 2002 and Ye *et al.*, 2009.

2 As argued before, entrepreneurial farming is grounded on the acquisition of commodities in order to convert these into more money. Land is one of the key commodities that needs to be acquired. Sun, however, 'constructed' his own land through the labour available in the family.

3 There is some confusion here (as is often the case in the countryside). According to some sources the farmer has a property of 400 mu. However, he only declared 326mu when claiming his food subsidy. This, according to some informants, was for fear that some of his possessions might be taken back by the state.

4 This income was realized and recorded in 2005. It was, of course, *produced* in the preceding year (2004). This also applies to the other years. However, here, and in the other tables, we follow the logic of the record book. The exception is Table 5.3 where incomes are directly related to the costs incurred in production.

5 This is a folk concept used in many parts of the world. It simply refers to the difference between monetary income and monetary costs (i.e. the difference between the inflow

and outflow of money). For a general discussion see Ploeg, 2003, Chapter 2 and also Yong and Ploeg, 2014.

6 China has also undoubtedly had periods when farming did not realize any 'clean part' that benefited the peasant producers. At times all the surplus produced was taken away or taxed by ruling elites. The farmer whose records we examined also went through such periods, especially in the initial years during the reclamation of land, when the surplus levels were very low. Today, the augmented 'clean part' represents (just as it did in the past in, say, the Netherlands) emancipation and progress.

7 As argued by Fei, 'for those who work on the land themselves, the wage is not a real item of expenditure; it is a part of their standard of living.' (Fei and Chang, 1945:154).

8 There are not only two full-time labour equivalents from the family working on the farm, but also two families (the one of the farmer and another of the wife's brother) supported by the income generated by the farm.

9 Of course, the farmer only works part-time, since he is also active as village leader. His wife is working full-time on the farm and her brother also makes a considerable contribution. Together the input of family labour equals nearly 2.0 labour units. Another point of reference is living expenses. According to Table 5.2 these are, on average (for the 2005–2007 period) equal to some 21,780 yuan per year, which is more or less identical to the 'calculated salary' that we have included in Table 5.3.

10 Other important elements of this flow are food for the family's consumption (vegetables, pig meat, chicken meat, etc.), straw and soy used for heating the house and cooking as well as fertilizing the land.

11 In his study on peasant agriculture in Yunnan, Fei and Chang (1945) arrived at a similar conclusion (see page 50 on pig-rearing and page 64 on the importance of non-commodity exchanges).

12 'Buying' occurred here in the sense that he contracted the land use rights from the village committees for a 30–50 year period.

13 The biography of Zhao is documented extensively in Ye, 2002 and in Ye *et al.*, 2009.

14 As a matter of fact Zhao has already invested some 300,000 yuan in building a brick factory for his son.

15 The land was acquired through transactions with the village committees, but at that time it was still a useless resource. It was at best an asset. It only became a resource (and part of a wider resource base) through the hard work of the farmers involved (draining wet lands, levelling, manuring and planting trees).

16 It is not however just one block, but scattered pieces. This is because only some villagers were willing to lease their land or part of it. This is a striking difference with the cases we will discuss in the next chapter.

17 He goes not only to the market to sell their own products but also to observe and chat with traders. He receives market information continuously on his mobile telephone.

18 Here we follow Wang (2009) who distinguished the 'consumptive use of hired labour' (by women and elderly in the countryside in order overcome the lack of available family labour) from 'productive utilization of hired labour' in capitalist farms. This is close to the recent argument by Ye and Pan (2015): 'in reality the dichotomized boundary built on the presence or absence of hired workers has been blurred. Family farms are showing huge diversities in their ways of production organization [. . .]: they can farm self-owned land, state-owned land or rented land; their production can be labour-intensive, technique-intensive or capital-intensive [. . .]; they may produce most of their own needs, only a small part, or none without ceasing to be family farms; they may hire from one to a few labourers without becoming anything like capitalist enterprises.'

19 Their estimation is that, on average, diversified vegetable production may result in a GVP/mu of 5,000 yuan. Their input use is estimated as being around 25 per cent of GVP. Thus, the labour income is some 3,750 yuan/mu. With 80 mu this generates a yearly income (in good years) of 240,000 yuan. From this the workers need to be paid, the land, and, of course, Xiao and Wang as well.

20 Nor can it be understood in terms of 'accumulation by dispossession' (Harvey, 2003). While there definitely is such 'accumulation by dispossession' in China (as elsewhere) the particular cases discussed here cannot be understood in this way. There is no accumulation (understood as investing surplus value in order to obtain more surplus value) here, nor is there dispossession (not even in the Xiao/Wang case since the land is leased to them on a voluntary and short-term basis). Equally, there is no accumulation *by* dispossession. The wealth produced by the Xiao/Wang farm is not based on the takeover of the wealth of others, but on the productive use of the acquired resources.

21 This also requires structural preconditions, such as a fully working and stable land market, widely available commercial credit and a wide acceptance of entrepreneurial repertoires. None of these conditions are generally present in China's countryside.

22 A single acquisition of a plot of land (especially when funded with one's own savings) does not necessarily result in the land being commoditized: when it is used according to the local cultural repertoire it functions as a use value, as a resource (Appadurai, 1986).

23 Land that cannot be worked by the family, due to migration or some other reason, is temporarily assigned to relatives, neighbours and/or friends in order for them to care for it. This is done on a reciprocal basis. No rent is paid.

24 The iron ore is taken out of a mixture of sand and water with electromagnetic power. The remaining mix of 'washed sands' is piped to a huge basin where the water is slowly drained away.

25 In this respect it is telling that the Party Secretary (who is the owner of the factory in the village) does not want to process the material from the fields in the valley.

26 We have put 'land market' here in inverted commas to indicate that it is a market that is far from 'perfect'.

27 Only when many of those who are migrating are willing to sell their use-rights, would there be opportunities for repetition.

28 Only when magnitude is combined with other features such as ongoing expansion, high commoditization, high debts to assets ratio, the elevated scale of farming, etc., does it become a significant characteristic.

29 In turn, this is due to entrepreneurial farming being a financial operation taking place in situations where there is a squeeze on agriculture.

30 As a result of which relations of mutual competition and the yardstick of competitiveness are introduced.

31 If we take into account the presence of 395,000 non-family farms in China (see Chapter 6, note 1), it could be argued that there are, in China as a whole, some 482,000 larger peasant (or family) farms. This is just 0.25 per cent of all farms in China.

32 农业部首次对全国家庭农场发展情况统计调查结果. www.sd.ifeng.com/zt/sdjtnc/jdjtnc/detail_2013_06/21/919015_0.shtml, 2013/06/21

33 陈明鹤. 土地流转与家庭农场的关系探讨：以辽宁为例. 党政干部学刊，2013年第8期, 45–50.

6 The rise of capitalist modes of farming

Jan Douwe van der Ploeg and Ye Jingzhong

The previous chapter focussed on entrepreneurial farming; this chapter discusses three different mechanisms currently used in China to construct new, corporate farm enterprises. These are the centralization of milking in new, large units (an operation that followed the melamine scandal), the creation of co-operatives for horticultural production (which goes together with the introduction of new technologies) and the contractual shift of usufruct rights of the land from peasant producers to newly created, large private farm enterprises[1]. We also discuss the shifting distribution of benefits and costs for each of these mechanisms. The chapter concludes that it is highly questionable whether these large–scale, capitalist farm enterprises are making any meaningful contribution to agricultural growth, rural development or food security.

Some theoretical notions

In *Making Capitalism in Rural China* Webber rightly observes that 'capitalism [. . .] is a specific way of producing and distributing goods and services' (Webber, 2012:9).[2] And he immediately adds that 'there is a host of *other* ways of producing and distributing goods and services' (ibid., italic added). In this respect Webber refers to 'peasants and small traders who are independent commodity producers; cooperatives that integrate ownership and work in various ways; and state-planned and controlled organisations that are not in business to make profits [. . .] and there are many others' (ibid.). This position differs from those who view capitalism as the central nervous system of society as a whole, a view that implies that *all* activities (all production and all marketing) should be understood as capitalist. In Webber's view this vision is problematic, since 'in most places, even in "capitalist societies", all these [other] ways [e.g. peasant production, co-operatives, state enterprises, etc.] are present: they may provide preconditions for, be subservient to the needs of, or be completely independent of capitalist producers' (Webber, 2012:10).[3]

The relations between capitalist and other ways of producing and marketing are governed by 'the economy', understood as 'a system of rules, formal and informal, governing and unifying the allocation of labour within and between different forms of production' (ibid.).

Thus different and mutually competing 'systems of valuation' (Webber, 2012:19) can, and generally do, operate within the same social and economic context. Our argument here is that in Chinese agriculture there are essentially three such systems of valuation. These are peasant agriculture, entrepreneurial agriculture and capitalist agriculture. They are contrasting ways of producing and marketing. Equally, there are contrasts in the ways in which each contributes to the generation of social wealth, the production of different class structures and different patterns that regulate access to (and ownership of) the means of production.

In the past China has experienced other systems of valuation (such as the people's communes, production teams, state enterprises). A few remnants of these remain. Currently, the agricultural sector is dominated by peasant agriculture, which arose phoenix-like from the ashes of the radical breakdown of the commune system at the end of the 1970s and beginning of the 1980s. However, there are also pockets of both capitalist agriculture (which emerged in the 1990s and 2000s) and entrepreneurial agriculture (although at the moment this is more an intention than a reality).

The interrelations between these three ways of farming are far from stable. It can be argued that China is currently experiencing a complex and contradictory process of transition that is continuously reshuffling the interrelations between the three ways of farming. However, the likely outcomes of this reshuffling remain unclear. Key government policy objectives, such as strengthening food sovereignty, enlarging the domestic market, reducing poverty and avoiding increased levels of carbon dioxide emissions will all be heavily influenced both by the dynamics and the outcomes of this transitional process and the resultant balance between the different ways of farming.

Analysing different ways of producing and marketing agricultural products necessarily involves two interconnected lines of inquiry: one that considers the social aspects, the other the material aspects. In terms of the social relations of production Bernstein (2010) has summarized four key questions: Who does what?; Who owns what?; Who gets what? and, finally, What is done with the surpluses? In terms of the second line of inquiry into the material aspects of processes of production and distribution, the main questions are: What are the main resources?; How are they developed?; How are they converted into products?; and How are these products channelled to consumers? In more operational terms this raises questions such as: Is all the land being used or only part of it?; Is water spoiled or used efficiently?; and Is production per unit of land high and increasing or low and stagnant? When it comes to distribution there are also key questions to be asked: Are consumer prices close to the prices paid to producers or is there a large (and widening) gap between the two?; and What is the material infrastructure that makes products flow through space and time and towards specific needs (while ignoring others)?

The two lines of inquiry are closely related to each other. Theoretically speaking, the social relations of production constitute the processes of production and distribution just as they regulate the distribution of the social wealth produced

(Poulantzas, 1975). From a methodological point of view this implies that we can study the nature, effects and social significance of different sets of social relations of production through (the comparative study of) the ways in which they mould the processes of production.

Introducing capitalist relations of production through centralization

Following the melamine scandal in 2008, the state introduced new rules that stipulate that milk destined to be channelled to the dairy industries can no longer be collected from individual dairy farmers but only from specific and officially recognized 'cow raising areas' (often referred to as 'districts' [*shao qu*]). These 'cow raising areas' are, in practice, newly built (and highly subsidized) farm enterprises that are privately owned. They provide 'shelter' for the owners' cattle as well as for the cattle of all dairy farmers in the surrounding area. Within these newly created units primary dairy production (milking, feeding, selection, etc.) and the subsequent transport of milk to dairy factories are tightly controlled. It is assumed that this will prevent any fraud with milk and dairy products.[4]

Until 2008 China's dairy farming had a dualistic structure. Alongside many small peasant producers, there were large capitalist farms that specialized in milk production. Some of these were converted from state operated enterprises that existed during the previous period of a state-led economy, others were the result of foreign investment and yet others were triggered by newly emerging market opportunities (such as the demand for organic milk within the 'embassy district' of Beijing). However, in 2008, 75 per cent of China's milk was produced on small farms (Webber, 2012:37).

In areas with considerable dairy production the new regulation that followed the melamine scandal provoked major changes. A county such as Yan Qing, in Hebei province, had some 2,000 households raising dairy cows – taken together some 30,000 heads. All these farmers had to bring their dairy cows to a 'cow raising area' (or to change over to breeding and fattening only).

The centralization of milking in newly created units is not, as such, a new phenomenon. In the north of Portugal the creation of jointly managed milking parlours (SCOM) was a major innovation that gave small farmers access to milking facilities[5] with optimal hygiene conditions (thus allowing them to obtain a better price for their milk); such facilities could not have been financed by individual farmers (Cristóvão *et al.*, 1994). Another scheme emerged in northern and central parts of Italy, where it was called *stalla sociale*, which can be translated as the 'co-operative stable' (Sandro, 1981). All the cattle from participating farmers (mostly smallholders and old farmers) were pooled together and housed in newly constructed, large stables. Here not only the milking, but also cattle-feeding, selection, veterinary care, fodder collection, etc., were centralized.[6] Participation in both these schemes was voluntary; that is a major difference from the current Chinese experiment. The new cow raising areas in China are imposed and they are obligatory. Peasants who do not join can no longer sell their milk.

Figure 6.1 Peasants feeding their cows in a cow raising area

In organizational terms the cow raising area consists of one large paddock (divided in smaller sections where individual herds are kept), linked to one large milking parlour. Alongside the paddock and the milking parlour there are the usual facilities for manure storage, silos for roughage and refined feed, offices, etc. The unit is organized as a privately owned enterprise. The owners are usually a couple with considerable experience in dairy farming. They often have the largest herd within the unit (up to, say, sixty dairy cows) and one of them (nearly always the woman) operates as the manager of the enterprise as a whole. The paddock also provides room for the herds of other local dairy farmers. Examples of the structure of these enterprises include two cow raising areas that we visited in 2010. One had 296 dairy cows from thirty-eight households and the other had 400 dairy cows from eighteen households (see Figure 6.1). Each participating farmer remains the owner of his/her own cows. Each one takes care of his/her own herd – so the cows are fed and cared for by the original owners.[7] The milking however is centralized and is done by specialized workers, contracted by the enterprise. Veterinary care and interventions are also the responsibility of the enterprise as a whole. The enterprise may also assist with breeding decisions. Thus, a complex pattern of relations exists between the dairy farmers and the enterprise. The dairy farmer relocates his cows in the cow raising area and may also deliver roughage (mostly stalks from maize). The farmer does the daily feeding of his/her herd but often using (some) feed and fodder provided by the enterprise. For its part the new enterprise does the milking, stores the milk and is responsible for delivering the milk to the dairy industry. When calves are born, the enterprise might deliver them to the farmer. Manure is sometimes allocated to the farmers, although some enterprises sell the manure. The largest slice of the value of the milk goes to the farmers: with the milk price at 3.30 yuan per litre, 2.85 is paid to the farmer and 0.45 yuan is kept by the enterprise for 'managerial assistance' (2010 prices). For the owners of a cow raising area this might yield a total of more than 600,000 yuan per year.

Apart from these relationships there are additional relationships that link the enterprise with the wider environment:

- Workers are contracted from the nearby village, especially for the milking.
- Roughage is bought in, often from large catchment areas (grass might even be bought in north- eastern China) and refined food (concentrates) is bought from industries in Beijing or Tianjin.
- The savings of the managing couple are used to finance the cow raising area, although these are supplemented with considerable government subsidies (on average, 190,000 yuan annually) and bank loans.
- The site is rented from the village committee.
- Pregnant heifers are bought in order to replace old or badly producing cows.
- And finally, there is the flow of milk from the enterprise to the dairy industry and the associated reverse flow of payments for the milk.

Apart from the scale of operation there is one important and quite decisive difference between this new constellation and the peasant units in the villages. While peasant farms are grounded on a non-commodified and self-owned resource base (that might very well be extended in a step-by-step way), the cow raising areas or districts are essentially based on a complex set of commodity flows. The peasant units are based on a stock of permanently available resources; the newly emerging enterprises are based on commodity flows (the same phenomenon that we encountered in the previous chapter –the difference here is that it is not only the main resources that become commodities, but also labour). Labour is mobilized through the labour market, feed and fodder is always bought in (there is no self-provisioning here) and replacement cattle are bought in (rather than raised in the enterprise itself). In short, the newly emerging enterprise is highly commoditized (as opposed to the peasant units that are based on a non-commoditized, or far less commoditized, resource base). As a consequence, the notion of profit-making becomes a guiding beacon in daily management.

It has been argued that the concentration of dairy cows in one single cow raising area contributes extra value – notably through more hygienic conditions in the new, large milking parlours – which might translate into better prices and improved prospects for the dairy farmers involved. On the other hand, the separation of the cows from the original unit (and daily care) might have counterproductive effects (it may reduce the productive lifespan), just as the dependence on commodity flows might increase cost levels. Probably the result is not so much an increase in the total value produced – what we see here is, above all, a *redistribution* of value (and probably a decrease in the total value produced).

At the level of the newly emerging enterprise there is an increase in production. At the level of the regional rural economy as a whole, however, there only is a reallocation of production (from many dispersed and small dairy farms to one single, large enterprise).[8] It is questionable whether these new enterprises will prove to be as stable as the initial peasant units. Together with this reallocation there is a redistribution of value (through the 0.45 yuan/litre management fee) and possibly, at specific points, a loss of value. Peasants living far away cannot bring the fodder they have available to the new stable – it will get lost. The reduction of longevity equally represents a loss.

The managers of the new enterprises sometimes refer to their farms as 'the head of the dragon' [*Long Tou*]: examples that should be followed. And of course, taken in isolation they look impressive, far more impressive than the much smaller peasant units – the more so since they can operate complex and expensive mechanical devices. However, the 'dragon head enterprises' (or 'flagship companies' to use another expression) are propelled by and depend upon the efforts and sacrifices of others.

Over the years we have visited several of these cow raising areas. During one year we were accompanied by an experienced Dutch dairy farmer, to have an extra set of trained eyes to observe and interpret the many details within each farm, and of different types, which together often provide information about several important features. One impression that emerged repeatedly is that the appearance of the cattle is poor. They are meagre, badly developed and, as the data shows, their longevity is indeed limited. This might be an inheritance from the previous situation (of individually held and managed cows), it may be a temporary effect of grouping cows together (this often increases the prevalence of diseases), but it may also be due to poor management. Even if the new 'managers' might have previously had sixty dairy cows, a jump towards 400 animals represents a completely different situation, with new requirements that might very well go beyond the capacities of these managers. Be it as it may, for the moment the new districts are far from an illuminating or inspiring example of productivity, efficiency and animal health.

This impression is confirmed by the available literature. Webber (2012:33–34) refers to the Zhao Jun farm located in Helingeer county in Inner Mongolia – a province with a long-standing tradition of dairy farming. This farm had 180 cows, one-third of them dairy cows (the others were pregnant or calves and heifers). Despite the Holstein Friesians (a high yielding breed), three milking turns a day, and artificial insemination (with semen bought in the USA), they only provide about 6,000 kg of milk per year. This is low, very low given the technologies applied. Another indicator of technical inefficiency is that the farm employs fifteen workers (including a full-time supervisor and a full-time college educated vet) to keep the farm running. That amounts to four milking cows per person – most peasants perform far better!

Webber also analyses (2012:36) the Wang farm owned by He Hua Limited Company and located in Beidaolaban. In 2006 this farm had 155 cows in milk (plus 150 young animals, 150 dry cows and 3,000 mu planted with corn for cattle feeding). It employed 120 people (equal to eighty full-time workers) which is the equivalent of just under two milking cows per full-time employee. Nonetheless, there were plans to expand the operation to 1,500 dairy cows (!).

Although they formally operate as co-operatives, the newly created cow raising areas clearly show the contours of a capitalist enterprise. Except for the minor contributions of the small dairy farmers themselves, labour is *wage labour* which is controlled by a new management layer. Once shaped and controlled by smallholders, the process of dairy production is now virtually completely expropriated by and oriented towards other interests.[9] This is underpinned by huge investments of capital that require a return. In short, the cow raising areas – at least many of them – are capitalist farms in disguise. The same applies to the newly created horticultural co-operatives, which we discuss below.

Horticultural co-operatives: another example of centralization

Since ancient times the city of Beijing (and many other cities in the north-east of China) has been surrounded by huge areas dedicated to horticultural production. The classical design of the greenhouses includes an earthen wall that stores heat. This means that the winter temperature within the greenhouse can be 2 or 3 degrees Celsius when outside there are frost levels of up to minus 25 degrees Celsius. This permits year round production without any additional heating. To catch the maximum heat (during summer), the greenhouses are south facing. They have plastic covers which are shielded, during cold nights, with a heavy 'blanket' that helps to avoid the loss of the heat accumulated during the day. In these greenhouses, that usually cover 1 mu of land (15 mu = 1 hectare), a wide range of vegetables is grown.

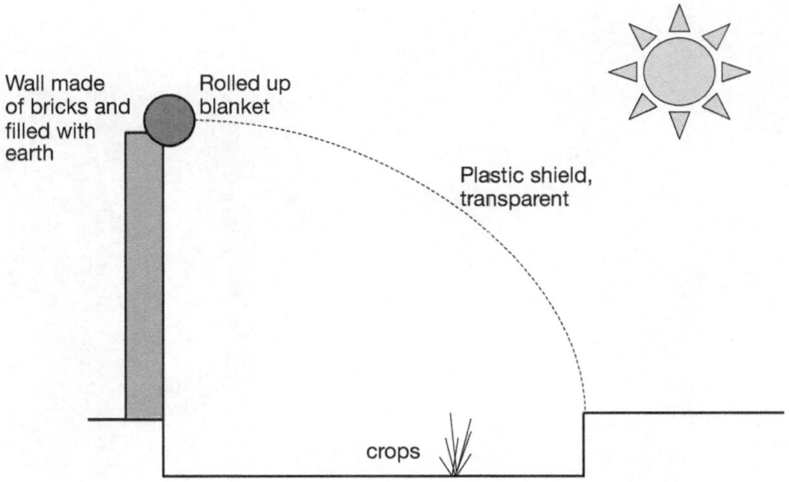

Figure 6.2 The architecture of a greenhouse

Currently, greenhouses are increasingly grouped together in 'co-operatives'. This is being strongly encouraged by the Ministry of Agriculture which is looking for more 'scale', as it calls it. These co-operatives imply another concentration or centralization of production. One manager plans the production in, say, 100 greenhouses that are grouped together. The peasant owners receive a rent for their greenhouse. Some of them might be contracted as salaried workers by the co-operative. It is not very clear whether or not there are mechanisms for profit sharing and, if so, how they function.

Nearly all the newly created co-operatives have constructed new greenhouses, applying a completely new technological design (see Figure 6.3).[10] What is lost is the ingenious system for harvesting and storing heat, which characterized the 'classical' greenhouses.

It is said that the new co-operatives are a promising solution for situations where many villagers have migrated to the industries in the south-east of China. The co-operatives are presented as an interesting mechanism to prevent greenhouses laying idle. However, it is also clear that the organizational and technological adaptations are creating a new pattern which is highly dependent on flows of fossil energy that enter the enterprises as a commodity – just as the labour force is now entering them through a commodity relation.

From an analytical point of view, both the cow raising areas and the horticultural co-operatives represent a partial takeover (albeit neither definitive nor formal) of some of the main resources (sometimes of all the resources) used in agriculture. Peasants lose their autonomy (in as far as it is grounded in the direct control over their resource base) and they also lose control over the labour process. At the same

Figure 6.3 Classical greenhouses (on the left) and new greenhouses (on the right)

time these two new models represent an imposition of capitalist relations of production – at least partly. Many of those engaged in dairy farming and in horticultural production are now doing so as *workers*. They enter the process of production through wage relations. Combined with this there is another change: the process of labour and production is no longer designed and controlled by peasants (reflecting their specific rationality). Instead it is now controlled by managers who orient the organization and development of production to obtaining high rates of return on invested capital (and/or on objectives such as gaining market share, economic power, etc.). One consequence of this (but only one of a wider range) is that the mechanism of labour investment (discussed in Chapter 4) is made inoperative: peasants cannot use it anymore, within these new, entrepreneurial constellations, to improve their resource base. Improvements need to be bought.

Whether these enterprises will be able to maintain the once highly intense levels of production is an open question that can only be answered through further observation. First, high intensity levels are no longer a primary objective (as they are in the peasant economy where increased intensity translates into improved incomes); production is primarily oriented towards profits. Second, the labour input is no longer a given (as is the case in the peasant economy). In the new enterprises, the labour input might be decreased in order to improve efficiency and profitability. The quality of labour will probably also be diminished through the newly emerging wage-labour relations. Third, the food security of the smallholder families is worsened.

Horticultural co-operatives and cow raising areas are both mechanisms that are meant to re-link peasant production with downstream markets through new capitalist enterprises that control the linkages between production and the market. These are referred to, not only by the managers, but also in official policy discourse as 'dragon head enterprises' (see also Schneider, 2013). This concept is central to policy schemes, introduced in the late 1990s, to promote vertical integration (Ye and Pan, 2015). Many of the new dragon head enterprises were founded by private urban capital but the state has also made considerable contributions. During the 2000–2005 period alone, central government invested 12 billion yuan at the national level to support dragon head enterprises (Huang, 2010a). Local governments followed this lead with similar efforts to support local-level dragon heads. Nationwide, the number of dragon heads engaging household farmers in contract farming rose from 8,377 in 1996 to 58,186 in 2002 (Niu, 2006). In 2001, in one single coastal province of China, Shangdong, 1.4 million hectares (ha) of cultivated land – about 20 per cent of Shandong's total arable land – was devoted to contract farming and 15 million rural households were participating in contract farming (Huang, 2010a). The rapid growth of contract farming is also a part of China's 'hidden agricultural revolution' under which structural changes are occurring in the domestic food regime without significant changes in the means of production. Huang (2010b) refers to this process as capitalization without proletarianization.[11] Alongside this process there is also the emergence of capitalist enterprises that *directly* organize and control agricultural production. This will be discussed in the next section.

Renting land, building large capitalist farms

Although there is a lack of exact data, the number of extended, capitalist farm enterprises has grown rapidly over the last decade. The mechanism through which they are established is as follows. Single entrepreneurs with access to sufficient capital contract the usufruct rights over the land of villagers. Negotiations over these rights often take place at the level of a village as a whole or at production team level. Thus, more or less complete 'blocks' of land might be obtained. The contracts for leasing the land may cover different time spans (from, e.g., 3 to 30 years). They also stipulate the yearly lease to be paid and other rights and duties. In this way it is possible to construct agricultural enterprises of a considerable size. Some of the people renting the land to the new enterprise may also get employment as salaried workers. Conventional thinking assumes that these new enterprises will operate more efficiently than the far smaller peasant enterprises. In addition, they may well invest in new technologies (often beyond the reach of peasant producers). This will lead to higher yields and total production and more wealth will be generated. The search for profits (returns on invested capital) is seen as the powerful engine driving such enterprises forward.

It is also assumed that the overall benefits are mutual. The peasants not only receive a yearly payment for their land, there are also employment opportunities. In addition, it is thought that this model strongly facilitates labour migration. Since the land is now cared for by the enterprise the household members can leave without having to return for the peak agricultural periods. Sometimes it is even claimed that agriculture as a whole is considerably strengthened by this shift of land from peasant farms to capitalist enterprises: it is a shift that moves agriculture from being mainly run by women towards a new, male dominated agriculture. In the opinion of some this figures as a major improvement.

The Liang enterprise: an example

The Liang enterprise is located in Huangjue village[12] in the province of Sichuan. Liang is the owner. He is from a neighbouring village and has peasant origins. He became rich through construction works and especially by buying the right to demolish areas earmarked for urban reconstruction and then reselling the bricks, iron, tiles and whatever else remained. This generated considerable profits, which were partly invested in several large pig and poultry farms. The farm in Huangjue village was started in the year 2010 and was in its fourth year of full operation when we visited it for the first time. The farm covers nearly 500 mu,[13] which was obtained through a set of carefully coordinated land transfers. Four hundred mu of this land is used for vegetable production. The rest is used for producing trees (ginseng trees, but also the slowly growing taxus tree). The enterprise also has a unit for pig production. There are 1,000 sows; the piglets are sold.[14]

In 2013, during the busy season, up to 150 workers were employed. In 2014 this decreased to 100–120 workers. One of the managers who guided us around the farm assured us that 'the employment problem is now resolved in the village

Figure 6.4 Female workers in the Liang enterprise

. . . the villagers now earn more than in the past from their work and their land'. In practice the villagers get 800 jin of rice for each mu of land they lease to the enterprise (2 jin = 1 kilogram). At the moment this equates to 1,040 yuan. Wages range from 40–50 yuan per day. Most workers are women, older than 50 years and contracted on a daily basis.

It is remarkable that the nearly 500 mu is in one single block. This concentration of land implies that more than 200 households all simultaneously accepted lease contracts. This is in stark contrast to a neighbouring farm of 80 mu that has many 'holes', i.e. plots where owners do not want to reallocate the use right of the land. It is even more remarkable since the contracts stipulate a lease period of at least 15 years. The manager says that indeed some 'extra explanation was needed'. One important factor explaining the construction of this single block is that 'the stubborn peasants' who didn't want to relinquish their land were offered pieces of land equal to their initial property at the periphery of the block.

The farm produces a range of different vegetables and also some flower varieties. The yearly production is around 1 million yuan from 400 mu, an average production per mu of 2,500 yuan. There is considerable use of inputs. The manager explained to us that this farm had the potential to be more productive than neighbouring peasant farms because the latter 'hardly apply any technology'. By way of example, this farm 'mixes pesticides' (according to advice from contracted specialists). We also noted widespread use of a growth hormone designed to enlarge fruits and flowers and that the workers did not use protective clothing.

Yet, so far the vegetable unit has only produced deficits. In 2011 there was a deficit of 2 million yuan; in 2012 it was more than 1 million and we were informed that it once again showed considerable (but unspecified) losses in 2013. The bad results are related to various external factors:

> in the first year we were not familiar with the soil, in the second year flooding destroyed 100 mu of tomatoes . . . the money was literally flowing away . . . . Later on, in the second year we faced a drought. If all goes well this year we should at least be able to balance our sales and costs.

Considerable investments have been made in this farm. Four million yuan in 2011, 2 million in 2012 and at least 1 million in 2013. This raises two particular questions. First, how can one explain the seemingly contradictory situation of such high investments and ongoing losses? And second, why would anyone continually plough such large sums into a loss making enterprise?

According to the farm manager the owner (or 'boss' as the manager likes to call him): 'does not care much about this. His aim is to improve the wellbeing of the local people here, to help them . . . . He is determined to bring many benefits to this village.'

It is hugely difficult to produce different vegetables on a large scale. The logistics are particularly problematic. They require a meticulous planning that covers time, space, labour inputs, soil fertility, market expectations and many other variables. Once a crop is harvested the land needs to be worked immediately in order to plant the next crop. Crops need to be rotated in order to control the presence and spread of pests and diseases. Cropping schemes need to match the availability of labour and water. And so on and so forth. Evidently, the management of the Liang farm is not up to meeting these complex tasks.[15] As Miss Lue, a member of the village committee (which in general is highly favourable of this type of enterprise) observes: 'This farm is too big to manage well.'

It is also telling that a former worker at Liang's farm concluded, on the basis of her experiences there, that 'it could be done far better', which led her to start her own 80 mu vegetable farm enterprise.[16]

These ongoing losses could lead us to conclude that (to date at least) the Liang enterprise is functioning badly. However, this would be too easy. If profit-making were the immediate objective then selling the use right on the 400 mu to another entrepreneur would be the easiest solution. Yet: 'The boss will absolutely not sell this farm. He says "as long as my trees grow here, I will stay here"'.[17]

Maybe there are other objectives completely different from, but as valid as, short-term profit-making. There may be an expectation that the value of land (and/or the price of crops) will rise considerably in the future. The owner could be anticipating regulatory changes (i.e. the need to combine pig production with the availability of land to dispose of the manure) that will make the vegetable farm a more viable enterprise. Or there may be other factors that we are not aware of.[18]

In general terms two sets of questions need to be asked in situations like this. The first set concerns the *developmental* dimension: Do these newly emerging enterprises generate extra social wealth (or do they simply redistribute the existing social wealth in a different way)? And is the emergence of these new enterprises inevitable, or are alternative trajectories possible? The second set of questions quite clearly raises *class issues*: labour conditions, wage levels, relations of negotiation, the presence or absence of labour unions, etc. (Bernstein, 2010).

Developmental issues

Compared to the production previously realized in the fields now controlled by the Liang enterprise, there has, at least at first sight, been an increase: from some 1,950 yuan/mu to 2,500 yuan/mu (see Table 6.1). However, the figure of 1,950 yuan/mu only includes the production of maize and rapeseed. On top of this there was much non-marketed production, including vegetables, fish (from a fishpond), chickens and one or two pigs. If these are taken into account it might be concluded that the Liang enterprise hardly realizes any extra production on the land it controls. This is confirmed if we compare it to production levels in other parts of the village that have access to sufficient water where 2,400 yuan is achieved (again, non-marketed products are not included).

If, however, we compare the Liang enterprise to small-scale vegetable growing in peasant farming then the underperformance of this large enterprise becomes evident.[19] In the small plots meticulous care is given to the crops and intercropping allows for constant flows of different marketable products. It is possible to produce up to 10,000 yuan/mu. The realization of such exceptional levels is evidently not feasible on the much larger scale of enterprises like Liang's. Nonetheless, this difference (of 7,500 yuan/mu) shows that the 400 mu controlled by the Liang enterprise is not realizing its full productive potential. This is clearly illustrated in Table 6.1 which also includes data on mixed farming. This is derived from a 20 mu farm that contains, alongside 10 mu of tea, walnut trees (eighty), 200 orange trees, 100 pigs, maize production, and trees for wood production. This is one of those farms in which peasant farming is being, as it were, 'reinvented' (see Chapter 7). It achieves on average an annual net income of 100,000 yuan – making a net income of 5,000 yuan/mu per year.

In reality, capitalist enterprises can only achieve higher yields than peasant farms if the former make use of technologies that are not accessible to peasant farmers and/or are able to cultivate high value crops that are beyond peasants' reach.[20] This is also illustrated in the section of Table 6.1 that looks at the case of specialized ginger production. In Huangjue village there is a 50 mu farm operated by Li, who comes from a neighbouring village. There is no tradition of ginger production in Huangjue village – especially on a large-scale. Ginger is typically a high value product: it can fetch up to 11 yuan/jin. However, large-scale production requires a huge investment, up to 13,000 yuan per mu. This means that it is out of reach for most peasant producers.

Yet, specialized ginger production can result in very high benefits: up to 33,000 yuan per mu which roughly translates into a gross profit of 12,400 yuan/mu. There are, though, at least three problems that stem from such high profits. The first, of course, is that these profits will not stay in the village.[21] They are largely siphoned-off, and will be removed to other places (largely because of the capitalist organization of farming). Second, in the case of ginger one needs also to take into account that cultivation cannot be repeated more than three times on the same plot. After three crops the pressure of disease and pests is too much. Rental contracts are therefore only for a three-year period. After that the ginger enterprise will move to another location. These ginger farms are a form of 'hit-and-run agriculture' (Ploeg, 2008). The third problem is that ginger is extremely sensitive to disease and even slight variation in climatic conditions. If a bad harvest occurs not only is the potential production lost – but also the very high costs incurred. This is a very big risk to face, even (or maybe especially) for a capitalist farm. In 2012 there were two ginger farms operating in this village. One of them has since gone broke.

Drawing all these different comparisons together one can only arrive at one possible conclusion: Liang's enterprise does not contribute to the production of social wealth. Its production levels are less than (or certainly not more than equal to) the ones previously realized in these fields and they are far below the potential levels of specialized vegetable production. Equally, they are far below the very high profits that can be realized from speculative hit-and-run agriculture (i.e. from ginger). The economic performance of Liang's farm is very poor: so far it has only produced deficits.

The presence of Liang's enterprise has several effects on the village economy. The villagers receive 1,040 yuan for each mu of contracted land. For an average smallholding (2.5 mu) this makes 2,600 yuan. Twenty working days/mu were needed to achieve the previous level of production (1,950 yuan/mu), i.e. fifty working days for the average household. If the farmers were paid for this work at the current rate (45 yuan/day) this would provide them with 2,250 yuan over the year. When the Liang enterprise is busy it employs between 100 and 120 people (in 2014). One hundred and sixty households have contracted their land to the

Table 6.1 Comparative overview of levels of production per mu in Huangjue village

Peasant agriculture on the lower parts (rice and rape seed) (GVP/mu)	2,400 yuan/mu
Peasant agriculture on the higher, more sandy parts (maize + rapeseed) (GVP/mu)	1,950 yuan/mu
Full-time peasant farming (chicken + vegetables) (Lv Jianlin) (GVP/mu)	7,500 yuan/mu
Liang enterprise (vegetables on 400 mu) (GVP/mu)	2,500 yuan/mu
Miss Xiao's Enterprise (vegetables on 80 mu) (GVP/mu)	2,500 yuan/mu
Vegetable production on small family plot (GVP/mu)	8,000–10,000 yuan/mu
Minus inputs (i.e. net income)	3,000–4,000 yuan/mu
Minus inputs and all labour paid	1,000–1,500 yuan/mu
Mixed farming (20 mu) net income (all paid costs deduced)	5,000 yuan/mu
Specialized ginger production (GVP/mu)	33,000 yuan/mu
Minus inputs and labour	12,400 yuan/mu

enterprise and it seems very improbable that the enterprise can give all these households at least fifty days employment per year.

Taking everything together, it might be concluded that the 500 mu of land that is now concentrated in the Liang enterprise contribute significantly less to the village economy than before. If there are any benefits, these may reside in a reduction of drudgery for women[22] and more flexibility for men and young women to engage in labour migration.[23]

Apart from contributing to the village economies, enterprises like these are also expected to contribute to the national economy and to food security at national level. Yet our analysis (especially the data in Table 6.1) shows that neither of these expectations are being met. The Liang enterprise does not contribute to either of these goals. Nor does it coexist with peasant farming: it has *replaced* peasant farming. In doing so it does not increase total agricultural production, nor does it enlarge the social wealth produced: it only produces losses. And, if compared to emerging alternatives, such as ginger production by new agricultural entrepreneurs, specialized vegetable production or mixed farming in 'reinvented' peasant farms, the Liang farm is a significant step backwards. When it comes to food security, it is noteworthy that the Liang farm produces less than others do – or less than what is possible. Thus it negatively affects both social wealth and food security which could have serious medium- and long-term implications.

There is another aspect that should be mentioned here. Capitalist farms, such as the Liang enterprise, represent a *selective* process of growth. Only fertile flat land, available in large plots, with irrigation (or at least thereof) is subject to land transfer. All other plots are ignored. Compared to the Chinese peasantry's meticulous use of even the smallest plots, wherever they are located, and their impressive capacity to convert marginal lands, through labour investments, into highly productive fields, the capitalist farm enterprise is highly selective: it will only use those resources that promise immediate and high returns on the capital invested.

One cannot explain the emergence of capitalist farms such as Liang's enterprise in terms of their technical and/or economic efficiency. Nor can it be explained in terms of their capacity to produce considerable (or even any) short term profits. As we have seen, there are no profits and both technical and economic performance are poor.[24] For an explanation of the emergence of these capitalist farms we must look at another, completely different concept: that of *control*.[25] The owners want control over large amounts of agricultural resources, as this will enable them to obtain large shares in different markets. Large market shares, in turn, translate into economic power and this may help to rebalance, somewhere in the future, the terms of trade and to create 'profitability' through skimming the market. This control might be used to convert agriculture into an attractive *financial operation*. Such an operation might even include speculating on changes in government policy and associated increases in the value of land. It might equally entail speculating on future food scarcities and rising prices. Or it may be a way of converting farms into a vehicle for tapping into subsidies (the Liang farm received at least 700,000 yuan for building the stables for housing pigs and there is considerable, but unverifiable, local gossip about other subsidies).

Such well-established control might also be used to underpin other financial operations that stretch far beyond the farm boundaries. It is possible to link an agricultural enterprise to an industrial enterprise, a trading company or a construction firm. This can be done by grouping them together in a holding (although far more complex solutions are also feasible). Such constructions allow the shifting of profits from industry, trade or whatever towards agriculture. In many countries, including China, this can bring considerable tax benefits. When invested in agriculture (or other 'green' activities) profits are exempt from tax liability or subject to much lower rates. At the same time it is possible to shift losses from the agricultural branch to the other activities. This again can translate into tax reductions. By co-ordinating these mechanisms, the gains obtained through the tax reductions realized by the holding as a whole may greatly exceed the losses from the agricultural enterprise. Thus the capitalist farm becomes an element of a far wider financial operation and as such does not have to make profits. On the contrary, in the context of the financial operation as a whole a loss making farm can even be quite 'profitable'.

In times of high financial insecurity constructions such as the one discussed above may also be used to channel capital to 'safe havens': as long as land is not losing its value it is a safe 'parking place' for capital – especially when longer-term price increases for the land and/or subsidy flows (associated with 'greening') are expected.

It would be somewhat premature to make the claim that Liang's farm enterprise is involved in such financial machinations. Although we had the chance to talk at some length with Liang himself in spring 2014, we do not really know his strategy. There are, however, some interesting indications of what might be the underlying logic. Liang's enterprise appears to be co-owned by partners whose main activities are in construction. They frequently visit the farm and they complain loudly about the many millions of yuan they have to pay in taxes. Another important indication is that Liang's professional résumé clearly shows that he has mastered the art of financial engineering, the precise skill needed for designing and running the sort of complex financial operations discussed above.

Class issues

Farms such as the Liang enterprise do *not* contribute to the creation of a solid labour class. It is the older people who are employed. They are the only ones prepared to accept the poor working conditions and low pay. Those able to move to the cities, where much better wages can be earned, will continue to do so. Interestingly, the rural enterprises of the type we discuss here are dominated by traditional patron–client relations (as exemplified in one of the earlier quotes, see page 118). There is a lack of outspoken capital–labour relations (including potentially a complete proletarization of the peasantry). Workers only work casually and seasonally. Their wages are supplemented with the rent they receive for transferring their land to the Liang enterprise. In addition, most households that supply their labour to the enterprise also maintain small kitchen gardens and tiny plots for rice or maize production. Of those involved, 70 per cent rent just part of their land to the Liang enterprise, keeping a plot to till themselves. Only 30 per cent have rented all their land, thus becoming solely dependent on wages.

In this system of patronage getting work is perceived as a favour. Work is distributed among a large group of potential workers through *guanxi* relations. Thus, workers try to 'please' the patron as much as they can. Beyond that the Chinese norm of *mian zi* ('losing face') is strongly related to one's performance at work. Badly performed tasks imply that the worker 'loses face'. This excludes typical resistance of 'the second kind', such as foot dragging, carelessness and a slowing down of work rhythms. Ironically, the capitalist farm enterprise partly relies, in this respect, on non-capitalist relations – just as it reproduces them.

Intertwinement

In this and the previous chapters we have described different development trajectories in China's still predominantly peasant agriculture. These descriptions show that rural villages in China are definitely not 'empty nests' lacking in endogenous dynamism. They also show that a further intensification of peasant farming is far from being an impossibility. Rather, intensification is continuing and is grounded on new mechanisms, such as embedded specialization, the reorganization of space and a 'reinvention of peasant farming'. This, together with other processes, makes these villages into highly dynamic places.

However, alongside the consolidation and growth of the peasant economy, there are other tendencies that clearly point in different directions. We have illustrated these with the centralization of dairy farming and horticulture into new and much larger 'dragon head' enterprises and with new processes of land acquisition and concentration that result in the formation of capitalist farm enterprises. These processes clearly reconstitute farming on a new basis: land and labour are converted into commodities; farming is increasingly grounded upon multiple commodity flows; and finally, the units of production become part of overarching and complex financial operations.

There is a clear intertwinement, if not interlocking (Long and Ploeg, 1994; Ploeg, 2003) of these different development trajectories. Peasants working elsewhere allows for pieces of land to become available (through the village committees or through direct negotiations and transactions) for those who want to expand their units of production. In the peasant economy 'nearby fields' are the most valued. Fields that are located far away can become interesting opportunities for newly emerging 'peasant entrepreneurs'. Village committees often are in dire need of money (the more so since local taxes and levies have been abolished). Thus, contracting the possession rights of extended and as yet unallocated areas to newly emerging 'peasant entrepreneurs' can be interesting for them. In this way the interests of village committees and newly emerging entrepreneurs can come into alignment. Their different prospects become intertwined if not solidly interlocked.

The same occurs with migrants who are not in the village often enough to care for their land. This allows for the construction of horticultural co-operatives. New policies, such as those arising from food scares (e.g. the melamine scandal) are creating new features in the landscape such as the (enforced) concentration of cattle into cow raising areas. Finally, the emergence of corporate strategies can be grounded on the interlocking of different projects and tendencies. There is a strong *ideological* underpinning to this: the view that it is only large farms that bring development, that investments only (or mainly) occur on large farms, that peasant farming is stagnant and that all rural people just want to migrate, tie these differing sub-processes and projects together.

While these trajectories are highly interlocked there are limits on the possibilities for them to harmoniously coexist. The newly emerging capitalist pattern needs resources: far more resources than the peasant type of development requires. It needs more land per unit of labour, just as it needs far more land for each productive unit. It needs more energy per unit (as illustrated with the new greenhouses). It requires more spending (because of the increased reliance on commodities flowing from upstream markets). It needs more 'favours' from local and regional government. All this implies that relations between peasant farming and capitalist farming may well become highly antagonistic. We are already seeing vehement competition between peasant agriculture and capitalist agriculture for material and symbolic resources – and the state finds itself in a complex inter-mediating position. The outcomes of this competition will not be neutral: they will critically affect the creation and maintenance of attractive rural livelihoods, good income levels and high levels of ongoing agrarian growth (*sannong*).

Land transfers: to whom and for what?

Although large corporate farms receive much attention in today's debates, they represent only a very minor fraction of China's farms and only contribute a very small slice of total agricultural production. Many of the very large farms (some of 50,000 mu or more) are located in Heilongjiang province. Typically, these farms are divided into hundreds, sometimes thousands, of smaller units that are managed by individual farmers (assisted by the technical staff of the 'overarching' enterprise and by contractors with specialist machinery). Productivity growth here is moderate compared to other Chinese provinces (Chen *et al.*, 2008). Be that as it may, the perspective of developing agriculture through the creation of large-scale farm enterprises, based on salaried labour, is increasingly gaining traction in state apparatuses and influential think tanks. The implied transfer of resources from peasant farms (or, more generally, from rural villages) to the new capitalist enterprises is framed in terms that juxtapose an assumedly stagnant and weak peasant economy (in dire need of benefits coming from elsewhere) with an assumedly vibrant large-scale capitalist agriculture able to generate many benefits for peasants and the nation as a whole (see Figure 6.5 for an overview). In this scenario the transfer of land is presented as allowing villagers to reap multiple benefits, while the new enterprises will function far better (in terms of investments, production and efficiency) than the previous peasant farms.

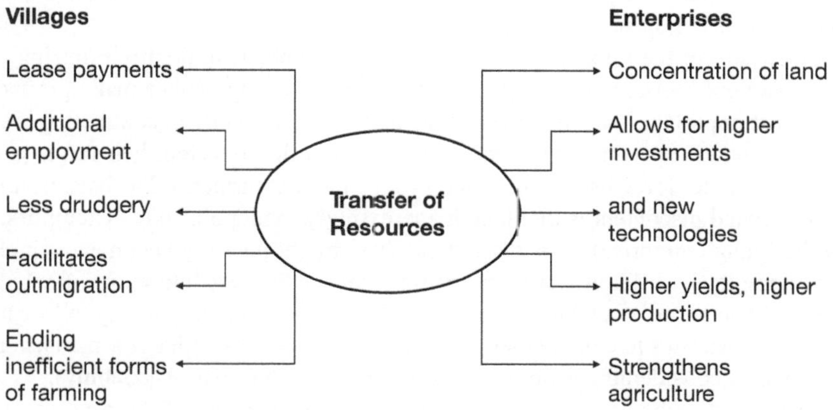

Figure 6.5 Overview of the most frequently mentioned benefits of land transfers

However, as this chapter shows, these assumed benefits rarely manifest themselves. Although it goes without saying that for some households, especially those consisting of elderly people, a transfer of land can be beneficial,[26] it is only occasionally that more wealth is generated. Most capitalist farms produce less per unit of land (or per animal) than peasant farms. They do not generate extra employment (instead, much productive employment is lost) and they do not introduce any upward pressure on wage levels. In the current scheme of things, capitalist farms do not contribute any additional value to the Chinese countryside. Peasant agriculture and capitalist farming are clearly competing systems of valuation and production. But it is not the system that creates the most value that is 'winning' the competition. The tendency is the other way. Capitalist farms are coming to the fore, even when their main 'value' resides in them taking control of the resources that may be decisive in speculative cycles in the near future. The production of food, let alone the achievement of food security, are not central objectives for the financial operations which are taking over a large portion of farming. Food production is, at best, a secondary objective. It only occurs in as far as it is supportive of other objectives. This should certainly give cause for alarm.

Is there resistance to these land transfers? So far this has not been the case in Huangjue village. The 'narrative' summarized in Figure 6.5, and actively transmitted to villagers through the village committee and the leaders of the different production teams (who operate as brokers in the process of land transfer) is evidently effective. It is the rents that are to be paid annually that are central to the relationship between the two parties, not the transfer of land as such (as a consequence rents in this village are increased by 100 yuan/mu every year). 'Stubborn' people can effectively be removed to the 'margins' as we explained above. The *guanxi* relations that regulate the distribution of work opportunities make those who rented the land to the enterprise refrain from explicit criticism.

The feeling that 'Liang is eating the country's money' is only expressed in closed night-time drinking sessions.

Elsewhere in the village,[27] though, there is discontent. Implicitly referring to the Liang enterprise, people say that 'large-scale farming cannot make profits'. The most important issue, however, is the increasing difficulty in contracting land faced by relatively young peasants who want to further develop their farms (see the next chapter for a discussion of such young peasant farmers). The background to the related discontent is that land is increasingly monopolized (as exemplified by the Liang enterprise). This discontent, then, might possibly become a vehicle for a completely different development trajectory in the near future; one in which the land becoming available is distributed in a (more) egalitarian way, allowing interested peasants to enlarge their own farming operations with a few mu (which can make a considerable difference within the single household economy).

At present such an alternative trajectory is blocked. Many of the peasants we interviewed indicated that it is impossible to have access to more land. They claim that it costs between 3,000 and 10,000 yuan per mu (rent plus inputs for crops to be sown). This money is not available; this is evidently related to state policy that makes subsidies available for large-scale enterprises but gives hardly any support to peasant farms.

This is how these interlocked trajectories are affecting short- and medium-term development prospects. The opportunities for peasant farming are blocked. In the longer term another problem might also emerge: peasants may find that the land, traditionally the cornerstone of rural social security mechanisms will not be available when labour migrants need it – simply because it has been rented out for long periods. Taken together this implies that the current combination of tacit agreement and open discontent might very well change to *resistance*.

Update

At the end of March 2015 we are suddenly confronted with a surprise. The Liang farm is laying barren. Liang himself disappeared in November 2014. Nobody knows where he went, not even the staff that have been managing the farm. Rumours have it that he mortgaged the farm and lent out the credit obtained to other people and other enterprises, at a considerably higher rate of interest.[28] However, something went wrong: those borrowing the money would not, or could not, repay their debts. It is also possible that the whole structure he created was considered by the authorities to be a fraud. Whatever the reason for his disappearance this latest twist in the story seems in keeping with the main elements of our analysis. First, farming was not the main objective of the Liang enterprise; it was secondary to more overarching *financial operations*. Second, it highlights that such constellations are inherently insecure and lack any continuity. Farming can be discontinued at any point and the farm deactivated.

In the meantime, the Liang enterprise has not paid the yearly rents for the land it was leasing. From a legal point of view this implies that the peasant families with usufruct rights are free again to do what they want with the land. Some are making use of it, others are not (partly because cutting and uprooting the trees that have been planted is too large a task). Others are leasing to other newcomers: Zhu contracted land for planting 'long beans' and Wang rented 150 mu of land for planting spring onions, tomatoes and egg plants (aubergines). The rents today are typically lower than before (850 yuan/mu) and it seems that some kind of hit-and-run agriculture is emerging. The choice of crops reflects underlying speculation as does the use of labour- extensive techniques. If the market situation is favourable, huge profits can be reaped – up to 10,000 yuan per mu for spring onions, for example.

However, a considerable part of Liang's former agricultural enterprise still lays barren and the stables previously filled with hens and pigs are now empty. Some of the areas planted with trees are being overrun by weeds. It is as if the empty fields and stables and the nearly asphyxiated trees are saying that so far this has been a huge failure – what to do now?

Notes

1 According to the Agricultural Census 2006 there were 395,000 agricultural enterprises that differ from peasant smallholdings in China. These might be co-operatives, state enterprises, corporate farms, joint ventures with foreign capital, enterprises of the army or whatever. Compared to the total for peasant farms (200 million) their number is small. However, their economic, political and symbolic importance goes far beyond their mere quantitative presence.

2 The wage-labour relation is central to capitalist production. The labour force enters the process of production only through the wage-labour relation. The means of production function as capital; they have to render a surplus value that is used to invest in order to produce more surplus value – a process that is known as capital accumulation.

3 Chen and Zhang (2013) argue that Webber's conception of capitalism is too narrow since his analysis is mainly confined 'to the labor supply part of [. . .] emerging capitalism in rural China' (2013:606). They argue that there are many other facets to the emergence of capitalism in China.

4 This new structure seems to be based on the assumption that individual dairy farmers were (and will be) the ones that tend to dilute the milk with water and melamine. As such the milking has been, as it were, expropriated and concentrated in a single place that allows for control.

5 Only milking is done in these co-operative milking parlours. The cattle remain in the individual farmer's stables and feeding is also done individually.

6 Most of these *stalle sociale* were created in the 1970s. However, they all disappeared due to major inefficiencies that led to bankruptcies.

7 This is probably a temporary phenomenon. For dairy farmers, anywhere in the world, it is always very difficult to part with their cows. Without the possibility of continuing to care for their own animals they probably would have strongly resisted the transfer of their cows to the new districts.

8 In some places this tends to be a selective process because a minimum of, say, ten dairy cows is required for access to the 'cow area'. This excludes smaller producers who face the prospect of closing down their dairy operation. However, in several places this problem is 'mediated' by permitting small producers to transport their milk from their small farm to the cow district to be mixed with the milk produced there. Such arrangements are typical of China and reflect local power relations and local leaders' fear of being denounced as 'bad leaders'.

9 This is not necessarily the case. In 2014 we visited a 'cow area' in Sichuan that was located in an old tea factory. Ten dairy farmers lived there, with their wives and children (at least during the week), keeping control of production processes and taking the main decisions by themselves. With an average of twenty cows these households are able to produce an income that ranges from 40,000 to 60,000 yuan/year. The cow area offers them large stables (with room for further expansion) that they probably could not construct at their original farms. Again, the exact configuration of the 'cow districts' or 'areas' depends very much on local circumstances and power relations.

10 The new technology is mostly used as an argument to tear down the existing greenhouses and replace them with new ones. The argument is that individual vegetable producers cannot make the jump to the new technological patterns (seen as representing progress) and co-operativization is needed in order to create the scale required for the new investments.

11 This remains a somewhat ambiguous concept and could lead in two opposing directions. One would be that peasant agriculture is increasingly subordinated to the dragon heads; the other would be that peasant farming enlarges its resource base with 'capital goods' (understood in the Chayanovian sense). Dragon heads may be, in certain circumstances, a vehicle that allows peasant farms to enrich their resource base. The ambiguity of many situations is probably due to the complex and contradictory processes of transition that are currently taking place in rural China.

12 In Huangjue village there are six large farm enterprises (data from 2014). Previously there was another one but it went broke and its land was reclaimed by the villagers. Of a total agricultural area of 4,056 mu some 25 per cent has been transferred to these large enterprises.

13 This total agricultural area of 4,056 mu includes the river bed where farming is possible when the water level is low. There is no fallow land. Those who migrate often rent their land to other villagers. Some families work the land from as many as seven other families. Of the total population (3,158 people/1,200 households), 200 men are involved in labour migration (that is, they work beyond the county borders, some going as far away as Sudan) and more than 100 women (mainly doing housework in Chengdu or working in restaurants in Guanzho). If those working nearby are included there are at least 1,000 labour migrants.

14 Liang also has three other pig farms in the village. These are run by old couples who work on a contract basis. Liang also owns and operates a construction firm.

15 According to the manager of the farm this also happens because the owner frequently intervenes and imposes decisions that turn out to be wrong. She argues that this leads to a lack of a 'sense of achievement' within the enterprise.

16 This farm (described in some detail in Chapter 5) covers 80 mu and achieves a more or less similar level of production (2,500 yuan per mu or perhaps slightly higher; see also Table 6.1). The owner, Miss Xiao, claims that she achieves positive results. She joins in with the work with the other labourers and lives on the farm. During peak periods some fifteen workers are employed; for the rest of the year, five or six. When one compares the two farms one gets the impression that Liang's is highly inefficient, employing too much labour and achieving poor results.

17 This is telling, the more so since at least 1 million yuan has been invested in planting trees here. The young trees are bought at between 400 and 500 yuan per tree. They can later be sold for much higher prices. The salient detail is that Liang is violating the law here. When contracting land it is formally forbidden to change the nature of land use (from, e.g., food production to tree growing). At the same time cutting (or removing) trees without the written authorization of the state is also illegal. However, in situations like this the law is one thing, power another.

18 Formally speaking, the owner pays the contractor for any improvements made to the land at the end of the contract period. The contractor is entitled to compensation. In

practice such payments may be far more than the peasant owners can possibly afford. Thus, investments might be a means of perpetuating access to the land.

19 Hu (2014:257) also reports for his area of study that 'the largest rented farm performed much more poorly than expected by the villagers [. . .] the rice productivity was only about 4,500 kg/ha, substantially lower than the village average'.

20 There is also a third possibility of course: that unfavourable circumstances (no market access, no credit, water being grabbed by others, extreme poverty, etc.) could induce yield decreases in peasant farms. In such a situation the yields from peasant farms might become lower than those from capitalist farms which do not have to deal with these negative influences.

21 The villagers receive a lease price that is slightly higher than the one paid by the Liang enterprise (1,200 yuan/mu). However, the fertility of their land will be far lower after the lease period.

22 Assuming that they do not work in the Liang enterprise. Then the drudgery will not be any less.

23 However, they would lack a fall-back position (in case of sickness, unemployment, etc.).

24 It would be wrong to assume that this is a typical Chinese aberration. Visser *et al.* (2014) show that more than 50 per cent of large landholdings in Russia have a negative profitability. Kuns *et al.* (2014) show the same is true of huge Swedish and Norwegian investments in Russian and Ukrainian agriculture. For a specific account of Ukraine see Lerman and Sedik, 2007 and Deininger *et al.*, 2013.

25 This was indirectly confirmed by the manager who informed us that the owner is 'not feeling upset about the losses' which is in considerable contrast with the manager who feels 'very frustrated'. He added: 'we do not know Mr Liang's way of thinking.'

26 Especially when their children are unwilling or unable to return. These elder people find it increasingly difficult to work the land. Official state pensions are only 60 yuan/month and then the rent for transferred land is a welcome supplement.

27 Huangjue village has ten production teams. Production teams one and two rented out their land to the Liang enterprise.

28 We know from other places that this mechanism is frequently used. It is possibly because interest rates for agricultural loans are much lower than those in other economic sectors.

7 Reinventing peasant farming

Jan Douwe van der Ploeg, Ye Jingzhong and Wu Huifang

State interventions and/or those of national and international capital groups are leading to the emergence of new farm types and ways of farming across China. Yet despite the presence of such new ways of farming, the vast majority of newly emerging patterns of farming are essentially endogenous in nature. They are the outcome of Chinese farmers reinventing farming. This reinvention is linked to the discovery that the countryside is a space where attractive incomes can be generated. It is also linked to the construction of new, and differently structured agricultural production units and entails a reshuffling of the interrelations between labour migration and farming. All this may have, in the near future, a significant impact on Chinese society.

This chapter describes the differences between what we call classical farming – which is highly dependent on labour migration – and the newly-emerging types of farming that generate levels of income that are comparable to or higher than those obtained through labour migration. We will pay special attention to the mechanisms that sustain the innovativeness of the newly emerging farming types. Finally, the strategies underlying these reinvented farms will be discussed.

From classical to new patterns of farming

The typical Chinese peasant farm – as constituted from the start of the 1980s after the introduction of the Household Responsibility System (HRS), the development of Township and Village Enterprises (TVE) and, especially, the beginning of the massive labour migration – consists of, say, 5 mu of agricultural land. Part of this land is dedicated to self-provisioning: to the production of the food needed to feed the peasant family. Beyond that there is a cash crop (sometimes more than one) that is sold in order to generate a monetary income. Usually, there is no sharp boundary between the two. Through barter the cash crop might be converted into food for the family; and the surpluses of food crops might be marketed. All this depends on a range of factors from the ecological conditions through to the family situation. A constant feature, though, is that the monetary income produced by the farm is far from sufficient to cover the needs of the family. In particular, what are known as the 'big expenses' (which are closely related with the reproduction of the family over time – education for the children, the construction of a house for the son, his wedding) cannot be covered by the operation of this classical farm. Hence, men (and sometimes women) engage in labour migration in order to raise extra income. The earnings from labour migration typically contribute the main share of the total monetary income of the family. That is not to say that farming is redundant. On the contrary, while for most Chinese peasant households the farm income is indeed supplementary, it is nonetheless indispensable.[1] One cannot do without farming.[2] When the men migrate (which is mostly the case), they usually come back once or twice a year to help with heavy agricultural duties – however, their role is additional; the women do most of the agricultural work. When they become older, the men try to find a job closer to home so that they can be at home more frequently. And finally they return to the village and re-engage in farming on a daily basis.

This traditional pattern is now being replaced by the emergence of a very different pattern. In the newly emerging forms of farming men are returning far earlier to the village. They often come back after 10 to 12 years of labour migration (at the age of, say, 30 years). They bring considerable savings with them that they invest in farming. And they themselves reappear in the fields, no longer in an additional role, but now as central actors. From now on they may also be involved in other economic activities, but these are supplementary, while farming becomes the main source of income – one that is not meagre but is thought to be acceptable (or even good to very good), at least equal or better than the income that can be gained through migrant labour. Underlying all this is a fundamental change: *agriculture is being rediscovered as, and subsequently reorganized into, a space where good incomes can be realized*. Pivotal to this is the development of novel ways of farming. Farming, of course, has always been there. But now it is being reinvented. It is being made into something new. Formerly, farming only produced a supplementary income. In the newly invented agriculture it generates the main flow of income. Associated with this there are many other differences (see Table 7.1 for an overview), some of which we will discuss here in some detail.

New products, new production methods

The search for new products (and services) and for new ways of organizing production (new techniques, new skills, etc.) and/or new ways to link to the markets are key aspects in this reinvention of farming. On the whole, Chinese farming always has been highly diversified (although to differing degrees in different regions). Now farmers are searching for products or services that contain a high value added per unit and which can be marketed in a way that ensures that this added value remains within the farm itself. Examples are specialized production of ornamental trees, the development of fruit yards (apples, plums, etc.), nurseries for afforestation, the production of free-range eggs, the building of agro-tourism facilities, the processing of agricultural products into prized specialties (e.g. sweet potatoes into glass noodles), breeding herds and introducing specialized pig and poultry breeding in places where this does not yet exist. The search also includes the methods of production (e.g. automated irrigation, new forms of pest and disease

Table 7.1 Differences between classical and new ways of farming

Type of agriculture	*'Classical'*	*'Reinvented'*
Perception of rural areas.	A space where one cannot possibly produce a good income.	An opportunity to make a good income.
Balance between labour migration and farming.	Labour migration delivers most of the income; agricultural activities are supplementary. Labour migration spans most of the man's productive life and is associated with a typical gender division of labour: men doing the important, income-generating work and women doing the secondary work.	Labour migration allows for capital formation. It is embedded in a shortened time-horizon. The man and wife work together in the renewed agricultural activities.
Man's role in farming.	Additional.	Central.
Magnitude of capital formation.	Reduced (and oriented at generational reproduction).	Large and strategic (and located in production).
Involvement in networks.	Only in networks that regulate labour migration.	In a wide array of networks.
Identities.	Peasant worker.	'Advanced peasant'. (see Figure 7.2)
Dynamics.	Routines dominate; dependent on externally provided innovations.	Novelty production is central.
Time perspective.	Cyclical.	Unfolding (and aiming at progressively improving one's own situation.
Investments in agriculture.	Mainly short-term.	Mainly long-term.

control, strengthening nutrient cycles through mixed farming), new ways of marketing and new forms of informal and formal co-operation. With these new products and methods, the potential volume of the farm can be enlarged considerably, and this is mostly done in a step-by-step way.

Getting bigger: what does it mean?

The reinvention of farming is resulting in a considerable increase in economic volume, which in turn allows for the increase in income. The amount of resources that are being used and combined in the new productive activities is also enlarged. In terms of land, however, the picture is less clear. Some of the new farms do indeed contract more, sometimes far more, land than before. Others, though, can increase their economic volume considerably without expanding their acreage. This is the case, for example, with the nurseries for afforestation, flower growing, free-range eggs, processing activities, agro-tourism and several other activities. This nuance may seem pedantic, but it is not without relevance.

In China there is considerable debate about farm-size. When perceived through the lens of 'modernization', China's peasant farms are far too small. They should be enlarged considerably in order to allow for savings, investments and growth. Thus, a primary objective of agricultural policy should be to enable farmers to accelerate expansion of acreage. What we see, instead, in the examples mentioned above, is an enlargement of the *economic* volume of the peasant farms without (necessarily) an increase in acreage: they produce more value added. Enlarging the acreage is not an objective in itself, let alone the primary objective. Producing more value added comes first and this may or may not require an increase in the land being used. But the acquisition of land is subordinate to the goal of increasing value added. As a consequence, agricultural policies should not focus on concentrating land into fewer, but larger, holdings; it should, instead, facilitate peasants' search for opportunities to generate more value added. Those farmers who need more land can easily obtain it with their own means without policy support.

An example

Between the ages of eighteen and twenty-eight Zhang worked in distant industries, until the outbreak of severe acute respiratory syndrome (SARS) in 2003 forced him to return to the village where he was born. He bought a lorry,[3] engaged in transporting rocks from the mountains to the local factories and started to work the land of his family. After a while he contracted more land. Every now and then he works in the local iron factory or operates a bus service. Sometimes Zhang performs as a musician. He is one of the young people in the village who is very much involved in reinventing farming. He explains this as follows:

> To create your own job is far better than working for others. Working for others is not consistent, it is not reliable. Today you have the job, tomorrow you may lose it. That is why I am starting some businesses now, raising livestock, planting trees, and now I am starting to try out egg production.

Table 7.2 summarizes the highly diversified nature of Zhang's agrarian activities. It shows that, alongside food production for his own family (a feature that is encountered in all reinvented farms), there is a strong emphasis on the production of apples and 'green eggs'. Together these different productive activities can be seen as composing a 'patchwork'. They are connected and, together, they compose the farm. This patchwork is colourful and is constantly changing: new patches are added, old ones are partly or completely covered over. The evolving patchwork is the outcome of step-by-step changes and it is never completed. It refers to a *modular* farm (Thomas *et al*., 2015). When looking carefully at the farm one can see that it is, indeed, *under construction*. Time and again new modules are added.

Figure 7.1 Zhang with a grafted tree

This was evident when, in 2012, Zhang spoke about his apples (which at that time needed 3 more years before coming into full production):

> I have 300 trees and this variety produces very big apples, each weighing 1 jin each. Vendors really like these apples and will come to the village to buy them. They can pay up to 4.5 yuan/jin. Sometimes they even contract the produce before I start the harvest. A tree can produce 80 jin when in full production. Thus these apple trees will render me a lot of money . . . 45,000 yuan perhaps, maybe even be more. This is far more than you can get through migrant labour, because there you have your costs as well [. . .]. But why didn't I plant more: 6 mu, instead of 3? That is simple. That is because this new activity needs skill, and it needs people to take care of the trees. Maybe in the future I will plant more, but I first want to see whether it works well like this and I probably have to develop my skills further.

On Zhang's farm there is currently 8.5 mu under cultivation. The farm as a whole is a bit larger (10 mu); there is still a part that has not yet been developed into land that can be tilled. Of the total 10 mu, 4.5 mu was allocated to his family in the 1980s. Since he returned Zhang contracted another 5.5 mu. The average peasant

Table 7.2 The patchwork that is Zhang's farm

Product	*Destination*	*Stage of development*	*Area in production*
Sweet potatoes	Only for family use		0.5 mu
Groundnuts	To be processed into oil, for family use only		0.8 mu
Maize	Partly for family use, rest to be sold or exchanged		1.2 mu
Vegetables	For family use		Interplanted
300 apple trees	For the market	Not yet in full production	3.0 mu
20 walnut trees	For the market	In production	0.5 mu
110 chestnut trees	For the market	In production	2.5 mu (est.)
'Green eggs'	'Nested market'	Currently raising chickens	

family here cultivates between 4 and 6 mu, so it can hardly be said that Zhang has significantly expanded his farm acreage. Yes, he does farm more land than the average peasant family, but the significant difference resides in the capacity to earn up to 45,000 yuan a year from the 3 mu plot planted with apple trees.

Long-term investments

This patchwork is underpinned by investments. These are long-term investments that are based partly on labour and partly on the savings achieved during labour migration. Zhang explains:

> Yes, the investments that I have made in planting trees, raising chickens and buying a lorry have all been with *saved* money. It is ours. It is mine, it is not borrowed, it is not from the bank . . . This is important, it means that I can take the risk [. . .] After marriage and especially when you have children, you have all the responsibility yourself, you have to get by on your own. You have responsibility for the family. This means that you cannot easily take a loan from the bank. So, planting trees and developing the farm would be impossible. But having your own money means that we can do this . . . .

There are considerable efforts now, just as there were considerable efforts in the past (during the period of migrant labour). These efforts are now coming together in the long-term investment that will generate a flow of income which will continue for several years, decades perhaps. As such this is a textbook example of the balance of drudgery and utility described by Chayanov (Ploeg, 2013). It is a balance that covers, in this particular case, extended periods: it links past, present and future in a meaningful way – the investment being the link between migration in the past and agrarian incomes that will be generated in the future. The balance also has a considerable spatial dimension, linking activities in distant industries and urban construction sites to hidden fields in remote villages.

The investments do not imply a definitive adieu to migration and/or 'economic activities' (as Chinese peasants call everything apart from farming). The investments are strongly rooted in, and dependent on, previous labour migration. And while these investments are 'maturing', other economic activities remain a key element. Beyond that, Chinese peasants, both the 'classical' and the 'renewing' types, don't like to have all their eggs in one basket. In the meantime, though, the balance between the different baskets is changing considerably.

As time goes on the total investments involved can become huge. This can be illustrated by the case of Li who now produces watermelons and mushrooms in greenhouses that cover 30 mu of land. He claims that his total investments (2014) added up to 250,000 yuan.[4] He started to invest in 2008 when the economic crisis and some family problems led him to decide to return to the village. With his savings he rented 1 mu of land on which he built a greenhouse. As well as caring for his mushrooms Li also worked in a factory in the nearby county town. The mushroom production turned out to be successful so in the next year he could plant up 2 mu of land. This was repeated, step-by-step, until Li was managing 10 mu of mushrooms and 18 mu of watermelons. The fact that his investments are self-financed gives him, as he argues, the 'strength' he needs:

> I do not have to look to annual profits or losses. What matters for me is the 10-year period. I may lose for 5 years and then win for 5, but in the end I will have more money [. . .]. My experience is that I can afford to lose up to 30,000 yuan in a single year [. . .] In any year I invest [in current expenses] up to 250,000 yuan. Now, if I only get back 220,000 yuan, due to bad prices or bad harvests, then I have a loss of 30,000. Even then I recoup 220,000 and that is enough to start again next year. In this way I can always continue. For those with debts that is impossible.

On the patchwork

Previously we referred to the layout of Zhang's farm as a patchwork. There are many different plots scattered over two villages with crops, trees, chickens and dogs. It gives an impression of messiness and the patchwork seems to be incidental. It looks as though it is supposed to be like this, but it easily could have looked differently too. It seems that the farm structure is extremely provisional. If we also take into account Zhang's four different vehicles, his computer and internet connection, his house, yard and musical instruments, and the fact that he bought a shop in a nearby town for his daughter, then an overwhelming impression of a patchwork, of a loosely structured portfolio of all kinds of economic activities, comes to mind. And if we return next year, one thing will be sure: the patchwork will again look different. The farm is, indeed, to a considerable degree provisional. It really does seem to be a 'crazy quilt' (as a patchwork is called in some places).

Observers trained in modernization theories would surely raise their eyebrows. For them this kaleidoscopic combination of crops, trees, animals, vehicles, an urban shop, music making, transport activities, experiments with chickens and trading is hardly justifiable. Modernizers would argue that a farmer such as Zhang should concentrate on one activity, preferably the most profitable one, and try to optimize the technical and economic efficiency of this selected activity.

But these first impressions and the immediate appearance might be misleading. One can very well argue that the patchwork is a rational response to market and climate induced risks (and/or a rational adaptation to micro-ecological differences). At the same time the patchwork also turns the peasant farm into *a laboratory that allows for learning-by-doing and for novelty production*.[5] The patchwork is a place where new products, services and methods of producing them can be tested. It represents a place where many small experiments are organized in order to show the best possible combinations of different inputs and input levels. The patchwork is an expression of the farm's development and dynamism. And the many experiments provide signals that are understood (and sometimes misunderstood) by the farmer about what goes well, what could be improved further, what should be different, and what should be discarded.

From isolated cases to a newly emerging pattern?

Alongside Zhang we have come across many other young – and sometimes older – peasant farmers who are reinventing farming. In Table 7.3, we summarize some key information about nine cases that we have studied in depth.[6] Taken together these cases provide the basis, we believe, for identifying a newly emerging pattern.

We believe that the instances summarized in this table are something more than just exceptional cases and that they go beyond the subjectivity of the peasants involved. Somehow these cases reflect 'subterranean' changes in Chinese society:

Table 7.3 Characteristics of farmers who are reinventing agriculture

Name	*Current age*	*Age when returning*	*Agricultural income (in yuan per year)*	*Type*
Zhang	39	28	(45,000) expected	Apple trees, 'green eggs'
Didi	49	29	25,000	Glass noodles
Xu Zhen Ming	46	33	40,000	Goats, glass noodles
Xiu, Sun	42	24	45,000	Goats, glass noodles
Ju-an	39	33	60,000	Chicken
Liu Zhentong	58	35	Not known	Agro-tourism facilities + free- range eggs
Xiu	66	45	30,000	Tree nurseries
Wang Liang Qing	40	28	100,000	Tea, pigs, maize, oranges, walnuts, eucalyptus
Li	44	34	120,000 (but highly variable)	Mushrooms, watermelon

they are hardly visible but very important. In the 1980s and 1990s the peasantry discovered labour migration as a promising new road that allowed for higher incomes and which, especially for young people, offered a step outside the dullness of village life. But in the 2000s and in the beginning of the 2010s the negative sides of labour migration increasingly came to the fore. It is, in many ways, a form of hardship. And from 2008 onwards it became obvious that employment in the large industries is inherently insecure. There is also a growing trend for industry to move from the coast inland and towards the west, in search of cheaper labour. As a result many labour migrants (and especially those who, as yet, do not have to face the 'big expenses' of secondary school education for their children, a wedding and providing them with a house) are looking for alternatives. Whereas people's eyes used to be focused on the cities, they are now also looking for opportunities in the villages – the more so since a vibrant rural non-farming economy is developing (J. Zhang *et al.*, 2006). Agriculture now offers new possibilities for making money[7] and obtaining (informal) loans from friends and family is very easy nowadays (because there is, as people say, 'now a lot of money in the village'). At the interface of these different 'subterranean' tendencies there is, we believe, *the reinvention of farming*.

When it comes to the relative importance of reinvented peasant farming, we have to be very clear. In quantitative terms the cases summarized in Table 7.3 represent only a miniscule part of the economy of the villages we have been studying. In strictly quantitative terms it looks like an almost irrelevant phenomenon. However, in *qualitative* terms we are probably witnessing the first swallows that herald a new spring. In this respect it is significant that stories are now being told, among labour migrants, about how 'one can now make a living in the village as well'.

At first sight it seems difficult to believe that we might be witnessing a new trend. It could be that these individuals are going, in a seemingly 'heroic way',[8] against the mainstream. The countryside is increasingly being drained of its labour force and, as impressive as the efforts of these outliers are, they remain seemingly insignificant. This, at least, is what most observers (provided that they noticed these exceptions in the first place) will think.

Here we argue that it might be different. The first point that has to be made clear is that these newly emerging peasant realities do not start out of 'heroism'. Although the projects of the reinventing peasants might at some moments take on an aspect of doggedness, strength and insubordination (Ye *et al.*, 2015), the cases we know absolutely do not start from such ingredients. Rather, they emerge from *incidents*. These might be incidents located in the places where migrant labour is heading (as with the outbreak of SARS in the case of Zhang described above). They might also be incidents (i.e. undesired and unexpected events) within the peasant's family: a son or daughter suffering a chronic health problem that requires the presence of the father, or parents who need a lot of assistance and therefore a more or less permanent residency in the village. In the cases we know there were repeatedly such *incidents* that provoked the early return and the subsequent 'heroic' search for an agriculture that provides an acceptable income (i.e. a reinvented agriculture).

But then, these same incidents represent, we believe, ever so many *interstices* (Holloway, 2010) in the standardized routines that lead the rural labour force towards the industrial areas of the eastern coast. We also believe that the frequency of such incidents will increase. In comparison with the past, migrant workers are less inclined to accept hardship and/or injustice. Industry moving inland from the coastal zone may well produce many new incidents. Migrant workers will not like the lower salaries associated with the inland move. Similar episodes might happen in the villages as women will be less inclined to solely take care of the children and parents-in-law, just as men are becoming less tolerant of the burden of being separated from their wives, children and parents. In short, we expect that industrial development and the associated town–countryside relations will increasingly generate interstices that result in growing numbers of incidents. In turn these incidents will provoke the question of whether there are alternatives to the well-established routines, triggering creative processes that might indeed result in a much broader reinvention of peasant farming. These processes will even get amplified since labour migration hardly ever results in what migrant workers call 'a decent job': 'I definitely do not want to work away from the village any more. Here in the village I have freedom. Nobody is in control of me, there is no boss commanding me. Here I have decency.'

Some more numbers

In Sanggang, a village with a population of 680 inhabitants (including migrant workers), there were (in 2012/2013) some twenty men who returned much earlier than expected from migrant work in order to actively engage in farming (and other local activities). They represent some 10 per cent of all households in the village. The distribution of the early returners over the age-groups is telling. According to Zhang (who became a member of the village committee in 2014 and therefore knows the local statistics):

> 'There are no young men of less than 25 years in the village. They are all engaged in labour migration or at university. Nearly all the men older than 50 are back in the village again (or are working and living in nearby townships, allowing them to frequently be in the village). [. . .] What is new is that around 5 per cent of the young men between 25 and 30 are back in the village. About 30 per cent of those aged between 30 and 40 years old have returned and even more (40 per cent) of those between 40–50 years are returnees.

This suggests a barely noted but nonetheless undeniable acceleration of return to the village. Comparable data is mentioned in the PhD manuscript of Meng Xiandan. Within the group of couples in which both man and woman live in the village (as opposed to the group of 'left-behind women' whose husbands are absent because they are involved in labour migration), 90 per cent of the men are active, in one way or another, in farming and 14 per cent are active solely in farming (Meng, 2014).

Patchwork and rag bag: the ingredients of novelty production

As indicated before, the farms of Chinese peasants, and especially those of the reinventing peasants, always look rather chaotic. There seems to be no notion of order: the farms are loosely structured and appear like patchworks. Everything seems to be contingent, if not emergent. This overall impression is further strengthened when one enters the houses of the 'reinventors'. They look like small workshops with ten activities going on at the same time. They also look like a store house. Nothing is thrown away. Even the meaningless (especially the meaningless, it seems) is stored here. With no disrespect many of these houses look like lumber sheds. But they are a lumber shed that relates to the farm as a 'ragbag' does to a patchwork.

Similar ragbags are encountered within the fields, or at the margins of the fields and/or in the stables.[9] When walking with Zhang through his fields with apple trees we stopped at one point. It was muddy, there was water, a tiny pond, an unpaved road (or something that came close to a road), dirt and a pole for electricity. In short, there was clutter. For Zhang, though, it was different: 'I could make a little reservoir here and have an electrical pump so that I can more easily irrigate the field and the hill behind it – there is electricity nearby.'

At that time the little pond was just part of the mess around the field.[10] It was part of Zhang's ragbag. But as with many other elements in the ragbag it has the potential to be converted into something useful. Zhang tells us he will probably construct a little house here on this field. This will be used to guard the field when the harvest is almost ready. He will also keep a watchdog here. As he is telling us this, we suddenly understand why he has two big dogs in his house in the village. And why the son of Zhao came along with his dog to mate with one of Zhang's dogs. At first we thought this was 'just another part of the chaos'. But these dogs are meaningful elements of the ragbag. They are meant to be added, later on, to the patchwork. The ragbag is a reservoir of elements that are, at this very moment, perhaps useless, but they might become highly useful in the future, that is: within the *evolving* patchwork.[11]

Figure 7.2 might help to explain the dialectics of peasant innovativeness (a term coined by Osti, 1991) or novelty-production (as the same process of generating innovations is called in the STS tradition,[12] see Rip and Schot, 2001). Taken in isolation both the patchwork (of fields, crops, animals, trucks, etc.) and the ragbag (the pole, the dogs, the pond, etc.) are relatively stable (although at the interfaces of the partly overlapping patches many unexpected events may occur). However, considerable dynamism may emerge when these elements are actively combined. This basically occurs by considering each of the single patches and asking: 'Which

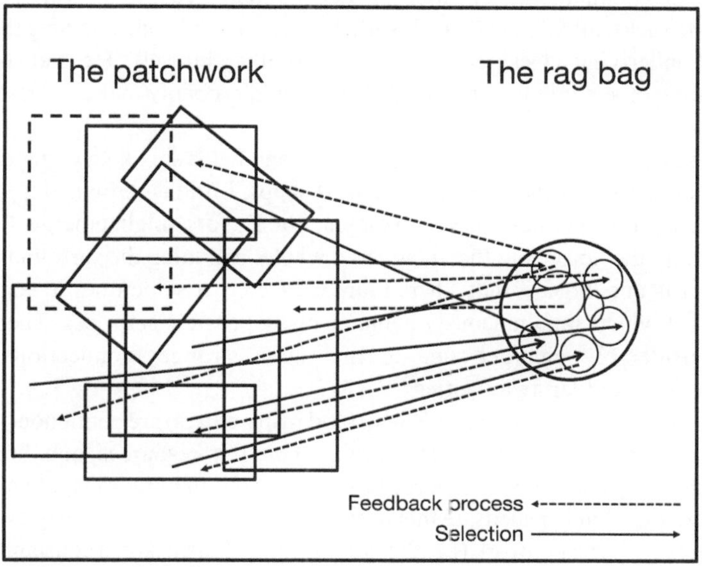

Figure 7.2 The intertwining of patchwork and rag bag

of the many available rags could *fit* in the patchwork (or be linked to it)?'. This is a *selective* process: it departs from the existing whole made up of different patches and asks whether specific rags may be integrated in the historically evolving patchwork and, if yes, to what degree. Specific rags can be used to create new, productive connections between different patches and/or improve the productivity of specific patches. Etc., etc.

The selection of rags and their subsequent integration in the patchwork is driven by conscious feedback: it goes from the patchwork to the ragbag. But the opposite direction is followed as well. It starts with the rags and the leading question is: 'What can this particular rag contribute to singular patches or to the patchwork as a whole?'. That is, the rags are used to *explore the potential* of patches that are, so far, unseen or not understood.[13] The starting point is not a specific patch (as in the first process of selection); here the starting point is a specific rag and all the patches are screened to see whether their performance can be improved because they entail possibilities not seen so far.[14]

Telling landscapes and farms

Our grandfathers (both on the Chinese and on the Dutch side) had, as did every peasant of that time, their lumber sheds.[15] Whatever they found, they kept in this shed. It was often completely unclear what most of the found items could be used for. But to throw something away was thought to be the most stupid thing one could do. For it might become useful, although one did not know how, when or where. Our grandfathers were fond of their lumber sheds; there were pots, and crates, and sacks, all filled with well- sorted items. The only one knowing the order, and the underlying classification, was grandfather himself. He had difficulty in admitting others into his little shed. They would probably make a 'mess'(!) out of it.

Chinese peasant houses and the margins of many fields are like our grandfathers' proverbial lumber sheds. They are like workshops. Little is thrown away, a lot is kept: waiting for a possible, as yet unknown, use. But one might emerge. In a way, it is not only the house, but the village as a whole, including the surrounding eco-system: it is one large, jointly shared lumber shed that sometimes looks terribly chaotic but which simultaneously contains many possible surprises. The mess is now considerably enlarged due to the much stronger connections[16] with neighbouring (and far away) cities.[17]

What Chinese peasants such as Zhang (and many others) are continuously doing is to tie different rags together, thus turning them into resources, and then trying to enlarge and improve this resource base by slotting in other rags, other objects taken from their lumber sheds. In this way the patchwork is continuously expanding and changing. The patchwork (i.e. the farm) is never finished, it is constantly being invented. It is always under construction. The patchwork and ragbag make the Chinese peasant farm into a field laboratory (Stuiver *et al.*, 2003). The patchwork thus contains many traces of previous decisions, of previous assemblages. In the case of Zhang's farm, the chestnut and walnut trees reflect, and remind him of, his return from labour migration in 2003 and his decision to make a living in the village and stay there. The apple trees tell another story (about expectations, investments, drudgery and utility). The different varieties, some new, others traditional, equally tell their story (about how to link to traders and street vendors). Understanding this story (i.e. exploring the 'biography of things', see Appadurai, 1986) is not easy, but once it starts it makes farms into an intriguing and promising whole. Thus, the patchwork (i.e. the richly chequered farm unit) is full of memories. It is, as it were, a spirited whole – a living organism. If you study it carefully, it starts to tell you the story of those who constructed it, the story of their ragbag and patchwork – and how the two interacted, mutually enriching each other.

It is well known that processes of genetic improvement rely on heterogeneity. A wide range of varieties is needed in order to combine, cross and select and then again repeat this procedure. This applies to animal breeding and plant breeding and applies equally to agriculture as a whole. A highly diverse agriculture that interacts with a changing environment will generate a range of responses to these changes. And the variety of responses allows for comparison and evaluation: so one can identify the most appropriate and satisfactory response(s).

The presence of the many lumber sheds contributes to the heterogeneity of Chinese agriculture and provides many resources for its incredible dynamism. As such, the story of Chinese agriculture is, to a large degree, the story of the lumber sheds on which it draws and therefore the story of those who created, filled and ordered these many lumber sheds.

The peasant-like nature of reinvented farming

The actors who are currently reinventing farming operate very much in peasant-like ways. They are 'entrepreneurial' (in the Schumpeterian sense) in so far as they do their best to find, use and further develop market and technological opportunities. The discussion (above) on novelty production is a case in point. These peasants are not *buying* a new technology and then implementing it as a ready-made *turnkey project* – instead, they themselves *develop* their own novelties and then improve them in a *step-by-step* way. They are very eager to test, use and further develop new possibilities. In this respect they are, indeed, very 'entrepreneurial'. However, their farming activities (plus the associated processing and marketing) are not organized according to what is known as the *model* of entrepreneurial farming (as specified, e.g., in Ploeg, 2008).[18] Instead, we are witnessing a highly dynamic and increasingly prosperous *peasant* economy. While reinventing and subsequently reconstituting farming, the renewing peasants maintain the basic rationale of the peasant way of farming – they stay at arm's length from the entrepreneurial rationale (or model). Table 7.4 summarizes some

Table 7.4 Features of the reinvented peasant farm and the entrepreneurial farm

	Reinvented peasant farms	*Entrepreneurial farms*
Relation with financial capital	'Let your own money circulate'	Use loans, credit from banks
Dependency on subsidies	'Meet all expenses from one's own wallet'	Expenses partly paid by the state
Relation with labour market	Own labour central, possibly with additional hired labour	Hired labour central, entrepreneur takes the role of manager
Nature of production	Highly diversified	Specialized in one or two remunerative activities
Type of resources used	Strongly grounded on locally available and self-controlled resources	High dependency on external inputs
Knowledge	Developing own knowledge and skills through learning-by-doing	Buying knowledge, contracting engineers and agronomists
Temporal horizon of innovative process	Step-by-step; slowly evolving over time	'Turnkey' project; required knowledge is bought-in
Nature of innovative process	Novelty production	Dependent on externally developed innovations
Process of growth	Modest scale increases; emphasis on labour-driven intensification	Jumps in scale increase
Risks	Building security and resilience	'Gambling'; highly risk-sensitive

of the basic contrasts between the two models[19] – building upon the empirical examples we have encountered in rural China.

Underlying the differences between reinvented peasant farming and entrepreneurial (and/or corporate) farming there is, we believe, a clear – although far from explicit – strategy. Using qualitative research data (derived from extensive interviews) we will try to reconstruct this strategy in the following section.

On strategy

The strategy of the renewing farmers is grounded in a *double denial*. They do not want to continue with labour migration and neither do they want to fall back on the 'classical' peasant agriculture they left in order to engage in labour migration. The way out is to reinvent farming or, as described in the following interview fragment, to *develop* farming: 'In my mind it is all about income *in the future* or income *now*. When you want income now, you have to engage in migrant labour. If you want income in the future you have to develop farming . . . .'

One thing is certain, it is impossible to generate enough income through 'classical' farming (as we have defined it here). Even considerably enlarging the acreage will not do:

> The maximum land that a couple can work (without machines) is 20 mu, but then you cannot have another economic activity. But the grain price is low and you need a lot of fertilizer, etc. So if you plant it all with maize it will render you 20,000 yuan gross income and that means 10,000 net income. In migration you can earn net 20,000 yuan. So, farming is not economic.

Hence, 'classical' farming does not present a way forward. The same, though, is true of labour migration. It renders an income, but it also represents hardship. It too is not a desirable way forward. Such a new road forward has to be *constructed*. Farming needs to be reinvented. If classical farming is the thesis, and labour migration the antithesis, then reinvented farming is the synthesis. It combines elements that were once strictly separated: the promise of a good income and being located in one's own rural village. This is illustrated in Figure 7.3 with Chinese words.

The '*xianjin'* position in Figure 7.3 is actively distanced from the other positions. In order to exist as '*xianjian'* it needs to be different, both materially and symbolically, from both classical farming and labour migration. This implies creativity. That is, according to one of our respondents (who has himself developed an important rural industry), it is all about 'living in advance' [*chao qian wi shy*: consciousness of being in advance, being ahead of the present].

At the strategy level, the specificity of reinvented peasant farming is stressed through a range of polarized notions. One set has already been introduced here. While classical farming and labour migration both centre on 'income now' (the difference being that in classical farming it is lower than in migration), reinvented farming centres on 'future income'. This is associated with another polarized

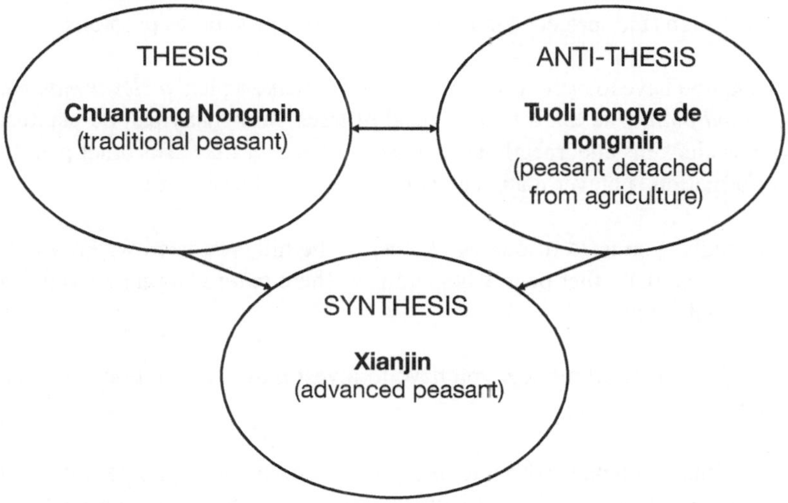

Figure 7.3 Thesis, antithesis, synthesis

notion. Migrant labour produces an income that is 'certain' (although this security might be threatened in many ways); in reinvented farming income is uncertain. Another key difference is about having (and developing) resources rather than having none at all:

'If you work hard in the village you can earn more money than in migration because we have more resources in the village.'

Resources are a distinguishing element. As such mobilizing, combining and developing resources is central to the strategies of the reinventing peasants. There is a Chinese saying: 'You cannot kill a person without a sword' (meaning, you need resources to do something).

Having resources is not only an issue of access. It assumes the capacity to envisage potential where others only perceive the absence of any promise: 'Others saw just a bare hill, but Zhang saw a garden for fruit trees' (a village leader talking about Zhang's vision).

Seeing such potential qualities requires courage (*yong qi*). This depends, just as the strength of the associated expectations (*xi wang*), on social networks and the information that is gathered through them. The family is also crucial here:

> If the family supports you, you can do something new; if you are just by yourself things will get very difficult; and of course your wife, she is the most important . . . when your wife is not complaining you know that you possibly are on the right track.

Expectations and courage need to go hand in hand because: 'you never know for sure whether you are on the right track, you can only know that afterwards.'

Nonetheless, our respondents specify several elements that contribute to making the new design (the 'project' of a reinvented farm) as solid as possible:

> First, you have to have wisdom, you need to make a plan, *a clever man makes a good plan*,[20] he knows which kind of trees, how many can be planted, he plans chickens underneath these trees . . . A clever man asks other people for information, a clever man is always looking for knowledge . . .
>
> Secondly, you need resources. Access to the hill, you need water, you need *guan xi* so that other people help you . . . [here *guan xi* might be understood as social capital].
>
> Thirdly, you need money, you have to invest in the hill, in trees, in a pump, in pipes . . . .
>
> And finally you need labour to help you; this might be in your family, or from friends in the village, both are okay, other labour might be a problem.

Other elements that help to make for a solid design (a solid project) explicitly concern the way it relates to the 'outside' world:

> Distrust the nice stories and the promises of others, be as careful as you can. Others might easily cheat you. And then, do as much as possible by yourself. And rely on your own resources, resources that can become your own. That is the most reliable. You should not become dependent on strangers' resources, resources controlled by others. And you should try out everything in small steps, first make an experiment on 0.3 mu for instance, before you go to 1 mu and then to 3 mu. It needs to be done in small steps . . . .

The relation with downstream markets plays an equally important role in the specification of the new resource base and the related farming activities. The key question here is whether or not to follow the market. Following the market is typical for entrepreneurial and corporate farming: this directs such farmers to the products that attract the best prices. By contrast peasants involved in the reinvention of farming typically show anti-cyclical behaviour. The stories abound:

> The Li brothers from Leshan saw [in 2012] spring onions attracted very good prices, so they planted large pieces of land with it [in 2013]. But then the prices dropped strongly. The brothers lost more than 200,000 yuan so they left. They had rented the land for 3 years but stopped paying so the land returned to the peasants. If you follow the market you will always be too late. Peasants will do it differently.

The creation of new resource combinations (and new types of farming) and developing these further introduces a new time perspective that is associated with the notion of *progress*. It is about moving ahead; about actively constructing a promising future. When asked about the objectives that inspired the creation of his new farm, Zhang says:

> The most important rule? Well-being in the *long run*, and this is not only about money but generally speaking . . . it is also about having a healthy body and a happy mind . . . and leaving a legacy for those who follow [. . .]. It is also about feeling proud. And for the rest it depends on your *ambitions* and your capacities. [. . .] Persistence is a quality, if you have chosen for something then you have to stick to this choice, you cannot change overnight and then again. There needs to be *continuity* [. . .]. Also *it takes time* to earn money from the market. Ten years ago I started to plant trees, now they earn me a lot of money. You have to see the profit in the *long run*. . . . [emphasis added]

All this underlines the fact that the farm is, indeed, permanently under construction (as discussed before). It is continuously unfolding. In more analytical terms, it is not only about production but as much (if not more) about enlarged reproduction. Another respondent said:

> Yes, we are *building* [. . .]. Of course, it is for the money, but secondly, you want also to leave a legacy for the next generation. So our farm is not just for the money [*zhing qian*: earn money]. And if somebody wanted to buy the business . . . Well, if the price is high enough . . . But that's a joke, I want to keep it, it is my dream . . . . [emphasis added]

And another villager, Didi, when explicitly asked the same question, said: 'Sell the farm? *Never, it is the root of the peasant, we cannot sell it.*'

This is a clear statement that was echoed by nearly all our respondents. However, some ambiguity remains. Didi added:

'At best we can rent it [i.e. the farm] to other people . . . .'

Be that as it may, development (constructing progress) versus staying more or less in the same situation (as is the case in classical farming and migrant labour) is an important pair of polarized notions that help to delineate, and to define, reinvented peasant farming. This clearly goes together with investments. Productive investments are central to reinvented farming. They are not used in labour migration and they are only secondary (and limited) in classical farming. Labour migration is mostly oriented towards *consumptive* investments (the 'big expenses' such as a house for the son and a prestigious wedding). By contrast, in reinvented farming, *productive* investments that result in capital formation (in the Chayanovian sense) are strategic:

> Others plant crops, I plant trees. I invest in trees [. . .]. Yes, my trees are my capital . . . You see, staying in the village and planting trees is thinking about the future, it will give income every year. Migration is also getting money every year, but it will stop when you retire from it. Hence, be in the village and making investments is more clever.

Box 7.1 A summary of the main strategic notions underlying the reinvention of farming

- Future income is seen as more important than current income.
- Being creative is important: develop activities not yet done by others.
- Don't feel afraid about 'unsure' income.
- Emphasis on resources (rather than on immediate income).
- Be knowledgeable and goal-oriented.
- Mobilize your social networks.
- Make a clear plan and try to be persistent.
- Don't follow the market.
- Time needs to represent progress (instead of repetition only).
- Enlarged reproduction is as important as increased production.
- Emphasis on capital formation (productive investments are crucial).
- Enough is enough (volume of activities defined by 'capacity' and 'conditions').

Together, these ingredients (see Box 7.1 for a brief summary) are combined in order to constitute strategies that result in making reinvented peasant farms. They are strategies that aim at 'earning an income in farming'; they orient young peasants who return from labour migration in the reinvention and reconstitution of farming.

Many of the ingredients in Box 7.1 come together in the response that Wu Liang Qing (see Table 7.3) gave when we asked him whether he considers himself to be a peasant or a business man. He argued that the difference is clear:

> businessmen have more money . . . I am a peasant. What I want to be is a *new peasant*, a professional peasant.[21] In the first place a new peasant has an open mind, he will find new ways to earn money, new ways to make a livelihood. Secondly, he has an open eye for technology and thirdly he can earn 1,000 yuan from a piece of land where normal peasants only earn 100 yuan.

This last element is especially telling. It shows how the search for new ways to earn money results in a considerable growth in the total social wealth produced. This is the fundamental difference with corporate and entrepreneurial farming.

Is it possible that the new strategies flow into entrepreneurial or corporate farming?

The strategies that result in the reinvention of peasant farming contain inbuilt limitations (like the automatic devices that can control and reduce the speed of cars) which may potentially hinder shifts towards entrepreneurial or corporate farming. The first of these devices is the lack of access to (or the unwillingness to access) the capital market. Formal bank loans are rarely available for people from the villages. They are constrained to operate with their own savings: 'We peasants don't have much money, so we have to invest step-by-step, and therefore you need to have ambitions. [He makes a joke] We need to have our own 5-year plan.'

The same point is often phrased in another way (see also Table 7.4): 'It has to be our own money that circulates, we cannot go beyond that.'

This answer was given by a pig breeder when asked why he chose to work with twenty sows and not with, say, thirty or forty. His answer was that the costs of acquiring these sows,[22] of building the stable and, especially, the acquisition of the necessary feed and fodder[23] should not exceed the savings that he and his wife obtained during labour migration. And once the farm is in operation, the selling of fattened piglets goes to financing the feed and fodder, medicines, insemination, etc. and providing an income for the family. Thus the breeder's (own) money indeed keeps circulating – and dependency on external funding (e.g. bank loans) is avoided.

A second inbuilt 'device' is related to learning and dealing with risks (features that are very important in the strategy of the reinventing farmers; see also Table 7.4): 'The farm needs to remain manageable. If something goes wrong there needs to be time available to get things right again. If not you stay in trouble.'

A third one relates to identity and social control. Zhao (who we met in Chapter 5), for example, contracted quite a lot of land. This often makes him the butt of people's jokes. While laughing, people ask him whether he will become the 'new landlord'. As difficult as it is to assess the precise meaning of such joking and nick-naming, we got the impression that it is also sending clear signals: 'do not become too big, do not exaggerate'.

Finally, reinventing farmers define an upper ceiling – derived from the amount required to live well (and to be able to meet the 'big expenses'). Going beyond that level, just to accumulate for the sake of accumulation, makes little or no sense within their world-view.

> Enough is enough . . . How far you develop depends in the first place on the capacity of people, if you can take care of 100 chickens then it does not make sense to have more than 100, then it stops. Secondly it depends on the economic conditions: if you have for instance one son who is already married and has a granddaughter and the son and his wife can get enough money by themselves, then you feel no pressure anymore, then you can invest the means you have in something new.[24]

These elements may well hinder a shift towards entrepreneurial or corporate farming – however, they do not completely exclude such a shift. Chinese agriculture is evidently going through a multifaceted transition (the outcomes of which will only be known several decades from now). In the associated turmoil there are possibly many developmental trajectories. In order to shape such patterns peasants, entrepreneurs and businessmen alike define their own (mutually contrasting) strategies. They do so by combining strategic ingredients (such as those summarized in Box 7.1). They may also combine these ingredients with notions derived from different settings. Earlier in this chapter we introduced Li, the grower of mushrooms and watermelons. He developed his farm in a typical peasant-like manner, following several of the strategic ingredients summarized in Box 7.1. In this way he too was reinventing the peasant farm. But slowly he started to combine these strategic ingredients with another one derived from industry[25]: '*having other people work for him in order to become rich*'. The possibility cannot be excluded that some of these reinventing farmers will, in the end, move towards entrepreneurial and/or capitalist ways of farming, especially in those farms where (temporary) workers are employed.

In the short run, though, this is not very probable. As Li explains:

> My dream is being a boss, but I cannot realize it now really because I am in farming, I am doing agriculture. I have to do the work myself, if not I hardly know how much time certain tasks take and then I cannot monitor the workers. I have to set also the standard, how the work is to be well done. Beyond that, if I would not work I would have to hire a manager to do what I do now and then I have to pay the manager and then I earn less income. There is no logic in that.

He continues: 'If after some more years I have earned enough money I may invest in another business, in making electrical artefacts for instance.[26] Then I can really have other people working for me and make me rich.'

This then is the irony of peasant life in present-day China (and especially of peasants who reinvent farming in order to earn better incomes). Being a successful peasant may bring them close to the dream of getting really rich by having others working for them. However, in order to make this dream come true they need to leave agriculture. For all kinds of reasons, Chinese agriculture hardly allows for capitalist and entrepreneurial models to be applied successfully. As Webber argues: 'The barriers to household producers who seek to become capitalist farmers are large.' (2012:40)

Conclusions

The reinvention of peasant farming is, we believe, crucial for China. It tends to construct a well-paid working population, strongly rooted in the countryside, that is able to deliver enough food to the cities while, at the same time, meeting new urban needs such as supplying agro-tourism facilities, regional specialties, high quality products and well cared for landscapes. The reinvention of peasant farming makes agriculture into a vibrant part of the rural economies (J. Zhang *et al.*, 2006) and it might, indirectly, strengthen the negotiating position of migrant workers engaged in industry – simply because it offers improved fall-back positions (as described by Brox, 2006 in the case of Norway).

The reinvention of farming (and even the longing for it) runs counter to attempts to expropriate and take over peasant agriculture (as exemplified by the creation of milk districts, horticultural co-operatives and the introduction of large-scale capitalist farm enterprises). It might be argued, therefore, that reinvention strategies are a main line of defence against a savage expropriation of Chinese peasant agriculture. In this respect reinvention represents a social struggle. It is a struggle that is occurring in the fields, in the houses, in the families and in the villages. It is a struggle encapsulated in the many new practices that result from the reinvention of farming.

Update

At the end of April 2015 we again met Liu Zhentong, the farmer who had been busy developing an agro-touristic farm on a hillside that is densely planted with exotic trees. To tell the truth, we had always been a bit sceptical about his enterprise, which seemed to be overambitious – but he continued to take steps forward and was continually making adaptations to his original plan. When we arrived we had a big surprise. A young man was painting an iron bow that was to become part of the arch that covers the entrance of the agro-touristic farm. He is Liu's son, 30 years old, well-trained and who, until recently, had worked as an engineer. He speaks English, is married, and he and his wife have one daughter. He quit his job in Baoding and decided to work together with his father, invest his savings in the farm and help to develop it further.

Sometimes it is difficult to believe that young people will come back to the villages to continue the farm work their parents are engaged in. This is especially the case if the young couple have good jobs that pay well. In reality, many of them won't return. One sees this in Sanggang and neighbouring villages (Liu's farm is located in one of these, Nangdu Gang). Maybe this returnee, the highly educated son of Liu Zhentong, highlights one of the most important features of the reinvention of farming – especially when it results in daring and courageous projects. It makes the return, and involvement, of youngsters into farming more probable.

Notes

1 It is indispensable for a variety of reasons and these may differ and change over time and from place to place. It may be related to the quality of food, with having a fall-back position in times of crisis, or having a form of pension for when one becomes old. It may be gender related (the women having their 'own' money for petty expenses). Having and keeping land is also deeply rooted in Chinese culture (see Chapter 3).

2 This is the case in the 'classical' farm type that can be found all over the country. This is not to deny, of course, that there are also many farmers who move for long periods and with their wife and children into the cities. They leave their land with relatives, neighbours and/or lease it to a third party. Sometimes this is a temporary solution, sometimes it is permanent.

3 Now he also has a minibus, a car and a three- wheeled tractor for farming.

4 Li refers here to the yearly expenditure on rent for the land, the annual renewal of plastic for the greenhouses, together with seeds, pesticides, water, wages, etc. These add up to some 9,000 yuan/mu.

5 A novelty is a new idea, a new practice or a new artefact that promises to work (although its 'working' is not yet well understood). A novelty differs from an innovation in that the latter is ready and well understood. A novelty is still under construction. For an extended discussion see Wiskerke and Ploeg, 2004.

6 The first seven cases are from Sanggang and Nangdugang village in Hebei province, the last two are from Huangjue village in Sichuan province.

7 This partly relates to the changes in agricultural and rural policies. See Ye *et al.*, 2010.

8 As we indicate later on, fellow peasants and village leaders also refer to the courage [*yong qi*] that underlies the construction of the reinvented peasant farms.

9 A typical innovation is housing for hens (that produce eggs) also being used simultaneously as a 'laboratory' to test the performance of pig production by having two sows in it as well.

10 Six months later, it had indeed been reconstituted and now feeds the irrigation system of the farm. But Zhang is still checking whether there will be enough water throughout the year. If not, the situation may change again.

11 The combination of ragbag and patchwork, of a lumber shed and a highly diversified farm, leads to unexpected interactions. It shows the uses and meanings of different elements that had previously gone unnoticed. It provokes a rethinking of the already established resource combination and may highlight potential that has so far gone unnoticed.

12 Science and Technology in Society.

13 The ragbag is a reservoir of materialized suggestions. It is increasingly enlarged through the use of the internet which allows peasants to connect to, and find suggestions available from, elsewhere. Nearly all the respondents have access to an internet connection.

14 The combination of patchwork and ragbag strongly resembles the heterogeneity of fields and the 'gene bank' of Peruvian potato producers, as described by Brush *et al.*, 1981. See also Ploeg, 1993.

15 This feature is evidently related to the space available in rural areas. In urban settings (especially those of today) such lumber sheds are far harder to find.

16 Travelling is an important ingredient in the biography of nearly all the actors mentioned in Table 7.3. For a theoretical discussion see Ye *et al.*, 2009.

17 This argument could be extended to the typical townships of the Chinese countryside. Here everything can be bought, everything can be repaired and everything can be changed into something different. That is, the ragbag is definitely strengthened by the chaotic offer of products and services available in typical county towns. It is not incidental that they typically look like 'frontier cities', i.e. places constructed and reconstructed at the shifting agricultural frontier.

18 As yet, the entrepreneurial model has not been clearly articulated in China. The factors that hamper a clear articulation are (1) the absence of a fully developed land market (land in China is not a free floating commodity); (2) the lack of access to the capital market (through formal credit) for small and medium size entrepreneurs; (3) the lack of any explicit policy that declares the largest part of the agricultural population to be redundant; and (4) the cheap price of labour. This last factor makes it attractive to employ a waged labour force and this moves these farm enterprises into the category of capitalist (or corporate) farms (see also Chapters 5 and 6).

19 Not all the features shown in Table 7.4 will be encountered in all reinvented farms. Yet our empirical analysis shows that reinvented peasant farms repeatedly display specific combinations of these features. Taken together the features shown in Table 7.4 compose an 'ideal typical model'. The same applies to the entrepreneurial model. And of course, on the ground there is no lack of complications and contradictions: some

farms combine many features from the left-hand column with some from the right-hand column. And vice versa. At the end of this chapter we explore some of these complexities.

20 Other respondents speak, in the same vein, of 'knowing where you are heading'. They often relate this to the need for persistence. Once you have a plan, you should not change it overnight. Persistence is needed, problems need to be faced.

21 This statement strongly resembles the notion of *Xianjian* discussed above (see Figure 7.3).

22 In fact this respondent started by buying just two sows and then bred the others until reaching twenty sows.

23 This is partly produced within the farm, partly bought from other farmers and is partly concentrates bought from traders.

24 In this respect it is interesting that reinventing peasants often use a particular 'window' in time. When they are around 30 years old they may already have amassed considerable savings, but at the same time the 'big expenses' (higher education, house and/or a wedding) for the children are still a long way off. Hence, their savings can be invested and will (probably) generate returns before it is time for the 'big expenses'.

25 When working in a factory in the county town he followed a course that included a video presentation in which a business man explained how he became rich. According to Li he said that he observed other people who became rich: 'I noted that they all had a lot of workers, they hired lots of people. That is the secret: a big guy gets money from other people working for him.'

26 Li has a nephew who is considering this possibility and may well join him.

8 Rural women

Glimpses of empowerment?

Xiandan Meng, Sabine de Rooij and Jan Douwe van der Ploeg

Although Chinese women are formally equal to men, in real life their situation is light years away from such equality. This applies especially to rural women. Many of them feel themselves to be just an appendage of their husbands. Rural women think of themselves as 'nobodies' (Meng, 2014:131). This lack of a positive identity is strongly interwoven, albeit in a highly complex and slowly changing way, with both the livelihood strategies of peasant families and the reigning cultural repertoires that confer meaning to the details of everyday life.

The main livelihood strategy of peasant families in China is to engage in both farming and other economic activities (see Chapter 2). Since the mid-1980s, structural and institutional changes, together with rapid economic growth, have opened up opportunities for rural people to escape poverty and hardship and improve the well-being of their families. The increase in employment in urban areas has led to rural outmigration becoming a common practice in rural households. Although both men and women leave the countryside to work in urban areas, the migration patterns are highly gendered. Migration is often temporary, either seasonal or circular, yet many women return to the countryside earlier than men (upon marriage or pregnancy) and for longer periods. In line with traditional rural gender roles, women stay in the village when it is necessary to take care of the children or elderly. In addition, they become the main farmers as their husbands stay away for longer, only returning home a few times a year (see e.g. FAO, 2011; Brauw *et al.*, 2012). The increasing participation of women in agriculture is referred to as the feminization of agriculture (Lastarria-Cornhiel, 2006; Kelkar, 2009; Deere, 2005). In theory, (long-term) male absence from home and the farm and the corresponding feminization of agriculture potentially offers a space for the empowerment of rural women. Female empowerment refers to the increasing ability of previously disempowered women to make choices that challenge existing power relations (Kabeer, 2005). However, as this study shows, male labour migration mainly seems to increase the workload of the women involved (see also Asis, 2003 and 2004; Hugo, 2000). Being alone on the farm for extended periods, having to do all the work and take the daily decisions is often experienced as a heavy burden. Alongside their solitude rural women have to deal with extra drudgery and this affects their well-being. Compared to women who work on the farm together with their husbands, the wives of long-distance workers feel lonelier and less happy and they judge their work situation as being less in line with what they desire. Yet, some of them manage to make the most of this difficult situation and see it as a challenge and an opportunity to gain empowerment. They are building on strengths they take from specific living conditions, qualities, work experiences and networks. This makes them feel a 'somebody' instead of one of those people nobody cares about.

The feminization of agriculture

The feminization of agriculture is an ambiguous concept.[1] According to Schutter (2013) it takes different forms and is dependent on the mode of farming. He observes '[three] different modalities through which agriculture is "feminized"' (2013:5). The first way, occurring in smallholder farms, is when husbands take paid off-farm jobs, at daily commuting distance or more remote locations. Either way, this implies that their wives have to do (most of) the farm work. When the farm produce is primarily used to fulfil the family's daily food needs, this model of farming is known as subsistence agriculture. In this case women's farm labour is not considered as 'work' but as part of caring for the family. If, however, the farming women start to produce for the local market, a different pathway to feminization occurs. Their work will be recognized as 'work'. But to be successful here, female farmers must overcome many gendered obstacles (e.g. access to land, credit, education, technology, services and cultural norms) and try to find ways to reconcile farm work with care tasks within the family. The third way to the feminization of agriculture is when women become workers in more capitalized forms of agriculture.

In this chapter we use data gathered from research on Chinese peasant families (Meng, 2014). These peasant women produce – to varying degrees – for both the market and their own consumption. A central question guiding this research was whether the feminization of peasant agriculture is empowering women. The Women's Empowerment in Agriculture Index (WEAI), a tool developed to measure the magnitude, nature and significance of women's engagement in the agriculture sector, was used as a framework to analyze the data. The WEAI highlights five domains:

1) autonomy and control in agriculture;
2) ownership, access to and decision-making power over productive resources;
3) sole or joint control over use of income;
4) time use (allocation of time to productive and domestic tasks and satisfaction with the time available for leisure activities); and
5) leadership in the community (membership in economic or social groups and comfort in speaking in public).

It also measures women's empowerment relative to the men within their households (Alkire *et al.*, 2012). Kabeer (2005) emphasizes that 'a process of empowerment often begins from within' and that this requires a sense of self-worth and agency. Kabeer stresses the importance of being aware of the link between people's feelings of self-worth and how they are seen by others and by society as a whole. She further reminds us not to ignore the impact of beliefs and values that support and justify gender inequality (Kabeer, 2005:14–15).

In this chapter we examine different dimensions and aspects of women's empowerment. We briefly pay attention to associated changes in agricultural production. Has the feminization of agriculture had a negative impact on food production (the size of cultivated land, the kind of crops grown, inputs used, output levels)? This question is relevant because it is frequently claimed that rural–urban male migration flows and the subsequent feminization of small-scale agriculture are associated with a decrease in agricultural productivity and that this can even pose a threat to food security (Tamang *et al.*, 2014).

Methodology

The analysis underlying this chapter builds on research undertaken in the period 2010–2012 in Qiaobei, a typical agricultural village in Jiangsu province. This province has high levels of outmigration.[2] In Qiaobei village there are 389 mainly male migrants from a total of 478 households. Most of these labour migrants work far away from home. The research included both quantitative and qualitative methods. The latter entailed life history research, participatory observation and interviews with key people within the village and township. The quantitative research involved face-to-face surveys of 100 farming women, all below the age of 55 years. An extensive description and discussion of the outcomes of the research can be found in Meng (2014). For the purpose of this chapter, the respondents are distinguished into three categories.

1) Respondents whose husband is a long distance worker who returns home only occasionally. These women are often referred to as 'left-behind' or 'staying behind' women. We refer to this category as the LD subgroup (LD: long distance). 39 per cent of respondents belong to this subgroup.
2) Respondents whose husband works within daily commuting distance, in a nearby county or town, and who comes home every evening. We refer to this category as the DC subgroup (DC: daily commuting). This subgroup accounts for 36 per cent of the women of Qiaobei village.
3) Respondents whose husband works in or around the village, or just on the family farm. We refer to this third category as the VIL subgroup (VIL: village). It covers 25 per cent of the sample. From this 25 per cent, 18 per cent have husbands who work in or around the village as well as on the family farm. The remaining 7 per cent only work on the farm. They seem to most resemble the people who are reinventing the peasant farm, who were discussed in Chapter 7.

Some general characteristics of the survey respondents

The women participating in the survey were between 23 and 55 years old with an average age of 45. The majority had more than one child (67 per cent). Only the respondents born in Qiaobei village lived close to their natal family (29 per cent). Compared to their husbands, the women had a lower educational status: the majority had no formal schooling and were illiterate (49 per cent) or educated to no more than elementary school level (26 per cent). This is in considerable contrast with the educational level of the men. The majority of their husbands had completed junior middle school (51 per cent), high school or a technical secondary school (23 per cent). Only 5 per cent of the men were illiterate. In this village, only a few respondents defined themselves as 'rich' (11 per cent) or '(very) poor' (24 per cent); the majority judged themselves to be 'in-between' (65 per cent). Almost half of the women had experience as a migrant worker (45 per cent). Practically all the husbands had paid work alongside farming (93 per cent). The average size of the land farmed by the households was 5 to 9 mu (15 mu = 1 hectare) with a variation from 1 to 20 mu. Rice and wheat were the main crops. A few also grew corn (10 per cent). Wheat and rice were often sold on the market (by 76 per cent and 64 per cent of the households respectively). The commercialization of rice occurred more often when the husbands were long distance workers (82 per cent of women in this group as opposed to 47 per cent of the women in the DC subgroup and 63 per cent of those in the VIL subgroup).[3] Breeding animals (mainly chickens, pigs and pigeons) was an agricultural activity on 39 per cent of the farms, done partly for self-consumption and partly for selling. The staying behind women were less involved in breeding activities (21 per cent as opposed to half of the women in the other subgroups). But when they were involved their main motivation was producing more nutritious food for their children (88 per cent). This is in contrast to women in the VIL subgroup of whom more than half mentioned selling as the main reason (54 per cent). On a few farms (all in the VIL subgroup) breeding was also a larger-scale activity (for instance 100 pigs; or 3,000 to 6,000 chickens).

Within the peasant families there is a clear gender division of labour, with women doing the '*inside*' work and men the '*outside*' work. These are powerful folk concepts that strongly relate to gender inequality. Women's work is confined to the inside – being outside is 'improper'. Being and working inside also generates invisibility. The outside is, by contrast, where men go, where commodities flow and money is earned. To go outside is daring, it requires strength and courage. In the not too distant past, the outside coincided with the agricultural fields (and with jobs elsewhere). Men worked in the fields while women stayed inside. Traditionally the work inside involved taking care of the household, children, grandchildren, the elderly and the vegetable garden. 'Inside' is where protection was organized and where one was protected. It was not about earning money, but about consumption. Overall, the typical inside was non-commoditized and was not governed by commodity relations. This again was associated with traditional gender roles: men controlled the money flows, women stayed at the margins.

Over recent decades this traditional gender division of labour has changed profoundly. With labour migration men have left the fields and the women have taken over. But remarkably, the folk concepts have remained the same, although they now apply to reshuffled empirical realities. Working *inside* now critically embraces working in the fields, while working *outside* applies more to the often very distant industries and construction sites where migrants go. Outside is still the space where the men work (wherever it is). The notion of inside has moved with the women to the agricultural fields (and market places, etc.). Together with this came a gendered redistribution of meaning and value. Once, working in the fields was judged to be the main and foremost activity of the household: it represented the (main) source of income and it was done (mainly) by the men. Now it is done mainly by women and thus its status has changed. It is work inside and of a secondary nature.

Theoretically it can be argued that agricultural activities ('inside') and migration work ('outside') are incommensurable entities. They are different from each other: yet both are needed by the household. But the reigning, and highly gendered, cultural repertoires construct a hierarchical relationship between them. The one (outside) is more important than the other (inside). Thus the dominant cultural notions render the work of men into something that is important, while women's work emerges as less important – because it is located *inside.*

Thus working inside now also includes farming, that is, work in the fields and sometimes also beyond (buying inputs, selling the output, gathering information and knowledge, etc.). For some women it also includes animal breeding. Women's work in the fields includes sowing seeds and transporting the seedlings to the fields, applying fertilizer, weed control, irrigation and doing part of the harvesting work in the peak seasons. In contrast to the women with husbands who return home every day, the staying behind women have to do that work largely by themselves. Their husbands come home only occasionally to do the '*heavy'* work (ploughing, sowing, transplanting) although some women have to do this as well. In contrast to this heavy work, women's labour is normally classified as '*light'*.

In addition to the farming activities, many of the women (41 per cent, on average) had paid work. They worked either in the home (e.g. sewing clothes, making toys, etc.) or elsewhere (running a small shop in the village, working in the watermelon plantation or in one of the local industries) or together with their husband (e.g. in construction, transport or trade). The farm women whose husbands worked in or around the village or just on the farm were relatively more involved in additional paid activities (64 per cent in the VIL subgroup as opposed to 25 per cent in the DC subgroup and 41 per cent in the LD subgroup).[4] They said they took on paid work 'to add to the family income' and 'to pay for daily expenses'. The income husbands earned was mostly needed to pay for a good education for their children (mentioned by 72 per cent of the women who said their husband was working 'outside'). The money was also used 'to create a financial buffer' (12 per cent), 'to pay for the son's marriage' (11 per cent) and 'to pay off debts' (5 per cent).

This gendered division of tasks was not questioned by the women (nor by their husbands). As the women explained, 'he can earn more money outside than me' and, equally important, they believe that the wife should be at home, that care giving is women's work. In fact, most women considered taking care of their children and/or grandchildren as their most important task within the farming family (79 per cent on average for all women).

Women's agricultural empowerment

The domain of productive resources

Land, including farmland, is public property in China but access to (agricultural) land is grounded in the legal right to use rural land. Under the law, women and men have equal land-use rights. Yet, the legal framework is at odds with rural social and cultural traditions and discriminates against rural Chinese women (Wang, 2013; Lin and Zhang, 2006). For instance, patrilocal residence gives marrying-in women the legitimate right to have land in their own name in their new home. But the 30 year land-use contracts, meant to sustain the stability of land-use rights, can make redistribution of land problematic, especially when farmland in the village is limited.[5] Therefore, women who can't be allocated land upon marriage may keep their land-use right in their natal village. This land is often not returned to the village for redistribution; her relatives take advantage of her share. As one of the women interviewed emphasized:

> Nobody wants to give back land when his daughter marries someone living in another place . . . a married daughter has land in her own name but she cannot take the land with her. The general view is that the land of a married daughter belongs to her natal family . . . . The wives of her brothers could meet the same problem after marriage and move to Qiaobei Village. From a long-term perspective, there is a balance . . . .

In case of divorce, widowhood and remarriage, access to the land can become problematic for women (see e.g. Ray Liaw, 2008; UNDP, 2003) and women's land might also be taken when they can't pay the fine for infringement of the one child policy. In fact, women's access to the use of farmland is uncertain.

It also happened to one of the women in this village:

> I do not have any land; my children do not have any land either. During the readjustment period, we should have been allocated at least three people's farm land. However, we wanted to have a boy, so I got pregnant again. But I have two daughters now. At that period of land readjustment, we did not have money to pay the fine for disobeying the one child policy. The village cadre was very angry and he did not give us the farmland we were entitled to. Now, my husband is the only one in my household who has land, around 1.5 mu. It is not enough for us to survive.

Most of the women who participated in the survey said they have land of their 'own' in their family (92 per cent). Yet it was not clear whether they were referring to land in their natal village or in Qiabei village. Anyhow, 'own' land doesn't seem the most important issue for them. What the women care about is whether their family has the same amount of farmland as the other families in their production team. If so, they seldom claim their land-use rights. What they can experience as a problem is that they do not always have as much land as they need or desire. Moreover, some of their land can be taken at any time by the village committee – for road construction, construction of residential buildings or the establishment of local industries and horticultural enterprises. One of the farm women said:

> I disagreed with the renting out of the land (2 mu) but the village leaders made this decision. The farmers had no voice in the negotiations with the watermelon plantation. I am afraid that the plantation also wants to use the rest of my land (1.5 mu) but we do not have the power to prevent this.

Another woman said:

> Now, there is almost no farmland in my family. Is that a safe situation? I am 51 years old, my husband is 54. How many more years can he work away from home to earn money? What can we do when he cannot migrate anymore? We are getting older and older and the bosses dislike employing elderly people. They think we are useless. We also don't want to rely on our sons if we can farm . . . Yes, the money you get from renting out is similar to the earnings from growing crops, but we don't have our own food now; the feeling is different. I do not want anybody to take more of my land but there is no way to avoid it if the government wants to use the land . . . We should be able to keep some land for farming. It is a kind of guarantee.

The women don't necessarily take enforced land reduction as a negative development. Women whose land was taken by the watermelon plantation have a lower work burden in agriculture, lower agricultural costs and an income from the rent. Some also work on the watermelon plantation or invest the extra time in paid work in or around the village. One woman said:

> It is good to sub-rent the farmland to the bosses; they can give us 1,000 yuan per mu per year which is almost equal to the earnings we got from growing crops. Moreover, we do not need to invest any time and money in it. We can also get some money by doing some extra job, such as making toys at home.

The staying behind women reported relative more changes in the size of land they farm: while 49 per cent said their farmland had reduced (as opposed to 37 per cent on average), 27 per cent (as opposed to 21 per cent on average) reported an increase. Those who had increased their land, farmed land from other family members. In some cases a reduction in land was the farmer's own decision and not enforced from above by the village committee. All the respondents said they negotiated the decision to reduce the farmed land with their husband and that the outcome was the result of common agreement, as were decisions about renting extra land.

Most of the respondents did not mention any problem in getting access to additional productive resources such as water and credit. The majority of women could access irrigation water (70 per cent) and those who mentioned difficulties either had farmland situated high up or land relatively far away from the village hydropower station (28 per cent). However, the privatization of the hydropower station was experienced as problematic. Ever since the village committee sold it to two villagers, the peasants have to pay 20 yuan each time they need to irrigate their land.

Taking a bank loan for agricultural production appears to be a rare phenomenon in Qiaobei village. Many women said they don't know if it is difficult to get such a loan (40 per cent); others said they didn't need it (21 per cent). One of the women explained:

> Farming inputs are divided into small pieces and that doesn't cost so much. For instance, when applying pesticides to rice it costs about 20 yuan per mu at a time. There is no need to apply for credit for farming. There is also no opportunity to expand the farming scale. There is not so much farmland available.

The limited need for bank loans is also the result of the local habit of peasants being able to pay for the agricultural inputs they have used after selling their harvest. This practice is based on relations of trust between peasants and sellers. A female peasant explained:

> If you don't have cash money you can get the fertilizer or pesticides without paying for them right away. The seller will record the deals and come to collect the money in the harvest season. They know you will have money when you sell the produce. There is no need to sign any kind of contract. We know each other; nobody will cheat or renege on the deals. Nobody wants to have a bad reputation. If you do not pay the money after the harvest, nobody will make deals with you next year.

Despite this most respondents were familiar with borrowing money (76 per cent, on average), mostly from relatives (93 per cent). Only a few women needed the money to buy agricultural inputs (9 per cent). The loans were primarily used to cover the costs of their children's education (36 per cent, on average). The staying behind women mentioned this more often (59 per cent). A loan to pay for agricultural inputs was more often the motive for the women whose husbands worked in the village or on the farm (39 per cent). Other reasons included covering the costs of the son's marriage and, linked to this, house (re)construction (26 per cent and 8 per cent respectively).

If a loan was needed, both wife and husband made efforts to obtain it: 43 per cent of women and 54 per cent of men. Staying behind women more often did this themselves (52 per cent as opposed to 45 per cent for the women in the DC subgroup and 28 per cent in the VIL subgroup). Most commonly men and women borrowed from their own respective families. Considering this, it is of no surprise that the women said they had no need for a credit organisation in the village.

Farm women have little knowledge about other financial services such as farm insurance or subsidies. For instance, less than half of the women knew how much agricultural subsidy they get. And though all the peasants in Qiaobei village bought farm insurance, less than half of all women said they or their husband had signed such a contract (45 per cent). Women do not acknowledge the importance of farm insurance and subsidies.

Basic knowledge of farming had been built up from childhood by only a minority of the women (33 per cent, on average). The most common way the women gathered this knowledge was learning-by-doing (56 per cent, on average). To keep their agricultural knowledge updated and to get new information, the women asked (or looked at the behaviour of) relatives, friends and other villagers (48 per cent, on average) and followed specific television programmes (17 per cent, on average). The husband was rarely a source of information, particularly not for the staying behind women (3 per cent as opposed to 12 per cent, on average). Additional sources of information were retailers/ wholesalers who sell agricultural inputs (fertilizer, pesticides, seeds, etc.) and machinery. Some of the women showed an interest in additional agricultural schooling. Women in the VIL subgroup were relatively more interested in information about seeds, fertilizer and pesticides, poultry breeding, planting techniques, marketing, agricultural technology and rural policy. Many women said they had no idea what kind of information they would need (40 per cent). The interviews also showed that information about the safe use of pesticides and herbicides is much needed.

Access to agricultural labour could be problematic for the farming women, although almost all staying behind women worked on the land without getting help (85 per cent) and more than half of them said they don't need it (58 per cent). Those women who need help (in the peak season) will ask family members to help them out but they are not always available. Hiring labour is possible but costs money and this can be an obstacle. When the women do need help it is mainly for using farm machinery (harvesting, ploughing) which is considered to be men's work. Even though the women could gain these (male) skills and would like to drive the machinery themselves, it is still difficult for them because the machines are tailored to male bodies. The biographies show that cultural norms can also be obstacles to getting help. It was said that staying behind women in need of extra labour are not always allowed to participate in the village's labour exchange system due to women's lower body strength. Villagers would also gossip about staying behind women if they are in contact with other men, even when these men are helping them out in the fields (e.g. with driving a tractor, ploughing).

The domain of agricultural production

Women's agricultural empowerment within the domain of agricultural production implies that women's farm work goes together with autonomy in production and having an input into productive decisions (Alkire *et al.*, 2012). That is to say they have sole or joint decision-making about the size of the cultivated area, what crops to produce and in what way (how much of which inputs to use and when, where to buy them, when to irrigate, etc.), when to sell what and for how much, to whom, how and where to acquire knowledge, etc. Data analysis shows that the staying behind women have a larger say in decisions about agricultural production compared to other women: half of them said that they take most of these decisions (51 per cent as opposed to 17 per cent of the women in the DC subgroup and 4 per cent in the VIL subgroup). In the comparative subgroups husbands had a larger say but joint decision-making is still common, in particular in households where husbands work in or around the village (56 per cent as opposed to 44 per cent in the DC subgroup and 23 per cent in the LD subgroup (see Table 8.1).

Most staying behind women who took decisions about agricultural production by themselves talked with their husbands about their farm work to some extent (75 per cent). But during regular telephone calls their children were the major topic and conversations were very short (a few minutes at a time).

Tables 8.2 and 8.3 show the division of specific farm tasks between husband and wife (who does what) and the decision-making associated with these (what to buy, when, how much from whom, what to sell, to whom, etc.) for the different subgroups. This explores the farm tasks that extend beyond work in the fields (i.e. buying inputs, selling outputs) and that assume additional knowledge, skills, access to markets and no mobility barriers.

Data analysis shows that the majority of women whose husband was a long distance worker were relatively more often responsible for the purchase of pesticides, seeds, and fertilizer (90 per cent, 84 per cent and 67 per cent respectively) and selling products on the market (46 per cent). When the husband worked closer to home these tasks were much more often carried out by the husband. The exception here is the purchase of farm machinery, one of the productive assets, and this is clearly a typical men's task in all situations. The fact that women are relatively less involved in the purchase of fertilizer is, as the women said, because the packaging of fertilizer is not designed for women (the packs being too heavy).

Table 8.1 Women's view on their present share in agricultural decision- making (%)

		Share of wife and husband in agricultural decision making		
Work location of husband	*Woman takes most of the decisions*	*Joint decision-making of husband and wife*	*Her husband takes most of the decisions*	*Total*
Long distance	51	23	26	100
Daily commuting distance	17	44	39	100
In or nearby the village	4	56	40	100
Total	27	39	34	100

Note: $X^2(4) = 20.82$; $p=0.000$

> It is too heavy for me to buy it and transport it by myself. Usually, my husband does it for me when he is at home. It is not perishable; you can store it at home. If it is not enough, then I buy just a little bit in the shop.

As can be seen, husbands who worked at a long distance were more involved in the selling of products than in the purchase of seeds, fertilizer and pesticides. Half of these men saw marketing of the produce as their work (46 per cent). The reason for this is that many husbands return home in the busy season to harvest the crops and to prepare the land for new crops. Before outmigration, selling used to be a man's task. So when he is home this pattern is resumed. The combination of her work in the family, a lack of knowledge about how to use the rice weighing machine or having no driving licence can also limit women's physical access to markets (to buy inputs or sell their produce). The introduction of electric bikes has been helpful in overcoming this latter obstacle. Those who can afford this kind of bike and also have access to one (some husbands use the bike to go to work) gain mobility and more autonomy.

Staying behind women mostly combine the purchase of inputs with decision-making about inputs (see Table 8.3). They decide when and what to buy and from whom. And though buying farm machinery is not a woman's task, many of these

Table 8.2 Division of farm labour within the household and location of husband's work (%)

Farms tasks	*Location of husband's work*	*Wife*	*Hus-band*	*Jointly*	*Others**	*Total*
Purchase of pesticides	Long distance	90	5	0	5	100
$X^2(6)=43.03; p=0.000$	Daily commuting distance	33	56	3	8	100
	In or near the village	20	76	4	0	100
Purchase of seeds	Long distance	84	8	0	8	100
$X^2(6)=43.39; p=0.000$	Daily commuting distance	31	58	3	8	100
	In or near the village	16	76	8	0	100
Purchase of fertilizer	Long distance	67	18	5	10	100
$X^2(6)=27.45; p=0.000$	Daily commuting distance	25	58	3	14	100
	In or near the village	20	76	4	0	100
Selling produce	Long distance	46	46	5	3	100
$X^2(6)= 19.09; p=0.004$	Daily commuting distance	25	50	25	0	100
	In or near the village	4	76	20	0	100
Purchase of farm machinery	Long distance	3	77	10	10	100
$X^2(6)= 5.12; p=0.529$; n.s.	Daily commuting distance	0	67	22	11	100
	In or near the village	0	84	12	4	100

Note: * mostly a relative

Table 8.3 Division of farm decision-making on the purchase of agricultural inputs within the household and location of husband's work (%)

Decision-making on the purchase of inputs and selling of the produce	*Work location of husband*	*Wife*	*Hus-band*	*Jointly*	*Others**	*Total*
Purchase of pesticides	Long distance	87	10	0	3	100
$X^2(6)=46.92$; $p=0.000$	Daily commuting distance	28	56	11	5	100
	In or near the village	12	80	8	0	100
Purchase of seeds	Long distance	85	13	0	2	100
$X^2(6)=41.85$; $p=0.000$	Daily commuting distance	31	56	8	5	100
	In or near the village	12	76	12	0	100
Purchase of fertilizer	Long distance	69	23	5	3	100
$X^2(6)=24.52$; $p=0.000$	Daily commuting distance	28	58	8	6	100
	In or near the village	16	76	8	0	100
Selling the produce	Long distance	33	41	23	3	100
$X^2(6)=10.23$; $p=0.115$; n.s.	Daily commuting distance	22	47	31	0	100
	In or near the village	4	68	28	0	100
Purchase of farm machinery	Long distance	0	15	75	10	100
$X^2(4)=20.37$; $p=0.000$	Daily commuting distance	0	56	33	11	100
	In or near the village	0	64	32	4	100

Note: * mostly relatives

women decided jointly with their husbands about such 'big' expenditures (75 per cent). When husbands return home every day only a third of the women had a voice in these things (33 per cent DC subgroup and 32 per cent VIL subgroup respectively).

All in all, these figures indicate a widening of the world in which the staying behind women operate. When husbands work far away, the boundary between inside and outside work clearly shifts and this also pertains to decision-making.

Women's work in the vegetable garden deserves separate attention. The vegetable garden is a typical women's domain; women decide what to grow, when and in what way.

Growing vegetables is of importance to the peasant family. It feeds the family, it saves on daily expenditure and the products function as a gift to maintain *guanxi*. The home plot is further used to grow the rice seedlings; this is important work because a good harvest depends on the quality of the seedlings. This is also women's work, particularly when the husband has migrated out. What is of interest is that, while women have increased the use of pesticides in staple food production in recent years, they don't use these in their vegetable gardens.

Time allocation

Within the family, cooking food and cleaning are chores that need to be done every day and are considered women's work. The women put, on average, 3.2 hours per day into this work. The number of daily hours spent on agricultural production varied and depended on the season. In peak periods, when crops are harvested, new seeds need to be sown, and fertilizer spread, it amounted to, on average, 10.5 hours per day (standard deviation 2.17). Women who bred animals spent 1.1 hours per day, on average, on this activity, mainly on cleaning and feeding (standard deviation 1.6). In the low seasons agricultural work (mainly irrigation and spreading fertilizer, pesticides and herbicides) took much less time, on average, 2.7 hours per day. When adding the time needed for cooking, cleaning and agricultural production we found that women worked, on average, 13.7 hours per day in the peak seasons. Taken together, they spent, on average, 5.9 hours per day for agricultural production, cooking and cleaning. The figures show that the hours staying behind women spend on agricultural production and the household differ greatly from the hours of the other farm women. Animal breeding extends the working day, and this involved in particular the women in the VIL subgroup (see Table 8.4).

It is obvious that farm women work more hours per day than the hours already mentioned above. Managing the household includes more than cooking and cleaning. Women also need to take care of (grand)children and/or the elderly, do the shopping, take care of the vegetable garden, and, for the women in the LD subgroup, also maintain *guanxi*.[6] In addition, 41 per cent of the women had paid work outside farming. Women in the VIL subgroup were more often involved in

Table 8.4 Number of working hours per day in agriculture, breeding, cooking and cleaning in busy and quieter seasons; VIL, DC and LD subgroups (%)

Type of work per season	*Hours per day spent on agricultural work, animal breeding and selected household tasks by women in the three subgroups*		
	LD	*DC*	*VIL*
In busy seasons			
Agricultural work	10.7	10.4	10.2
Cooking and cleaning	3.1	3.3	3.4
Subtotal	13.8	13.7	13.6
Animal breeding	0.6	0.7	1.8
Total in busy seasons	14.4	14.4	15.4
In quieter seasons			
Agricultural work	1.6	1.7	1.8
Cooking and cleaning	3.1	3.3	3.4
Subtotal	4.7	5.0	5.2
Animal breeding	0.6	0.7	1.8
Total in quieter seasons	5.3	5.7	7.0

this than the other women (64 per cent as opposed to 25 per cent within the DC subgroup and 41 per cent in the LD subgroup). The time they spent on extra paid work is not taken into account here. As such this data does not give a clear answer on the differences in working hours of farm women in different subgroups. What is clear is that all farm women have a high workload, especially in the busy seasons. Almost all the women (91 per cent on average) said they had no leisure time and there is no significant difference here between the women in the three subgroups. Women whose husbands worked far away bore an extra responsibility because they had to deal with daily problems largely on their own. They reported feeling more tired, stressed and lonely since their husbands migrated out.

Domain of income

The main income in the average household was the income earned by the husband from his work alongside farming (64 per cent). This was especially the case for households in the LD and DC subgroups (82 per cent and 72 per cent respectively, as opposed to 24 per cent in the VIL subgroup). In families where the husband worked around the village or on the farm, the income sources were more varied. Along with the income of the husband (25 per cent), women mentioned income from farming and animal breeding and from the son (36 per cent each) as the main sources. Women's share in the total income (from farming and/or extra paid work) was hardly ever indicated as the main income (by 2 per cent, on average).

According to most women (71 per cent, on average) there is no distinction between how their income and that of their husband is spent. Those who did see a difference said that their own income is primarily spent on daily or 'small' expenses such as food, water, energy, school fees for the young children, maintaining *guanxi* and farming.[7] The husband's income is used for 'big' expenses; daily expenses are only paid for with this money when strictly necessary. 'Big' expenditure primarily relates to ensuring a good future for their child(ren), especially the son. In this way the parents are investing in their own future; they expect their son to take care of them during old age. Important 'big' expenditure cited by the women included higher education of the child(ren) (34 per cent) and the marriage of the son (32 per cent), including a new house for him and his future wife (24 per cent) and household appliances (11 per cent). In practice, even 'things that cost more than 1000 yuan' or 'things that cost more money than the usual daily expenditure' can be referred to as 'big'.

In general, husbands had more control over 'big' expenditure (on average 49 per cent of men, as opposed to 6 per cent of the women, decided about such expenses) but these decisions were also quite often negotiated between husband and wife (44 per cent, on average). This applies to all the subgroups. Most women took decisions related to 'small' expenses by themselves. Interestingly, almost all the women mentioned daily spending as a major household expenditure (92 per cent, on average).

Farm women's beliefs and perceptions of the gender division of labour in agriculture and within the family

Women's beliefs and perceptions of their farm work and the gender division of labour within the household and the farm impede women's agricultural empowerment. Most respondents had a traditional opinion about the division of tasks between husband and wife within the family. The starting point is that taking care of the household and family members is women's work. And ever since earning money became important, and most husbands took a paid job, farm work has also become women's work *('*it is common in the village'*)*. Only agricultural tasks such as ploughing, harvesting and seeding, driving farm machinery and (large-scale) animal breeding remain men's work. Many women see their work as of a lower grade than men's work. The distinction they make between 'light' and 'heavy' work which coincides with women's and men's work, reflects this.

For some of the respondents, doing farm work and decision-making is not a self-evident combination. But many of them didn't question this. Two thirds of the women supported the statement, 'it is all right when the wife does most of the farm work while her husband decides about everything.' Even among the women that made most farm decisions by themselves, a group which contained relatively more left-behind women, the majority (59 per cent supported this statement. The women who disagreed most with this statement were those who took farm decisions jointly with their husbands (39 per cent as opposed to 23 per cent, on average).[8] Joint decision-making happened most often in the VIL subgroup. The view that a husband should take agricultural decisions even when he does not do the farm work is in line with their general recognition that the husband is the 'head of the household' (92 per cent). Taking decisions is part of that status.

Many respondents didn't think very highly of their work in agricultural production. They consider it work of the second order. Only a few mentioned farming as their most important task (17 per cent, on average) and the majority saw the role of agricultural production in their family as a supplementary economic activity (65 per cent, on average). Few women felt proud to be a farmer (21 per cent, on average). Moreover, in spite of all the farm work they did, almost half of them saw men as better farmers (45 per cent, on average). Here the migration status of the husband plays a key role. Staying behind women were more often convinced that women are better farmers (23 per cent as opposed to 6 per cent of the women in the DC subgroup and 4 per cent in the VIL subgroup).[9] At the same time, they also more often believed the opposite and said the husband was the better farmer (51 per cent as opposed to 42 per cent of the women in the DC subgroup and 40 per cent in the VIL subgroup). As such they less often held the opinion that husband and wife have similar farming qualities (26 per cent in the LD subgroup as opposed to 53 per cent of the women in the DC subgroup and 56 per cent in the VIL subgroup)[10] (see Table 8.5).

Farm work was not much liked by many respondents: not even a third said they liked it (31 per cent, on average). The staying behind women were most negative about farm work: only 13 per cent said they liked it and more than half said they

Table 8.5 Who is the better farmer: wife or husband? (%)

Work location of husband	*Who can farm better, husband or wife?*			
	Wife	*Husband*	*No difference*	*Total*
Long distance	23	51	26	100
Daily commuting distance	5	42	53	100
In or near the village	4	40	56	100
Average	12	45	43	100

didn't (56 per cent as opposed to 17 per cent of the women in the DC subgroup and 12 per cent in the VIL subgroup[11]).[12] It seems that more agricultural decision-making power does not automatically result in greater work satisfaction. In fact, if staying behind women had the choice, they would rather be a worker. Also half of the women in the other subgroups would have chosen this new identity (50 per cent and 48 per cent of the women in the DC and VIL subgroup respectively) but two thirds (67 per cent) of women in the LD subgroup would have taken this option. Being a worker has a higher status partly because of the higher earnings. Bearing this in mind, it is quite interesting that only 39 per cent of the women who said they would prefer to be a worker would abandon farming if there were (better) work opportunities elsewhere or if they had enough money/savings. The staying behind women more often made this choice (50 per cent). This, once more, is a sign of dissatisfaction with their current situation.

Despite this the fact that 68 per cent of all respondents (and 62 per cent of the staying behind women) said they cannot abandon farming shows that agricultural production is considered important within the peasant family. Almost two thirds of respondents (62 per cent) said that many villagers share this view.[13] Stopping farming is not an option because it provides food security for the household. This means that the role of staying behind women – as well as other women – is important (which does not imply that it is valued correspondingly). On the other hand, women's work for the family is also the main reason for them staying at home instead of becoming a worker. Beyond this, illiteracy, lack of skills, age discrimination in the labour market and health problems can block the way to paid work outside farming.

A majority of the women (69 per cent on average) said they would like to see a situation in which the husband and wife contributed equally to agriculture. The staying behind women were the most discontented about the current division of agricultural labour: 59 per cent said an equal contribution would be better, although this was currently the case for only a few (18 per cent). For the other women their current reality was closer to the situation they saw as better (see Table 8.6).

The income that can be earned by the husband is the decisive argument for those who think the woman should do most of the farm work. The husband's higher income is also a reason why more than half of the women believed that their own contribution to the family is subordinate to their husband's input (7 per cent as

Table 8.6 Farming women's view of the relative contribution of husband/wife to farm work; the current situation and what is considered a better situation; subgroups LD, DC and VIL (%)

The current situation[14]

Location of husband's work	*Wife does most of farm work*	*Husband and wife contribute equally*	*Husband does most of the farm work*	*Total*
Long distance	80	18	2	100
Daily commuting distance	31	64	5	100
In or near the village	0	80	20	100
Average	42	50	8	100

What is considered a better situation[15]

Location of husband's work	*Wife does most of farm work*	*Husband and wife contribute equally*	*Husband does most of the farm work*	*Total*
Long distance	36	59	5	100
Daily commuting distance	28	72	0	100
In or near the village	8	80	12	100
Average	26	69	5	100

opposed to 56 per cent) (see Table 8.7). The higher status they and other villagers awarded to paid work probably also plays a role in this respect.

Women whose husbands worked in or around the village or only on the farm considered that both partners made a more or less equal contribution, not only with respect to the family but also to agricultural production (see Table 8.7).

Table 8.7 further demonstrates that staying behind farm women undervalue their contribution to agricultural production. Whereas 80 per cent said they did most of the agricultural work (see Table 8.6), only 39 per cent thought their contribution was highest (39 per cent). The reasons why farm women devalue their own farm work is that they believe their husband has more agriculture-related knowledge and skills, that his work is heavier and that he takes more decisions.

There is a positive relation between women's view on 'who contributes most to agricultural production' and 'who is a better farmer' (see Figure 8.1). The more women see themselves as the better farmer the more they find their contribution to agriculture to be superior. There is also a positive relation between the location of the husband's work and women's view on who contributes most to agriculture. The staying behind women had a more positive view about their contribution to agriculture. A higher educational level also goes with a positive view on a woman's contribution to farming.

Figure 8.2 shows that women's opinion on who contributes most to agriculture has, in turn, an impact on their view of who contributes most to the

Table 8.7 Views of farming women of who contributes most to the family and agricultural production: wife or husband (%)

	Who contributes most to the family[16]				
Location of husband's work	*Woman*	*Equally*	*Husband*	–	*Total*
Long distance	13	33	54	–	100
Daily commuting distance	3	28	69	–	100
In or near the village	4	56	40	–	100
Average	7	37	56	–	100
	Who contributes most to agricultural production[17]				
Location of husband's work	*Woman*	*Equally*	*Husband*	*Son*	*Total*
Long distance	39	28	33	0	100
Daily commuting distance	25	19	56	0	100
In or near the village	4	44	48	4	100
Average	25	29	45	1	100

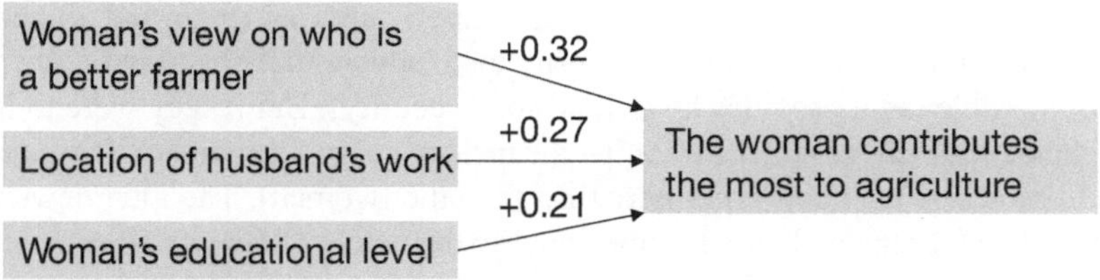

Figure 8.1 Women's view on their contribution to the household/family

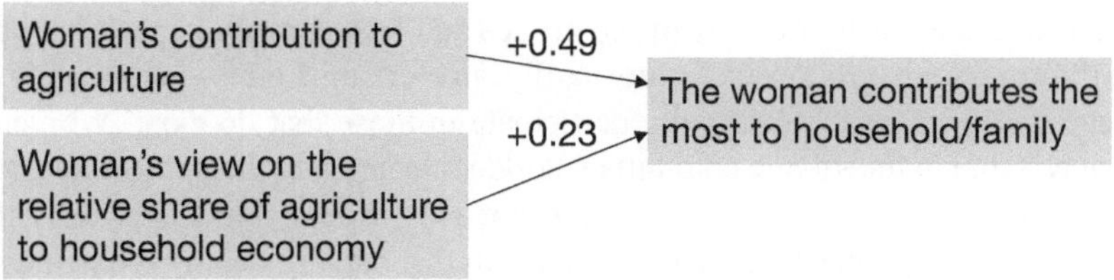

Figure 8.2 Women's view on their contribution to the household/family

household/family. The more they consider their contribution to agricultural production to be of relative importance the more they see their relative contribution to the household/family as important. The figure also shows that the more women value the contribution of agriculture to the household economy the more they see their contribution to the household/family as important. The same is true for their view on the relative share of agriculture to the household economy. The higher the share, the higher the contribution to the household is estimated.

Women's subjective well-being

Studies that aim to understand the social consequences of migration for rural communities, and the situation of those left behind, argue that outmigration has a negative impact on the well-being of the left behind women (see e.g. Mu and de Walle, 2009). These women have heavy workloads, suffer from stress and poor mental and physical health, feel lonely and insecure (see e.g. Wu and Ye, 2010) and have a relatively unsatisfactory and unstable marital relationship, (see e.g. Xu, 2009a and b). These factors all negatively affect women's development from a long-term perspective (see e.g. Yuan, 2006).

As already shown, the present research results support the finding that women whose husbands have migrated out are generally unhappier than women whose husbands work closer to home. They more often feel tired and more dissatisfied with farm work and with their overall personal situation. They believe their situation would improve if they could farm together with their husband instead of doing it on their own (see Table 8.6). Extra paid work outside the home would also improve their situation, giving them additional income and higher esteem in the village, but this is difficult to combine with women's work on the farm and for the family. Additionally, a lack of education and training, a lack of work opportunities and age discrimination are experienced as important impeding factors to finding outside jobs. A majority also believe that abandoning farming is not possible (as it provides household food security). But if they were to have the choice – which most clearly don't – the majority of women in the LD subgroup would rather be a worker (and more than the other women). The interviews also showed that the staying behind women have to deal more often with worries, stress and feelings of loneliness. Sharing such feelings with other women is not common. For some of them (21 per cent), membership of a local religious group in the village is a way out of isolation. Here they can meet other women, share feelings and even develop themselves. One of the women interviewed told us, for instance, that she has become a leader of one of the groups and has learned to speak in public. Apart from these groups there are hardly any organizations operating in the village and women do not usually participate in those that do exist. Women do not play a role in the village committee, nor do they attend village meetings. Most women (a surprising 89 per cent) were not even aware that there are village meetings or village elections. One mechanism that staying behind women use to take part in village life is maintaining *guanxi* relations. Normally, in rural China, maintaining social contacts is the man's territory. But after husbands have migrated out, it becomes women's task to attend ceremonies linked to marriages, births and deaths and to exchange gifts during traditional holidays (85 per cent of the staying behind women participated in these). Chatting with relatives (44 per cent) or playing poker (21 per cent) is another way to avoid loneliness.

The interviews show that some left-behind women cope more successfully with the hardships they endure than others. More education, experience with migrant work, participation in one of the local religious groups or living a short distance from their parents are all factors that help to improve these women's well-being.

The women who cope better have or create opportunities to communicate with others and are more able to ask for help or to help others. In this way they not only avoid loneliness but also feel more self-confident. By expanding their social networks they also become a 'somebody' instead of one of those people nobody knows or cares about.

Impact on agricultural production

Some scholars perceive women's increased participation in agriculture as a threat to food production (see e.g. Tamang *et al.*, 2014) while others disagree, arguing that productivity does not decrease and/or can improve when women have access to resources (e.g. Brauw *et al.*, 2012; Kelkar, 2009). The majority of women in our survey thought that households where the woman is the major farmer achieve similar production levels as households where husband and wife farm together (67 per cent, on average). With respect to their own farm, they said that yields had stayed more or less at the same level since their husbands migrated out (74 per cent) and a few (8 per cent) even mentioned improvements. They kept on working in line with the local farming style and almost all of them had maintained the same cropping scheme (95 per cent of this subgroup as opposed to 75 per cent, on average) and made use of the same inputs. What had changed was the quantities of fertilizer and pesticides they applied: two-thirds had increased the amount of fertilizer but this is also true of the farm women in the other subgroups (64 per cent of LD women had increased fertilizer use as opposed to 65 per cent on average). This is also true of pesticides. The main reason given for both trends was to increase productivity.

Conclusions

Rural gender studies undertaken elsewhere indicate that an increased involvement in the farm labour process in combination with a greater say in the many decisions that need to be taken (i.e. the feminization of agriculture) tend to empower farm women to some degree (e.g. Haugen and Villa, 2006; Safilios-Rothschild, 2006). This is particularly the case if farm strategies include multiple job holding (man and/or wife are also engaged in jobs outside the farm) and multifunctional farming (in which women have their own domain within the farm). These strategies also contribute to a loosening of traditional gender norms and changing work identities (e.g. Rooij, 1994; Ventura, 1994; Brandth, 2002; Bock, 2004; Haugen and Vik, 2008).

In Qiaobei village there clearly is, as yet, no such development. Although the role of rural women in the agricultural labour process and their say in everyday management decisions has increased substantially, their position remains subordinate. Using the WEAI tool, we have demonstrated that, on the whole, staying behind women do not experience more empowerment. Their access to resources remains more or less the same as before, the decisions about big expenses in the farm and within the household are dominated by the husband and their role in public village life is still minimal. Overall, male labour migration mainly has the effect of increasing the work burden of the women left-behind. Besides dealing with their solitude they also have to face more drudgery. Being alone in the farm, for extended periods, and having to do all the work and take the daily decisions is felt as a heavy load. Compared to women who work together with their husbands, especially the VIL subgroup, the wives of long term migrants (LD) are less happy. Their positions regress rather than progress.

This, we believe, is basically due to the dominant and largely unquestioned cultural framework in which notions such as *inside* and *outside* clearly demarcate unequal gender positions. This is further strengthened by notions such as *light* and *heavy*. Women engaged in farming do the light tasks (they only do light tasks), while men are capable of doing the heavy (important) tasks. 'Light' or 'heavy' are, of course, only partly intrinsic to the specific tasks (even a simple movement repeated the whole day becomes very heavy). It is more the fact that *women* are doing certain tasks that turns these into 'light' ones. And doing 'light' tasks means that the person doing them contributes less. The work done is, in synthesis, 'less important'. All this implies that men's contributions are perceived and valued as being far larger: they help to pay for the *big expenses*, while women's contributions are necessarily limited to *daily expenses* (the small ones). As long as such a cultural framework dominates, material changes (such as the feminization of agriculture) will *not* result in empowerment. Whatever is achieved by these rural women is systematically undervalued if not neglected. This is reflected in the sometimes very low self-esteem of rural women. While working hard, running a farm, caring for the family, maintaining *guanxi* relations, etc., etc. they are seen as – and feel themselves to be – nobodies.

Nonetheless, these same women (the female peasants of China) help to secure the continuity and growth of food production, and this would almost certainly be more the case if improved access to resources could be realized (see also UNDP, 2003). Together China's women peasants make the biggest 'agricultural system' of the world (see Chapter 1) function properly and develop continuously.

In the meantime, there are some glimpses of empowerment. Our analysis shows that those women who previously had migration experiences and/or have a higher level of education and/or participate in a local religious group are the ones who deal best with their situation and who are able to develop some agency, translating their own difficult role into a positive identity, a certain pride and a feeling of well-being. Working and/or living together (on a daily basis) with the husband is having the same effect.

There is an interesting theoretical aspect to all this. The more pluriactivity (not only at the level of the household, but also, particularly, at the level of the personal profile of the rural women) is a fully-fledged feature, the more the women like the different tasks in which they are engaged. This applies especially when farming is part of a larger portfolio of activities.

Notes

1 The phenomenon of the feminization of agriculture is interpreted in different ways: a) as a growing number of women participating in agricultural work; b) as an increase in the female to male labour ratio in agriculture; c) as the increasing role of women in both farm work and farm management. In this research the latter view is taken.
2 Between 2000–2005 the migration rate was the fourth highest in China (Meng, 2014:30).
3 Differences are statistically significant. (4)=10.10; p=0.039.
4 $X^2(2)=9.28$; $p=0.01$.
5 The last land distribution in Qiaobei village was in 1986, so for women who married a villager after that date it can be difficult to get land in their own right. For more information on rural women's land rights in China see e.g. Judd, 2007; Ray Liaw 2008.
6 Social relations build on reciprocal gifts.
7 This included relatively more staying behind women (41 per cent as opposed to 29 per cent on average; non-significant).
8 $X^2(4)=17.09$; $p=0.002$.
9 $X^2(4)=11,65$; $p=0,020$.
10 $X^2(4)= 11.65$; $p=0.020$.
11 $X^2(4)= 21.53$; $p=0.000$.
12 Young women (35 years or younger) disliked it most (71 per cent as opposed to 31 per cent, on average). Also women who had worked outside in the past disliked farm work more often than those who had not (42 per cent as opposed to 22 per cent; $X^2(2)= 8.02$; $p=0.018$).
13 The women in the LD subgroup were most negative about this; 31 per cent thought that villagers don't think farming is important. This compares to 22 per cent for the DC subgroup and 8 per cent among the VIL subgroup.
14 $X^2(4)= 44.78$; $p=0.000$.
15 $X^2(4)= 9.94$; $p=0.041$.
16 $X^2(4)= 8.87$; $p=0.064$; not statistically significant.
17 $X^2(6)= 15.44$; $p=0.017$.

9 The young and the rural

Sabine de Rooij, Wu Huifang and Jan Douwe van der Ploeg

This chapter argues that the growing weight of the urban (materially as well as symbolically) does not imply that the rural has a reduced relevance. It is not a zero-sum game. The growing importance of the urban is matched by a growing need for the rural. This point will be demonstrated by examining the views of students from rural and urban origins on farming, the countryside, the city and, especially, the differences between town and countryside. The students are from China Agricultural University (CAU), Beijing. Even if their work in the future is related to agriculture, they will most probably be located in a city and their professional trajectory will make them part of the urban population. Nonetheless, this does not mean, as this chapter shows, that they are indifferent to the rural.

There is a second, interrelated, theme discussed in this chapter. It concerns the nature of town–countryside relations. We show that young Chinese people perceive the two as antipodes. Yet, these two opposed poles are not ordered in terms of better or worse – or of being more or less attractive. The differences are perceived as multidimensional and the students make nuanced judgements. In some respects the rural is more attractive, in others the urban. This translates, in the end, into the need for a double transition: for the cities to become more rural, and the countryside more urban. That is to say, many services that have, until now, almost exclusively been available in urban centres need to be provided in the countryside. These students do not view the rural as being stagnant; rather they perceive it as changing – and think it needs to be further developed.

A third key theme of this chapter relates to the peasantry. The research makes clear that for the majority of these students any 'disappearance of the peasantry' is both unlikely and undesirable. In their opinion the peasantry represents a much needed continuity.

China Agricultural University is an interesting passage point. There are students with an urban background as well as students with a rural background. For many students with rural antecedents, CAU is a gateway that brings them to the city and, in the near future, to an urban job. Even if this job relates, from a substantive point of view, to the rural it does so from the city (the ministry, the university, the banks, the expert-systems, the food industries, etc.) and when they have graduated they will certainly obtain an urban *hukou* (residential permit) and live in one of the major cities. For students with an urban background CAU allows them to become knowledgeable, in one way or another, about the countryside, food and agriculture. In short, the students who participated in personal interviews and in the survey are part of a group that is simultaneously familiar with, yet distant from, the rural. What brings these different students together is that they all live in a kind of 'village', that is, the Western Campus of CAU. Fifteen years ago this campus was still located outside the city. Now it is surrounded on all sides by the continuously expanding metropolis. It is an undeniable expression of the scale and pace of the process of urbanization. One striking feature of the survey was the significant dissimilarities in some of the perceptions and attitudes of students with urban and rural backgrounds. This shows that it still does matter if one comes from the countryside or the city.[1]

Methodology

The analysis builds on information and data gathered from interviews and a survey conducted among CAU students in 2013 and 2014. In the interviews, open-ended questions were used to explore the students' views on the rural, its importance and function, the quality of rural and urban life and their direct relationship with the countryside. In addition, their opinions about and expectations for the future of agriculture, rural areas and agricultural and rural development policies were also investigated. These topics were discussed with ten male and ten female, – mostly Masters and PhD – students in social (the majority) and natural sciences, originating from both rural and urban areas.

In a second stage, a survey was conducted to explore these themes among a larger group of students. A list of thirty-nine questions[2] was first discussed with Chinese colleagues, then tested, adapted where necessary and translated into Chinese. The data was collected from a group of eighty-five male and sixty-seven female students from both rural and urban backgrounds, with a balance between the social and natural sciences and engineering. The majority (74 per cent) were BSc students.

To include a sufficiently large group of students with different disciplinary backgrounds several teachers in different departments were approached for their support. They were asked to distribute the survey and to invite the students to respond to the questions in the classroom after finishing their lecture. After the completed forms were collected a Chinese student entered the data in Statistical Package for the Social Sciences (SPSS) software (n=152). The home town of the students was used as the main variable to differentiate between rural and urban students. Starting with one of the rural–urban classifications used in China, a distinction was made between 'village', 'town', 'city', and 'metropolis'.[3] In the following analysis, only students originating from a village were considered to be rural; the other students are seen as urban. Students also were asked if they came from a farming family and if they saw themselves as 'someone from the countryside' or as a 'city dweller'. In some places these distinctions were also of significance and formed part of the analysis.

Background of the students

The students came from twenty-five different regions in China, though more were from provinces in the eastern part of China (70 per cent). Their origins varied from a village (40 per cent) to a town (19 per cent), a city (30 per cent) and a metropolis (11 per cent). Only a small percentage reported that their home town is located in a remote and isolated area (15 per cent). The majority came from a place near to a county town or city (46 per cent) or from a place within an urban centre (38 per cent). Almost a third of all students said that their parents (or one of them) are or were farmers (30 per cent). The male respondents more often had a rural background (46 per cent of males came from a village as opposed to 31 per cent of females).

Though less than half of the students were born in a village, more were familiar with the countryside as a place of residence. More than half of the students had spent their childhood and youth in the countryside (56 per cent)[4] for periods varying from less than 6 years (7 per cent) to 12 years or more (60 per cent). Looking back on their youth, most of the students had good recollections of this period and have primarily positive memories (81 per cent).

While all the students were now living in Beijing, not all of them saw themselves as 'Beijing city dwellers' (38 per cent). Quite a few students said they saw themselves as 'someone from the countryside', that is, as a rural person (44 per cent), while some couldn't make a choice (19 per cent). Rural students saw themselves more often as 'someone from the countryside' (70 per cent); 17 per cent didn't know how to respond. Generally, the more urban their roots, the more often the students saw themselves as 'Beijing city dwellers'.

The triple meaning of the rural

From the interviews with students, different images of the rural emerge. The rural clearly represents a many-faceted reality for them. Most respondents primarily associated the rural with agricultural activities and the associated features of small communities and community life. That is, as places where people have close relationships and help each other; or as places with a specific architecture and a specific history and culture (such as, for instance, local foods and recipes). The countryside was not seen as just a space where agricultural production takes place. The rural was understood, instead, as an organic unity of food production, peasants and village life. This is, precisely, *nongye*, i.e. peasant agriculture. This is clearly echoed in statements such as those that follow.

> For me, rurality is connected to paddy fields, large piles of rice straw, small houses or cottages, and the nice smell of food, different flowers and many nice people.

> I think of farmland, green grass, flowers, dogs, people who are familiar with each other, chatting with each other. I also think of children having fun and

> having a lot of things to explore. They go to the field, play together, and catch fish or crabs . . . .

The unity of agricultural production, peasants and villages clearly also has historical and cultural ramifications.

> I think of farmland, of the people who farm, and of history and culture. Rural people have a rural identity, they have [their] own customs.

It also translates into the rural being 'a way of life'.

> In rural areas, the relationship between people is very gentle. The emotional ties are very strong. I think, it is healthy for your body and soul to have contact with the soil and the nature. Rural people are more balanced. Compared to city people, their bodies may be more tired but their brains and emotions are very free.

At the same time the rural is perceived and defined as the counter-image of the urban. For better or worse, the rural differs from the urban. Most of the respondents constructed a diametrical opposition between the two. In this vein several respondents stressed the healthy environment, its beauty, quietness and the absence of stress. Here *nongcun* comes to the fore. This is illustrated in the following quotes, which clearly take the big cities as a point of reference.

> It is very quiet there, there are not many people, and it is not busy. It is also beautiful; there are rivers, mountains, nature, and fields, agriculture.

> In the countryside life is more natural, more human, more suited to human beings.

> In the countryside the environmental quality is better, it is clean, and it is also beautiful there. [. . .] You can live a quieter life in the rural area.

Not all the respondents saw rural areas as places that both organically unite food production, peasants and villages and also exist as a counter-image of the urban. Some respondents stressed both of these features while others put more emphasis on one or the other side of the equation. This illustrates a potential field of tension and conflict, as we indicate later when presenting data from the survey.

Table 9.1 summarizes the answers given to a question about the main functions of the countryside. It shows that many students perceive the rural as 'the place where the main activity is agriculture' (80 per cent). In fact, nearly all students believe that 'the countryside is there to produce food' (91 per cent). This links strongly to the importance students give to the issues of food security and food sovereignty: 92 per cent strongly agreed with the statement that 'it is important that China can provide its own food'.

It is also revealing that the majority of the students believed that agricultural activity 'is essential or increasingly important to maintain and develop rural areas' (78 per cent). Agricultural activities do not stand on their own – they are linked to peasants and villages. And therefore they should be developed. In this vein 97 per cent agreed with the statement that 'it is important that Chinese society cherishes its rural areas'. Clearly the rural is positively valued by these students for the combination and unity of food production, peasants and villages.

However, there is another side to the equation. Many students recognized that the rural has, or can have, multiple purposes. They believed that, along with food production, rural areas are a space 'to build up a livelihood and to further develop' (70 per cent),'to find solitude and peace' (69 per cent), and 'to preserve or recreate old landscapes' (66 per cent). They were also thought – though to a much lesser degree – as places 'to create growth and jobs' (38 per cent) and 'for recreation' (27 per cent) (see Table 9.1). Here the rural as a counter-image of the urban clearly appears.

In order to discover if there was a pattern in these answers, a factor analysis on these functions of the countryside was carried out. This analysis produced two factors: one centred on food production and the other reflecting the 'consumption countryside' – a counter-image of the cities (this factor has high loadings for recreation, landscapes, solitude and peace) (see Table 9.2). The finding that two factors emerge out of the analysis underlines that, indeed, the first and second notions of rurality (the organic unity of agricultural production, peasants and villages versus the counter-image of the urban) are independent dimensions of the rural. Some people combine the two, while others put the emphasis on one of them. This can give rise to contradictions.

It is interesting that the two factors shared high loadings for: 'build[ing] a livelihood and further develop [the countryside]'. This implies that there might be two different possible trajectories for developing the rural:

1) The construction of multifunctional countryside that allows for recreation, the maintenance of scenic landscapes, peace and quietness. Following Woods (2011, 2005) we refer to this option as 'the rural as a place of consumption'.
2) A countryside that mainly (or exclusively) has an agricultural function ('food production').

Table 9.1 Functions of the countryside according to students (%)

The countryside is there:	*Yes*	*No*
to produce food	91	9
to build up a livelihood and to further develop	70	30
to find solitude and peace	69	31
to preserve or recreate old landscapes	66	34
to create growth and jobs	38	62
for recreation	27	73

Table 9.2 Factor analysis on the assumed functions of the countryside

Functions of the countryside	*Factors*	
	Consumption countryside[5]	*Food production*
to produce food	–.16	.85
to build a livelihood and further develop	.51	.59
to create growth and jobs	.58	.25
for recreation	.63	–.17
to preserve as it is or recreate old landscapes	.74	–.03
to find solitude and peace	.63	.07

Notes: (1) Rotated Component Matrix[a] ; accounts for 52% of variance
(2) Extraction method: Principal Component Analysis
(3) Rotation method: Varimax with Kaiser Normalization. a. Rotation converged in 3 iterations

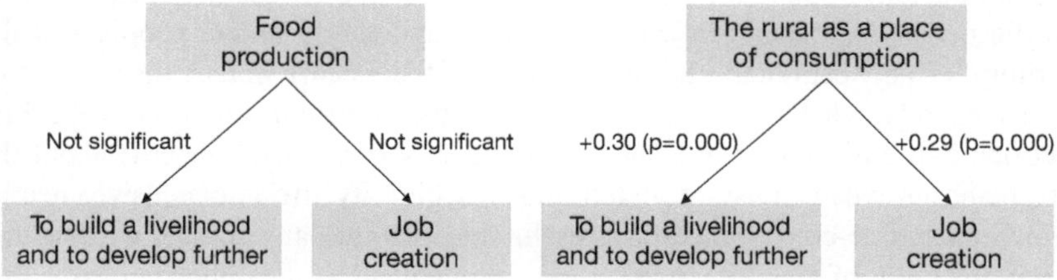

Figure 9.1 Assumed trajectories for rural development

In a next step, two variables were excluded from the analysis, that is, 'to create growth and jobs' and 'to build a livelihood and to further develop it'. A correlation matrix was then made to explore the relationship between these two functions and the two new components. The intriguing outcome of this exercise is that food production was not significantly associated with the variables 'creating growth and jobs' or 'building a livelihood and further development'. By contrast, the students saw a positive and significant association between 'more attention for old landscapes', 'recreation' and 'finding peace and solitude' and the development and creation of jobs (see Figure 9.1). Once again, this highlights that there are two contrasting trajectories for rural development. It also highlights that these contrasting patterns are likely to have different impacts on employment levels and well-being – at least within the perception of the students.

Rural or urban background: does it matter?

The survey results show that the students' background sometimes made a difference to their views. For instance, students coming from a village more often thought that 'relationships between rural people are easier and more sincere' than interpersonal relationships in urban areas (65 per cent as opposed to 48 per cent, on average, of the other students).[6] Students from a farming family were the most convinced of this (74 per cent as opposed to 47 per cent of students with parents in another profession).[7] Students were also asked to score the quality of rural social life on a scale from one to ten. Remarkably, 45 per cent of the students with a rural background valued this quality scoring it eight or more points – as opposed to 20 per cent of the other students.[8] Students whose parents farm were the most positive about this: 54 per cent gave eight or more points as opposed to 20 per cent of students whose parents do other work.[9] At the same time, being of peasant origin also seemed to influence students' perspective on the relationship between man and nature. They had a less romantic view: only 20 per cent thought that rural areas are 'a place characterized by an equilibrium between man and nature' (as opposed to 46 per cent of students whose parents did other work).[10]

However, in many other respects, the rural and urban students made similar judgements. The aspect of the rural most valued by both groups of students is its 'closeness to nature' (76 per cent gave ≥ eight points on a scale from one to ten). In addition, half of them highly valued (i.e. ≥ eight points) qualities such as (low) 'traffic density' (55 per cent), the 'quietness and safety of the people and the territory' (53 per cent) and 'the cost of living' (50 per cent). In fact, these qualities reflect exactly what the students lacked in their current situation – the big, metropolitan city. Indeed, the students frequently expressed concerns about the city being crowded, noisy, polluted and that big city life is expensive, hectic, anonymous, time-consuming and stressful and excludes any contact with nature. From this point of view, the rural is indeed imagined as the opposite of the city, i.e. as a counter-image of the urban. So it is of no surprise that nearly all students favoured the idea of bringing more of the rural into the metropolitan centres. According to them, 'metropolitan cities like Beijing need more green spaces where people can meet, relax, etc.' (95 per cent). They even think that 'urban agriculture/ horticultural initiatives have a future' (90 per cent).

The mirror side: the rural as place of hardship

At the same time, the research shows that the rural is also considered as a place of hardship and deprivation. Although these notions are, of course, relative (they are defined through comparison with the cities), the combination of farming and a difficult life draws on the strong historical continuity of Chinese society. For centuries, the countryside was the place for those who could not go to the cities and thus improve their lives. Of course, rural peoples' living conditions have improved – partly because they can now go to cities – but the rural continues to be associated with poverty, a lack of livelihood opportunities, poor living conditions, boredom and social strictures. Compared to urban areas, the standard of living still lags behind – and outmigration for work or education also has its social costs. It has impoverished family and community life and negatively affects the well-being of the left-behind rural children and wives (see Wu and Ye, 2014; Ye and Pan, 2011; Meng, 2014). Thus students' thinking of the rural also brought up mixed or negative thoughts and images. Farming is increasingly a less attractive livelihood prospect for rural youth.[11] The students interviewed expressed different aspects of these issues.

> This generation doesn't want to be farmers. That means being poor. The young generation wants to live in the city.

> When I think of rural China, I feel sadness. It might be beautiful and quiet but in the Chinese countryside you can only see elderly people and left-behind children. These children are cared for by their grandparents, they don't gather enough knowledge, and grandparents don't explain things to their grandchildren .... Last August, during the harvesting of the rice, I saw grandparents of over 60 years in the fields. They had to harvest the rice and take care of two grandchildren. That is too much for them.

The main disadvantages of the countryside, according to the students, are the low availability and bad quality of infrastructure and services. The lack and the quality of 'recreational and extra-curricular activities' (89 per cent), 'schools' (84 per cent), 'opportunities to make a living' (72 per cent), 'road links and public transport' (67 per cent) as well as poor or limited internet access (65 per cent) were considered to be the main drawbacks by most students. These aspects of the rural all scored five points or less (on a scale from one to ten). Similarly, more than half of the students judged communication systems in rural areas as poor (56 per cent). Students with rural and urban backgrounds did not think significantly differently about these aspects. The exception was the quality of the rural as a 'place for children to grow up well': the more urban the students the more negative their opinion of this. All in all, the students believed the quality of urban life to be higher. While they gave, on average, six points (on a one to ten scale) to the quality of urban life, the majority gave rural life a lower rating, of between one and five points (62 per cent).[12]

Clearly, cities are more attractive to most students because they offer much wanted opportunities and services (quality education and health care, more and better jobs, higher incomes, good public transport, shopping malls and leisure facilities) and more social opportunities. For many students these factors carry great weight. This explains why a majority did not see the rural 'as a nice place to live' (70 per cent) – or somewhere they would retire in the future (75 per cent). Some students express these negative views quite strongly. For example:

> Rural life is not attractive to young people; or only for a few people or those who have a job related to agriculture or rural development. Only young people, who have to be in the rural areas because of their work, will go and live there. Others want a high quality of life. Most young people migrate and leave the villages.

On the other hand, some students shifted the emphasis slightly by pointing to the possibility of improving working and living conditions in the countryside. One student said:

> I think that many people would stay in the countryside if they could have a job, if there were more and better job opportunities in the rural areas, where they could earn more money than they can today. [. . .] Our government needs to build factories there and give more subsidies to the farmers and horticulturalists. At the moment, rural life is boring for young people and there are no hospitals or schools . . . .

Overall, the interviews highlighted contradictory experiences of rural areas: they were not considered to be good places to live although they have many much wanted qualities that the city does not offer and were thought to have the potential to become better places to live. In this respect it is of interest that almost all the students judged it to be important to make rural areas more attractive for young people (88 per cent). As one student explained:

> Where I come from is a poor area but also an area where the environment is better than in the city, particularly in recent years because air pollution in the city has become worse. The differences in environmental quality between rural and urban areas have increased. In the rural areas people have a lot of space. In the city people live almost shoulder to shoulder. On the other hand, the chances of getting a better education are higher in Beijing and this is also true for other services such as health care. The rural areas have few resources; job opportunities and most kinds of services and facilities are scarce. For instance you can only find a few shops there. In the city this is all better.

Bringing more advantages of the urban to the rural – better infrastructure, services and amenities – could possibly change things and make the rural more attractive to the younger generation. This is reflected in the expectations for the future of rural areas. Students tended to be positive about this perspective (see Table 9.3). The prospect that was most often indicated as the most important is that, in the years to come, rural areas will be 'better places to live, where the economic resources are similar to urban areas, and other conditions are more equal' (40 per cent).[13] This choice was made by students with both rural and urban backgrounds. Even students who stated that 'rural areas have less access to resources than urban areas' believed that 'they will be a better place to live (in the future)'. Furthermore, quite a few students expected that, in the years to come, rural areas will 'offer the opportunity of living and working in perfect integration with the rest of the community' (29 per cent). The prospect that rural areas will become a place where one can 'live far from the urban chaos and where one can develop one's own life and work project' was considered very important by 22 per cent of the students. A similar number considered the role of information and communications technology in bridging the gap between rural and urban as being the most important. They thought that rural areas would be a place 'where all distances with urban areas and markets will be shortened thanks to the internet and where investments will be profitable' (22 per cent). At the same time, most students don't idealize rural areas. Only a few students believed that 'rural living

Table 9.3 Students' expectations regarding the future of rural areas (%)[14]

In the coming years, rural areas will be:	*Very Important*	*In-between*	*Less important*	*Total*
better places to live, provided the economic resources and other conditions are the same as in urban areas	40	48	12	100
places that offer the opportunity to live and work in perfect integration with the rest of the community	29	56	15	100
better places to live than urban areas even though rural economies will be relatively resource poor	26	54	20	100
places that offer the opportunity of living far from the urban chaos and where one can develop one's own life and work project	22	56	22	100
places where distances with urban areas and markets will be shortened thanks to internet access and where investments will be profitable	22	66	12	100
places that are too far from the rhythms imposed by contemporary lifestyles and destined to take a second place	15	37	48	100
worse places to live than urban areas, due to a lack of services	8	44	48	100
places where it will cost too much energy in the future to live and will not be sustainable	7	29	64	100
bad places to invest in as they will always be less favoured than urban centres	6	26	68	100

is the only way to make society more sustainable in ecological, economic and social respects' (9 per cent designated this expectation as most important).

The relative weight of positive expectations (that dominate over the negative expectations) allows for a tentative conclusion: the rural is, in the perception of these students, far from stagnant. It is an intrinsic part of important changes taking place in society. And for these respondents, these are not neutral: the rural is to be developed. It should be improved in many respects.

A factor analysis applied to these expectations reveals three factors – three different views on the development of the rural. One of these is a pessimistic view, the other two are more positive (see Table 9.4).

In the negative view (factor 1 in Table 9.4), rural areas figure as bad places due to the difficult economic and social conditions. In the first positive view (factor 2) the rural areas are considered as better places to live, provided that economic

Table 9.4 Students' future expectations of rural areas

In the coming years rural areas will be	*Factor*		
	1	*2*	*3*
places that offer the opportunity of living far from the urban chaos and where one can develop one's own life and work project	.06	.12	.71
places that offer the opportunity to live and work in perfect integration with the rest of the community	–.13	08	.77
better places to live than urban areas even though rural economies will be relatively resource poor	–.02	.76	–.02
better places to live, provided the economic resources and other conditions are the same as in urban areas	–.18	.75	.02
places where distances with urban areas and markets will be shortened thanks to internet access and where investments will be profitable	.01	.63	.32
places that are too far from the rhythms imposed by contemporary lifestyles and destined to take a second place	.59	.17	–.46
worse places to live than urban areas, due to a lack of services	.56	–.24	–.24
bad places to invest in as they will always be less favoured than urban centres	.75	–.10	–.03
places where it will cost too much energy in the future to live and will not be sustainable	.78	–.02	.22

Notes: (1) Rotated Component Matrix[a] (explains 56% of the variance)
(2) Extraction method: Principal Component Analysis (3) Rotation method: Varimax with Kaiser Normalization
a. Rotation converged in 5 iterations

resources are similar to those in urban areas and other conditions are equal as well. The standard of living is seen as comparable, even if economic conditions will not be the same. In the second positive view (factor 3) the qualities of rural areas are defined in terms of them being places that differ radically from the cities – especially in non-economic respects.[15]

Some students interviewed had quite optimistic views:

> I think that in about 10 years' time the rural areas and the villages will be more beautiful. The people will be richer and less labour will be needed there. People will have more and better educational opportunities and the health care will be improved. Maybe a lot of small cities will be built, small cities but bigger than villages and rural people will have better lives.

Some of them also expressed their willingness to become actively engaged in the improvement of rural areas:

> I think that in the future there will be jobs available in the rural areas. I want to have my own business, to be self-employed (be my own boss), I want to become farmer. That is why I study in this university. I like agriculture and I like electronic engineering. I want to do something with the combination, maybe also with ICT and the internet.

One of the students indicated that farming and agriculture will play a fundamental role in improving the rural area:

> It will be better than today. Agriculture will be improved and the farmers will earn more money. Their children will come back to the countryside.

And another argued:

> I think that young people from rural areas need to change their perspective. They need to bring their knowledge back to the villages. [. . .] Our roots are in the village. Many people have migrated out and that will lead to the rural culture getting wiped out. But we can restore the lost countryside with the help of the elderly people who are still in the village now. I will never be able to afford a house in the big city, [jokes] maybe just enough for a bathroom in the end. With the money we earn in the city, we will be able to build a house in the village and after some time we can go back to the village and live there.

From a space of memories to a space of disappointment

Generally speaking, hardship can be one of the decisive dimensions characterizing the rural; but for young people it is also a place that harbours pleasant memories:

> I associate the notion of the rural with going back to my home town, relatives, pleasure, and with entertainment. I have a positive image of rurality.

> We were free, we could climb mountains, we learned everything about the crops; we were having fun. And also we did a little homework; the tests at school were simple. Our life was free and easy and full of fun.

Studying at CAU does not imply that such memories fade away. They are often reactivated by being in Beijing. For many students who originate from a village, coming to Beijing was quite a change. A majority needed time to adjust (62 per cent) and some still found it difficult (5 per cent). Before actually moving to Beijing, living in the metropolis seemed attractive and this influenced more than one student to choose to study in Beijing.

> Beijing is the capital of China; this city provides more opportunities than other cities. There are a lot of people living here but there are also many universities, it offers a lot of culture, you can meet different people from different countries . . . . Living here means having more [opportunities]; you can get all kinds of information from abroad (USA, Europe), we have good public transport . . . . Beijing gives you also status . . . my parents are very proud when they tell other people that their son is studying in Beijing. It gives my parents a higher status . . . .

> Beijing has become a metropolis because it is the centre of the government. Therefore the public transport is very cheap here, the cheapest in China. You also find the best hospitals of China in this city and you get medical care for the lowest costs. [. . .] For children growing up here, it is easier to enter a good university. That is the reason why everyone wants to come to Beijing.

However, for quite a few rural students the reality is disappointing. They said life in Beijing had not met their expectations (60 per cent).

> Before I came to Beijing, I had high expectations of the city and what it has to offer. [. . .] I thought life would be stunning and that I could see many things and experience a lot. The reality was different.

> Before arriving here, I had a different view of big city life. But once being here, it did not meet my expectations. Beijing is too big. There are too many people here and the people in Beijing don't know each other. I live on the campus, I study. I have very few social contacts. I don't know how real Beijing people are, I don't know how they live. I don't know Beijing. Of course, I have been in the Forbidden City and I was on the Great Wall. But the city's soul is like other big cities, like [for] instance Shanghai; which has not got its own character. I don't know where to find the real Beijing. I am a migrant and maybe some years from now I will go back to my home town.

Without doubt, the city attracts rural youth. It seems like the place where personal emancipation is possible and which can offer them a way out of poverty. On the other hand, rural students cannot ignore the countryside. They take the rural with them to the city. It is the place where they grew up and it is the place of their memories. But then they discover that the city is also a space of disappointment. It is probably this combination that helps shape their views on the rural as something that needs to be developed, to be improved, made stronger.

In this respect it is probably revealing that some of the rural students interviewed talked as if they were migrant labourers:

> When young rural people migrate to cities to work there, they will earn money but in the end, it will be too expensive to live in the city – housing is very expensive – so they will return. Their parents and their friends live there and their parents are the most important for the children, they need to take care

> of their parents. And it is cheaper to build a house in the countryside. So children earn money in the city, build their house in the village and return after they have earned enough and they will find work in the village. Then they will have a better life. This will still be the situation in 10 years' time.

Almost all the students with rural origins travelled back to the countryside more than once a year (93 per cent). The most popular periods are the Spring Festival (the celebration of Chinese New Year), the summer holidays and the first week of October. Once home, they visit relatives, neighbours and, more rarely, old friends. Students from a farming family might also help their parents on the land. Returning to Beijing, two-thirds brought food back with them, mainly regionally specific food. Back at the university campus, they had regularly contact with their parents (81 per cent at least weekly), mostly by mobile telephone (95 per cent). In this way they were kept informed about what was going on in the family and in their home town. However, over time their ties with the village may gradually loosen.

> I go home every 2 or 3 months, I still have many friends in the village. But some are married and have children and we don't understand each other anymore. [. . .] But I still feel part of the community.

Most students of rural origin have some sense of belonging in their village. They said they feel part of the community when they visit their home town (88 per cent) and many saw themselves as 'someone from the countryside', that is, as a person with a rural identity (70 per cent). But of course, this does not apply to everyone:

> I do not feel very connected anymore. The villagers no longer see me as part of the village; they see me as a university student. The villagers are polite when they talk to me. But there is a difference in the way they talk to me and to my younger sister. With her, they gossip. She is part of the village. I am more of a stranger; I feel that by the way they talk to me, their politeness. But I do feel part of the family. [. . .] I feel urban and not rural. After 5 days being at home I want to return to the city. At home, I have nothing to do, I would become lazy if I stayed there too long. Also too many people come to our house and I am not so good at talking. [. . .] People want to know too much about me. They ask me if I have a boyfriend, when I'm going to marry and have a baby. These questions make me very nervous. My mother asks me these questions every day. She is really worried about me. It puts me under pressure. My father never talks with me about these kinds of things, but my mother and the neighbours do.

Looking for places to work and live: the rural and the urban matter

As already noted, the students recognized the relevance and importance of the rural – often in more than one respect. The question is whether, and how, they think of handling and including these concerns in their own work and lives after graduation. Bearing in mind that the respondents were studying at China Agricultural University, it is not surprising that quite a few students (60 per cent) were considering a job connected with rural issues. However, very few said they would like to get a job in a rural area (5 per cent). The students mostly wanted a job in a metropolis (55 per cent) or in a smaller city or town (40 per cent).

When asked whether they also would want to live in the place where they work, the responses change. Fewer students would also want to live in a metropolitan centre (42 per cent on average, compared to 55 per cent who said they would like to work there). More students found the idea of living and working in a smaller city or town, or living outside the city at commuting distance, more attractive than living and working in the countryside (59 per cent).

Place of origin made a difference to students' responses to these questions. Most students who came from a metropolis wanted to get a job in a big city (88 per cent) and live there as well (71 per cent). Rural students made other choices: far fewer said they would like a job in a metropolis (48 per cent) or to live there as well (35 per cent). They preferred to live outside the city at a daily commuting distance from their work (34 per cent), to work and live in a smaller city or town (22 per cent), or to work and live in the countryside (9 per cent). This means that 43 per cent of this group preferred the countryside as a place to live. Of the students originating from a metropolis only 24 per cent preferred to live outside the city at a commuting distance from their work.

When thinking about the kind of job they would like after graduation, the students we interviewed quite often mentioned a job in a university, as a teacher or researcher. Most students said they would prefer to work for the government or for an enterprise. Only a few said they would like to become a farmer, which is, as they are well aware, a difficult option.

> Many young people originating from the countryside actually want to return to the rural area. But before they can do this, they need to work in the city to earn money and to build a network. Then they can return and try to build something, to start their own enterprise. I think I would need 5 years to start my own farm.

Yet, these students did have quite specific ideas about the kind of farm they would like to build. Concerns about food safety and environmental problems inspired some of them to think about green, organic farming or urban farming:

> In the future when I have my own farm, I want to grow vegetables and fruits without using chemicals, pesticides and so on [. . .] I want to produce healthy food. Most people in China want to eat healthy food but where can they buy it? Many people don't trust the food sold in the supermarkets. Organic farming is the solution, the way to achieve food safety. All the friends in my network want this but we don't know how to achieve it. [. . .] The traditional farmers don't want to shift to organic production because they think it's too

> risky and they don't have a market channel. They don't know how they can market their products, how to create a market channel. I want to explore that issue and to help farmers to develop those channels so that consumers in the city can buy healthy products. I think that to produce healthy food is the main task of the students here in this university [. . .].

Another student outlined quite a different project:

> Our land policy changed and now we can rent land from other people. I would like to rent all the land in my village. I want to have a big farm, of about 2,000 mu. That is not very big. My ancestor was a landlord, he had more than 2,000 mu and also had land in the neighbouring village. My aim is to rent all the farmland and to create one big enterprise managed by me and to produce as sustainably as possible. That is to say, using the minimum of fertilizers, pesticides, and insecticides. I call this 'green production'. The soil is still too polluted for organic production; organic production also involves too many rules and is expensive. On a big farm I can guarantee a reduction in the use of pesticides and insecticides and generate less pollution.

Yet another student saw a promising prospect in multifunctional farming.

> My dream is get my own farm where I can invite parents and their children to learn how food is produced and also offer day tourism activities. I think this is a good way for farmers – but not the main development way in the future. The main mission for the farmers is to produce enough food that is also healthy. We have a lot of people living in China and we will have more people in the future so farmers will have enough work. As such it is clear that we will need the rural areas in the future.

Food safety, security and sovereignty: what type of agriculture?

Currently, food safety is a big issue in China. This is due to several recent food scandals. Among others, the milk scandal involving powdered baby milk adulterated with melamine is well known. Thousands of babies fell ill after drinking this milk and a few even died. There were further scandals: tainted pig meat, polluted cooking oil, contaminated mineral water, chicken legs contaminated with hydrogen peroxide as well as extensive areas of polluted agricultural land. Food safety can be put at risk in different parts of the food chain. During production (manifestly), but also at other points in the food chain: during processing and packaging, storage, transport and preparation. The latest food scandals in China have pushed these issues up the political agenda leading to stricter regulations, better monitoring of the food chain and harsher penalties for offences.

Food safety is also of concern to the students. Most of them (84 per cent) considered it a serious problem. The students interviewed shared several anxieties:

> There are all kinds of scandals about food. People are afraid of eating fruits because of the use of chemicals and we had the milk powder scandal. Food safety is an important issue, people feel it is unsafe to eat the food.

> I don't trust the quality of food very much, especially the food you can buy and eat in the city. If I could I would be a farmer, I would grow my own food. Now I must buy my food but this has passed through many hands (the food chain is long) and I don't know what has been done to the food in which parts of the chain. The television draws a lot of attention to unsafe food. But I have to take the risk; I have no other choice now. I don't think the government is entirely open about this issue, I think there are quite a few food safety problems in China.

As the students lived in the city and had to buy their food, the survey asked where and from whom they bought their food. Which selling point did they trust the most? Of course, being a student means having access to university canteens and the interviews showed that it was a daily routine for the students to go there for breakfast, lunch and dinner. In the survey we asked about a range of food outlets: the supermarket, directly from a farmer, a market or a street vendor. The results demonstrate that students mostly trusted the food sold by supermarkets (49 per cent) and then food which can be bought directly from a farmer (38 per cent). Food sold on markets and by street vendors was much less trusted (by 10 per cent and 1 per cent respectively). The students interviewed also mentioned organic food. They thought this could be trusted (though not all agreed with this) but none of them had ever bought organic food. The obstacles were the students' remoteness from the places where it can be bought (mostly at the outer sides of the city) and high prices.

> I trust the food which is served here in the campus restaurants. Also the food at home is safe although my parents use fertilizer and pesticides, but they don't use other poisonous substances. [. . .] Sometimes I buy food in the street, pancakes for instance, but no meat. And I do eat in the restaurant of the supermarket. I trust the food there.

There were different views on the causes of food problems (some students referred to too many small farms and some to the opposite, the production on industrial farms; the high use of pesticides and fertilizer; the outdated knowledge of peasants and inadequate information services; the long and non-transparent food chain, poor controls and lack of enforcement of regulations; cheating practices in different parts of the food chain; consumer demands and the free market). These different opinions resulted in different views on how to deal with the problems. This is demonstrated by the variation in students' opinions about the type of farm best able to guarantee food safety: is it the peasant farm, the large family farm or the industrial farm? It turns out that the students had most confidence in large family farms (36 per cent) but this is followed closely by industrial farms and

Table 9.5 Students' views on type of farm most able to guarantee food safety (%)

Farm type	%
Large family farm	36
Industrial farms	30
Peasant farm	26
Don't know	6
N/a	2
Total	100

peasant farms (30 per cent and 26 per cent respectively). So the differences in attitude towards the three types of farms was really rather small (see Table 9.5).

The students we interviewed expressed a range of opinions.

> The best products are produced by peasants, by smallholder farms. These farms are self-contained so they must make sure that their food is safe.

Whereas another argued that:

> food safety can also be improved by a reduction in the number of farmers. With fewer peasants, and more large farms, it will be easier to control the food.

Some students thought that government should play a central role in dealing with the food safety issue. They proposed more effective measures and improved instruments and additional or new policies. They mentioned designing a well-functioning control system, improved law enforcement, more financial support for farmers, training and the provision of information for farmers, more and better consumer information, scale enlargement in agriculture or the development of a certification system for organic produce. Others expected the pressure to come from organized consumer groups.

> The opinion of consumers is changing. They are paying more attention to their health and emphasize that a food safety policy is necessary. They will exert pressure so policy changes will take place.

> Some of my friends who study abroad talk a lot about getting their own farm plot once they are back in China. The reason is the food safety problem in China. I would like to have one as well. In the suburban area of Beijing it is always possible to rent a plot.

Food security is another major issue. As already mentioned, almost all the students thought that China should provide its own food. But, once again, their opinions about what type of farm is best able to guarantee food security were split.

In this respect they had, however, clearer preferences: industrial farms were the most trusted (41 per cent), followed by large family farms (34 per cent) and peasant farms (23 per cent). Nevertheless, the belief that peasant farms are essential with respect to food security was widespread (65 per cent). Peasant farms were deemed important in two ways: to cover the food needs of the population in general and to provide livelihood security for rural families: 'it gives rural people a safer future' (70 per cent). Considering this, it is no surprise that a majority believed that this type of farm will not disappear in the future (70 per cent).

Several respondents believed the highest yields are achieved by peasants and family farmers. They argued that these farmers work for themselves and are more careful. Moreover, having one's own plot means food security in times when income is uncertain and in old age.

> We cannot look to the industrial farms such as they have in the US nor to the type of farms we had in the past. The best option would be to combine traditional Chinese agriculture with an ecological way of farming. It is a labour intensive way of farming; in contrast to industrial farms they don't use [much] machinery. The traditional Chinese farmers farm very carefully. They don't waste any resources; they make sure that the resources are recycled.

> I think we should not depend on food from other countries as much as we do now. We import, for instance, a lot of soybeans. It is better to have food security at national level. We should be more independent. And family farming is the best way to guarantee this. When you have your own piece of land then you try to farm as best you can. Our small-scale farmers know how to farm in a good way and they strive for high production levels. I still remember my grandfather and his lesson of putting three seeds of maize in a triangle to assure a good harvest. When you have large-scale agriculture you have to employ labourers. The farm is not their own business, so they do feel not so responsible for the land and the products as family farmers. I do not believe that large-scale farms can do it better than family farms. Besides, you can expect many labour conflicts on farms with a lot of outside labour.

There was also support for a strategy that envisages the best way of securing food security through a range of different farms:

> To guarantee food security we should build on different types of farms. But the question is not only about food, it is also about rural people. At the moment, the peasantry is also important for the life in the cities. When industries are established in rural areas and more rural employment is created, then it will be possible to produce the food on big farms. For this to be realized you will need to find jobs for the peasants. But I think the peasants are needed to provide sufficient food. Big companies are steered by the global market and by making profit. So the danger is that they might not produce sufficient food. Peasants can guarantee a much more stable supply.

And what about peasants?

Peasant agriculture is an essential part of Chinese agriculture and many students believe that this type of farm is robust and will not disappear in the future. However, Chinese agriculture is in transition. Low incomes from farming, its low status – combined with young people's lack of farming knowledge and unwillingness to become farmers, rural outmigration, market liberalisation and economic globalization are all stimulating changes in agriculture. The students we interviewed observed contrasting developments.

> The peasants rent out their land and work as migrant workers in the cities or work for the new farmers. They can earn more money from their land this way than before. As a peasant they previously earned 100 yuan, now they are paid 150 yuan. Although the new farmers make 350 yuan, more than the peasants generated before, this new situation is also beneficial to the former peasants. The former peasants now also have extra income from their work in the city, or they have started a shop or sell something in the street, etc. They can also earn extra income by working on the land of the new farmers.

Students also told us how peasants are being replaced by agricultural entrepreneurs in several areas.

> The new farmers are often outsiders, from the town or the county. They are better educated, better skilled and they make use of ICT, a mobile phone, a computer, and other available tools. As a result they are better informed about the market. They are also changing the cropping scheme, producing more profitable crops (vegetables, fruits) on a larger scale and getting higher yields. They also build greenhouses. I have seen this in the eastern part of China [. . .]. These farmers contract the villagers to do the work. The villagers change from peasants into workers.

One student was not at all optimistic about the future of the Chinese peasantry:

> this transition is taking place at a high speed. Within two generations, I think, the peasants will have almost disappeared.

However, this view was certainly not shared by all the students. Some saw a future in which different types of farms – large industrial farms, farmers' co-operatives, family and peasant farms – will coexist, side- by- side:

> I think that in the future, large agricultural enterprises and peasants will coexist in Chinese agriculture. Maybe the large farms will dominate. Villages cannot refuse this trend. I don't know whether this is for the better or worse, I don't know. For now, it is good enough. But maybe not if it becomes too risky.

Others didn't think industrial farming had a future in China:

> Industrial farming doesn't fit with China. The natural conditions are not suited for industrial farming. We have a lot a land in mountainous areas and not in the plains. You cannot use tractors in the mountains and we have relatively little flat land. When industrial farms work under the same conditions as Chinese peasants, then their output is not higher because the peasants farm carefully. Besides, we can't just throw all the small farmers off from their land. It would cause social problems and the many small plots are very important for food security in China. We don't have the same national resources as the US; they can control the oil in the world. We cannot. So we can't guarantee food security in that way.

The students participating in the survey also expressed their opinions about future farm development. They were given four possible future farming strategies and were asked if they agree or disagree with these strategies ('fully agree', 'in-between', 'fully disagree'). As Table 9.6 shows, the students placed the highest expectations on multifunctional farms and the resilience of small family farms (46 per cent and 39 per cent of the sample respectively fully agreed with these strategies). *Nongmin* (i.e. peasants) remain as the cornerstone – albeit partly through developing multifunctional farms. The other two options for the future, 'fewer large farms using more technology' and farms 'managed by businessmen who aim at creating wealth for themselves and other people' were thought to be of less importance in the future (each fully agreed by only 15 per cent of the sample).

Table 9.6 Students' expectations about future farm development (%)

The way in which farms will develop in the near future	*I fully agree*
There will fewer, large farms using modern technology.	15%*
More farms will develop into multifunctional farms providing, besides food products, green energy, services to people (such as agri-tourism, educational farms, nature management) and employment opportunities.	46%
Small family farms will survive because they offer the elderly and other people food security and food safety.	39%
Farms will increasingly be managed by businessmen who consider farming to offer important opportunities for the creation of wealth for themselves and other people.	15%

Note: *Very few students (6.7%) of rural origin 'fully agreed' with this strategy - compared to 21.1% of students from urban origins ($X^2(2)$= 8.04; p- 0.018)

Conclusions

China is an urbanizing society. The sample used for this research project reflects, and represents, this process. The sample consists of young people from rural areas moving to the urban (and, later on, to urban jobs) and those from China's cities and metropolises. Despite this, the responses make clear, first, that urbanization cannot be understood as representing an adieu to the rural. The opposite is true: urbanization triggers a need for the rural. The more Chinese society becomes urbanized, the more it will need the rural. This occurs through several mechanisms and interrelations, many of which have been discussed in this chapter. The rural will remain relevant and of importance at the individual level as well as for society as a whole.

Second, the research shows that, at least among this specific sample, the rural is something that needs to be developed. Here the students envisaged the state as the main player (specifically in providing more facilities in the countryside), although some students also saw their own life projects as contributing to this. Given the issues that underlie this attitude (food safety, food security, the importance attached to an attractive and accessible countryside), it might be assumed that this interest (in the further development of the rural) will be shared by far more categories of Chinese society.

Third, the research findings strongly suggest that the *nongmin*, the typical Chinese peasant, is, in the perceptions of these young people – who are knowledgeable about such matters – far from being an obsolete phenomenon. They think that peasant agriculture will remain a key element of China's countryside, especially in terms of providing food security, employment and fall-back positions for migrant labourers and the elderly. They see peasant agriculture as most likely to develop through the adoption of multifunctional approaches – i.e. the delivery of services as agro-tourism.

Finally, it should be noted that many aspects of the rural are highly contested. The best way to equilibrate the balance of different rural functions (food production, attractive countryside, etc.) is subject to different opinions and may in the future be expected to generate friction and conflict – as is the case in many other parts of the world.

In short, China is urbanizing. This does not imply, though, that the rural has (or will have) less relevance. Rather, with continuing urbanization the relevance, value and importance of the rural is growing and being galvanized. However, how this should happen is, and will be, a subject of considerable contradiction, contestation and, possibly, social struggle.

Notes

1 The survey results showed only a few significant differences in the responses of male and female students. Most differences in their answers relate to (differences in) their urban and rural backgrounds.
2 Initially the questions were asked in English. Some of the questions derived from the research project, 'Young people's perception of rural areas', which was carried out

among young people from rural and urban areas in eight European Union member states. (Rete Rurale Nazionale' 2012). The reason for using questions from this research is to enable future comparisons between the views of young people in China and those in Europe on the rural.

3 The use of numerous and also changing definitions (by e.g. scientists, within and of different disciplines, policy makers and politicians) to determine what is rural and what is urban represents a problem here. It also demonstrates that the rural is a dynamic multifaceted concept. Land use patterns, distance to urban centres, population density, economic activity, the architecture and distinctive lifestyles and cultural traditions are among the criteria used to distinguish the rural from the urban (see e.g. Copus *et al.*, 2013). In China, administrative and statistical methodologies are the main approaches used to distinguish rural and urban populations (see e.g. OECD, 2010:40–1).

4 About a quarter of this group (23.5 per cent) had been taken care of by grandparents while their parents worked and lived elsewhere.

5 See Copus *et al.*, (2013). Generally, this term is used to describe areas where the rural economy is no longer dominated by the production of food and fibre, but where countryside public goods, environmental or cultural assets or local quality produce form the basis of 'consumption' activities enjoyed by urban visitors such as leisure, recreation, hospitality, etc.

6 Differences are statistically significant: $X^2(1)=4.05$; $p=0.044$.

7 Differences are statistically significant: $X^2(1)=9.28$; $p=0.002$.

8 Differences are statistically significant: $X^2(2)=11.69$; $p=0.003$.

9 Differences are statistically significant: $X^2(2)=18.32$; $p=0.000$.

10 Differences are statistically significant: $X^2(1)=9.66$; $p=0.002$.

11 This is a more general phenomenon that can be witnessed in other parts of the world. See e.g. White, 2011.

12 Students from villages were relatively more positive about the quality of rural life: 32 per cent gave six points or more (compared to 17 per cent, 13 per cent and 0 per cent respectively for those coming from a town, city or metropolis). Students with farming parents were the most positive 37 per cent (as opposed to 13 per cent of students whose parents did other work).

13 The students were presented with different future expectations for rural areas. For each of these they had to say whether in their view this prospect was 'very important', 'in-between' or 'less important'.

14 The expectations are put here in descending order of what was considered most important. In the questionnaire the expectations were put in a different order.

15 A correlation matrix applied to these factors and the place of origin showed a significant difference between students with rural and urban backgrounds with regard to the negative view. An urban background correlated with the negative view (Pearson coefficient -0.20; $p=0.014$). The two positive views were not significantly correlated with different places of origin.

10 Markets[1]

Jan Douwe van der Ploeg, Wu Huifang and He Congzhi

At the southern periphery of Beijing there is a huge market called Xin Fa Di that covers tens of hectares. It operates 24 hours a day but is especially frenetic between midnight and four a.m. This is when the restaurants, shops, street markets, small supermarkets and street vendors come to get their supplies for the coming day in order to feed the millions that live in Beijing. At first sight the market gives the impression of total chaos. There are thousands of trucks, small and large ones, three-wheeled tractors, carrier tricycles and refrigerated lorries parked up or trying to find their way somewhere. People seem to come from everywhere. Some (with lorries) come from far away. But others come from the areas around Beijing. These are mostly peasant-traders (see Figure 10.1). They harvest their produce in the evening, transport it to Xin Fa Di and sell it there to small traders who come in from Beijing and market the fresh vegetables to consumers the next morning. Every night thousands and thousands of these small peasant-traders arrive at the market. Together they deliver an incredible volume of produce. Alongside these peasant-traders there are also professional traders who come with lorries and big trucks, often travelling large distances from far flung provinces, who usually sell directly from the back of their vehicles (see Figure 10.2).

Careful observation soon teaches one that there are clear patterns in the apparent chaos. People have their own locations and different types of products are sold in specific spaces. Those who come to buy mostly know the people they will buy from, but before going directly to them they pass among other sellers to check the quality and prices. On our visits we talked with many traders; they are all very friendly and informative. But it is impossible to ask more than, say, four or five questions. After a few quick questions and answers they want to continue trading. Everybody is very busy and in a hurry. The products need to flow into Beijing and the sellers need to return home to prepare for their next trip.

Although it is dark, very dark in some places, and slippery from the rain, people move in a determined and focused way through this strange but fascinating place.

The basic principle underlying this market is, in itself, quite simple. There is an imaginary circle that surrounds Beijing (in practice this circle coincides with the fourth ring road). Traders and peasant-traders from outside may not cross this boundary with their merchandise. Equally the owners of restaurants, shopkeepers, street vendors, etc. from inside Beijing also may not cross this imaginary line and

Figure 10.1 A peasant-trader at Xin Fa Di in Beijing

Figure 10.2 A trader at Xin Fa Di

return with merchandise. Sellers and buyers can only meet at markets such as Xin Fa Di that are located on this boundary. These markets are obligatory passage points.[2] However, they are not obligatory passage points controlled by food empires as is the case in most of the Western world. On the contrary, places like Xin Fa Di are where tens of thousands of sellers and even more buyers meet. This is the type of market that was theorized by Adam Smith and other (neo) classical economists. They are markets in the classical sense of the word: just as they were known in the past. They are places where tens of thousands of people come together to engage in many, many transactions. Through these encounters transaction prices are negotiated, set, renegotiated and changed. Time and again. After spending some time at these markets you can almost feel the presence of the famous 'invisible hand'.

Xin Fa Di is a *market place*: a concrete place where sellers and buyers physically meet and engage in transactions that make specific products and/or services flow from one to the other. Concrete negotiations take place in such market places: negotiations that affect the flow of products just as, in their turn, the flow affects the negotiations. Taken together, the actors, the many transactions and the flow of products compose a whole that nobody really understands in every detail – and there is certainly no centre from which the different moves, actions and positions are orchestrated. There is no central direction, no pre-established script that makes people and things move. Nobody is in full control. Nonetheless, it is highly effective. Every morning meat, vegetables, fruits, flowers, herbs, drinks, fish, flagrances, traditional medicines, etc. are abundantly available throughout Beijing – all differentiated according to different quality and price levels.

Two years after we first visited Xin Fa Di (that is, the agricultural and food section) we learnt that it doesn't cover just a few dozen but 121 hectares. If one includes the 'service area' with garages, hotels, restaurants, etc. for those who attend the market, Xin Fa Di totals 230 hectares. Each year 14 million tons of agricultural produce are marketed here. This represents a total value of roughly 50 billion yuan per year (equivalent to around €5 billion). Eighty per cent of all the fresh, non-processed food and drinks (vegetables, fruits, grains, oil, wine, liquor, meat, eggs, etc.) consumed in Beijing passes through Xin Fa Di. All the imported fruits also pass through here. Xin Fa Di market currently has a staff of 2,000 people.

The area we first visited is where the peasants gather. Most of them come from the Da Xin district, the horticultural zone that provides Beijing with 'small vegetables' such as onions, cabbage, salads and the like. Overall, wholesale dealers provide the most produce to the market. The largest of them provides 1.9 billion yuan/year. There are twenty to thirty enterprises that each supply between 300 and 500 million yuan (trading value). These are followed by 3,000 smaller trading enterprises each delivering a trading value of some 50 million yuan per year. These wholesale traders come from all over China; they acquire the merchandise from local and regional peasants and traders at wholesale markets throughout the country and then they transport it to Beijing. Thus, the supply side is made up of thousands of peasants[3] and traders (who obtain most of their products from peasants all over the country). A richly chequered multitude of small traders (who supply the decentralized markets within Beijing that deliver food to specific quarters), street vendors, restaurant owners, supermarkets[4] and even consumers makes up the demand side. According to the city authorities Beijing consumes 16,000 tons of fruit every day and the same volume of vegetables. These needs are primarily met through peasant production. This is made possible by an extensively ramified network of trading relations, with Xin Fa Di as the hub within this network. The functioning of this market is strategic for food supply and security in Bejing.[5]

In current discourses about food provisioning, especially in the West, it is often argued that large cities need large farms to supply them. Xin Fa Di (and similar markets elsewhere in China) shows that this is not necessarily the case. Small producers and large cities can be very well connected through well-functioning *markets*.

The Xin Fa Di market is owned by an enterprise[6] also called Xin Fa Di. It also owns other markets throughout China. The Xin Fa Di market began in the early 1980s (after the 'opening' of China) as a small road market covering 15 mu (1 hectare) and was managed by some ten people. The land and the people belonged to the village of Xin Fen Di. In 2002 the villagers decided to develop the market further: it was decided to create an enterprise to own the market. The villagers are still the owners of this now enormously expanded enterprise. The village has become an investment group. Since imperial times this village and the surrounding area have been called Xin Fen Di which means 'new tomb area'. This was thought not to be a very appropriate name for a market. Thus the name was changed to Xin Fa Di which means 'newly developed area'. It is now an extended market place that feeds a metropolis of 26 million people.

Xin Fa Di is not unusual. According to data from the National Bureau of Statistics, at the end of 2008 there were 921 wholesale markets for farm produce, each of which had sales of over 100 million yuan. This represented an increase of 230 per cent since the end of 2004 (Ministry, 2010:61).

All these *market places* differ substantially from what today is referred to, in the West, as *value chains*. Value chains are extended and oligopolistic networks consisting of many inlets and outlets, and complex relations between the inlets, the outlets and the points of conversion. These are all governed by a centre that thus controls the network as a whole. Products are acquired at the inlets. They may be processed, packed, re-routed, stored, or whatever at the points of conversion and are delivered to final consumers at the outlets. A strategic feature here is that the inlets and the outlets do not function as markets. Although transactions take place, the conditions, volumes and prices are determined in, and by, the centre of control. Anyone who has tried to negotiate the price of products in a supermarket will know what we mean, just as every farmer who sells milk to the dairy industry knows that s/he has to accept the price offered by the dairy. It is a case of take it or leave it.

Often several value chains will be functioning at the same time in the same territory (different supermarkets coexisting alongside each other; or different food industries) and they might mutually compete. But the *architecture* of market places and value chains differs fundamentally (Gereffi *et al.*, 2005). In market places competition takes place at the level of (and even literally within) inlets and outlets. In market systems competition has shifted (if it exists at all) to another level, i.e. between the different chains as a whole.

In addition to China's many market places there are also several value chains (both Chinese and Western), just as there are, e.g. in Europe, value chains as well as market places. The difference resides in the balance. While value chains (consisting of the food industries, supermarkets, catering firms, trading companies, etc.) dominate in Europe, and market places only function at the margins, in China it is the other way around. The market places dominate (and they also include an extensive web of rural markets – see Figures 10.3 and 10.4), while market systems, at present, play only a modest role. In this respect it is interesting that large retail chains (e.g. Carrefour) also have to acquire their merchandise on and through market places such as Xin Fa Di.

The hegemony of value chains in the Global North has an important consequence: it widens the gap between the prices producers receive and those that consumers pay. This is shown graphically in Figure 10.5 (which is based on Italian data). In 2000, of every €100 paid by consumers for food, farmers received only €25.60. The share taken by the controlling parts of the chain equalled €68.40. In 2009 this 'marketing share' had risen to €72.90 while the farm share decreased further to €20.10.[7] Instead of *adding value*, these chains increasingly *appropriate the value* produced by farmers – they drain the productive systems which is the reason for referring to them as 'food empires' (Ploeg, 2008).

Figure 10.3 A rural market

Figure 10.4 Another aspect of rural markets (the distribution of seeds)[8]

Figure 10.5 Evolution of value chain of agriculture (in Italy)

Yet the amount that is received as *farm income* is far smaller than the farm shares suggest. We also have to deduct the cost of all the inputs used (seeds, fertilisers, fuels, etc.), the costs of contracted labour and capital costs. What remains as *real farm income* can be as little as €7.60 out of every €100 paid by consumers. In 2009 (a year in which the economic and financial crisis translated into an agricultural crisis) this decreased to €1.50. In the Netherlands similar relations apply. Of the total value added contained in pork, the processing industry took 32 per cent, retail 24 per cent and the delivery industries 29 per cent. For the farmers involved only 6 per cent remained (Hoste *et al*., 2004).

If we now turn our attention to China, we find a drastically different constellation (see Table 10.1). Although producer and consumer prices also differ from each

Table 10.1 Some consumer and producer prices in China

	Price paid to farmer	*Price paid by consumer (average of 35 main cities)*
Eggs	8.17	9.66
Beef	18.84	40.02
Mutton	23.19	47.30
Cabbage	0.96	2.32
Tomato	2.32	5.19

Source: Agricultural Price Survey 2013 a, b, c.

other, the differences are far smaller than those encountered throughout Europe. This data shows that China's agriculture has been far less colonized and squeezed by capital than is the case in the Global North. And this to a large extent is due to the omnipresence of *market places*.[9]

It is not, however, certain that this particular balance between market places and value chains will be maintained in the future. For instance, in recent years Xin Fa Di (the enterprise) has started to contract extensive areas of land and large numbers of workers throughout China in order to organize the supply to its own markets. In Southern Hainan, the Xin Fa Di enterprise has contracted, with the help of local government, 20,000 mu of land (1,330 hectares) that is used for vegetable production. One of the arguments used to explain this shift is that this is the only way to provide the 'complete control' needed to guarantee food safety. Thus the market place seems to give birth to new value chains. In China one never knows what will result in what. Neither does one know whether it is irreversible or not.

Another worrying phenomenon is the planned disappearance of the large 'Morning Market' in the north-eastern Haidian District of Beijing. Here tens of thousands of consumers come every morning to buy the ingredients for their daily meals. These are sold by traders who acquired their merchandise in the Xin Fa Di market. This part of the city is being redeveloped and more space is needed for new houses, larger roads and more green spaces. The market has been earmarked as the best site to redevelop. Whenever redevelopment occurs retail markets such as the 'Morning Market' nearly always automatically appear on the drawing tables of the city planners as presenting the most appropriate opportunities. This is partly because in the eyes of bureaucrats and authorities they represent confusion, chaos, anarchism and a lack of hygienic conditions. As such market places are dismantled; a hegemony of supermarkets will nearly always automatically follow – at least, when the outcry of the traders for a new market place remains unheeded.

So, on the one hand, market places are being threatened and are disappearing. It is paradoxical, on the other hand, that new markets keep emerging with a rhythm and speed that perhaps even surpasses the marginalization and disappearance of the old ones.

The construction of new, nested markets

Many of the market places now functioning in China were constructed in the years that followed the introduction of the Household Responsibility System (HRS). Before the beginning of the 1980s the state functioned as the distributive system: it imposed production quotas on producer organizations (communes, production teams) and distributed these to different consumer categories. From the start of the 1980s onwards the markets have increasingly taken over this role (for a succinct description see Ye *et al.*, 2010). However, the construction of such new markets (that physically materialized as interconnected market places) was not just limited to the 1980s. It continues today: every time new needs emerge, or new products come to the fore, *new* markets are created. This was the case with agro-tourism which emerged in the late 1980s more or less simultaneously in different places. One such initiative was in Sichuan province, where local peasants started to produce flowers and ornamental plants in order to sell the seedlings in the urban markets. This resulted in attractive, blossoming fields around the village that in turn attracted many visitors. One local peasant, Mr Xu, came up with the idea of offering accommodation and food to the many visitors. After discussing the idea at length with his family he started doing this. This resulted in the first '*nongjiale*', which roughly translates as the 'happy peasant home' (*nong* is peasant, *jia* is house and *le* is happiness) (Wu *et al.*, 2015).

The agro-tourism market is highly decentralized: there are many suppliers and many clients[10] looking for this new service. It is a market without a central focus: the internet offers the infrastructure for clients to find a location they like and the (usually informal) 'contracts' are directly established between the peasant family and potential visitors. Agro-tourism is now a huge market: 335 million tourists in 2007, generating a turnover of €5 billion (Ye *et al.*, 2010). Although the market is characterized by different levels[11] (it is, in other words, a segmented or differentiated market) it essentially remains, as Ye *et al.* argue, an example of a nested market. It is 'nested' in the value Chinese consumers attribute to staying in a peasant home (and enjoying the charms of the countryside) as well as in the capacity of Chinese peasants to deliver 'happy homes' and make other charms (i.e. local foods) accessible.

Ye *et al.* (2010) describe several other nested markets. Very frequently they turn out to be segments of wider markets, characterized by different prices, a different distribution of value added and different relations and dynamics. As in the case of the market for agro-tourism, these markets are also nested in social definitions of quality that are shared by producers and consumers and valorise distinctive products and services and the way they are produced.[12] Si *et al.*, 2014 provide information on other alternative food networks and especially on newly constructed farmers' markets. Further details can be found in Si (2014) and Schumilas (2014).

Alongside these huge nested markets, there are also many initiatives that are leading to the creation of small-scale nested markets. There are many different types. One particularly interesting case is the nested market for glass noodles (the processing of which we have already discussed in Chapter 4). Glass noodles are made from sweet potatoes and require a complex process of conversion (see Figure 10.6). They are very different from the noodles that are produced by food industries and commercialized through the main circuits.

Initially it was mostly migrant workers who, returning to the cities after Spring Festival, brought glass noodles with them as gifts for friends and patrons. Villagers also liked to give their children and grandchildren living elsewhere glass noodles. They were always a much appreciated gift.

This particular experience is currently being used by particular groups of consumers (in Beijing) and producers (mostly loosely organized in the village framework) to construct new, small-scale nested markets. One key factor in the emergence of these new markets is urban consumers' appreciation of fresh and genuine products like those consumed in rural households but which can rarely be found in the main urban markets. Another is the capacity of villagers to enlarge the production of such foods[13] and their need to obtain a good income from it. The third is building the connection between these two, i.e. connecting the urban and the rural. Together these three elements make newly emerging nested markets possible.

Making the connection is done in a seemingly simple way. In Beijing there are loosely structured consumer groups (in the beginning there was just one).[14] At the heart of the group is the consumer committee which functions as the outlet point. In the village there is a producers' committee[15] which functions as the

Figure 10.6 Processing sweet potatoes into glass noodles

inlet point. What connects the two, at least at first sight, is email and a standard form with a list of products. Consumers indicate how much of each product they want to acquire, the orders are bundled together and a small van delivers the bulk order to the outlet point, where consumers come and collect their individual orders. Figure 10.7 distinguishes these different flows with more precision.

The first flow starts with the participating producers drawing up an inventory of the products they have available, which evidently reflects seasonality and the like. Usually this is annotated with brief explanations about quality, product characteristics, etc. The producers' committee sends this form to the consumers' committee and it is then forwarded on to the individual consumers. They place their orders, inducing a second flow. These orders are then sent back to the producers' committee which distributes the orders between the producers. The harvested produce is brought to a central point (where packaging, weighing, quality control and slaughtering takes place) and then transported by two members of the committee in a van (which they own) to Beijing. This is flow 3: the delivery. The consumers' committee has already informed the consumers about

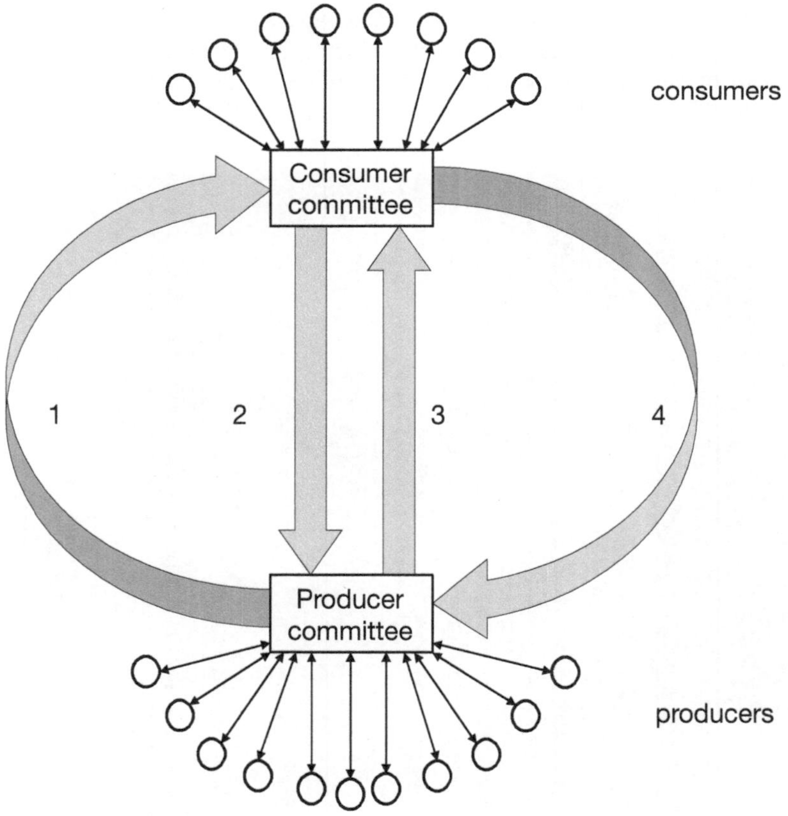

Figure 10.7 The flows within a small-scale nested market

the delivery place and time – where the consumers wait for its arrival, pay for the orders and take them home. This leads to the final flow, 4: the money goes back, usually with comments, compliments and/or complaints, to the committee in the village and the individual producers.

All this might seem an ordinary and simple part of everyday life. While that is true, at the same time there is something exceptional involved.

First, the flows summarized in Figure 10.8 represent completely new connections. People, places, things and qualities that were previously separated are being connected through this nested market. Fresh, pure and genuine countryside food of high quality that was previously only available to rural households now travels to the city, and the money that the urban people are willing to pay to access good food flows back to the village. Yet this is not all. Second, there are regular exchanges of information between the villagers and consumers – who are directly connected (in fact very well connected) – as they organise regular visits and discussions. Third, the products (as well as the payments) flow in a specific, socially defined way that is governed by the participating groups. These flows are not organised or orchestrated by a trader dealing in freely floating commodities. This certainly is a market, with voluntary transactions, prices, etc. But the transactions and the flows of products and money are embedded (i.e. nested) in a socially defined agreement between producers and consumers (see Ploeg, 2015a). This translates, in the fourth place, into the different flows having specific features. Here it is the producers themselves (from a particular locality) who inform consumers about availability. This is in marked contrast to the situation in a supermarket where availability is permanently secured. By contrast, in the nested market seasonality plays an important role. Another specific feature is that deliveries do not follow market opportunities but follow social agreements. In addition, the inlet and outlet points are democratically organized. They are not part of wider oligopolistic networks nor are they governed by external centres of command that orchestrate larger market systems.[16] This tends to lead towards an incipient form of food sovereignty.[17] This in turn relates to another difference: the inlet and outlet points (and the connections between them) do not represent or function as capital that is seeking to generate profits. The nested market functions through the participants own (i.e. self-owned and self-governed) resources. It is based on work done by the two committees, on having a van and drivers and more generally on the fields, knowledge and so on of the producers and the preferences of the consumers. Of key importance here is that the connections between urban and rural committees (these connections make up the 'heart' of the nested market) represent 'commons'. They are common property or, to use Eleanor Ostrom's concept: common pool resources (Ostrom, 1990; for an extended discussion of nested markets as common pool resources, see Polman *et al.*, 2010).

Finally, it should also be noted that the nested market can (at least potentially) change the social division of labour. Some capabilities that were previously exclusively linked to specific roles become more widely distributed. This can prove to be a source of problems, friction and dissatisfaction but it also has the potential to improve the quality of people's lives: to connect previously disconnected capacities. Let's illustrate this by taking the example of an organic vegetable producer with his/her own selling point (at the farm, in the local market, or wherever). In this case inlet and outlet points are identical, they overlap. The producer and the trader are one and the same. However, when considerable distances need to be bridged this evidently becomes impossible. This is where the trader normally enters the scene. But when the trader is replaced (albeit partially) by the type of nested market we are discussing here, then his/her tasks (exploring availability, making things and money flow, meeting consumer needs and translating them back to product selection, etc., etc.) are taken over by and shared among the participating consumers and producers (and especially the committees). This can considerably enrich their lives and empower them.

In the concrete case we have studied, a nested market that links Sanggang village in Hebei province with consumer groups in Beijing and which is supported by the College of Humanities and Development Studies, China Agricultural University, the building of such a nested market also involved a marked learning process. Many practical difficulties emerged,[18] new solutions were forged[19] and the nested market became increasingly institutionalized: it began to function smoothly according to a set of rules and expectations shared by both parties. At the same time, the electronic forms, the van, the direct communication between the committees and the constructed trust developed into a solid socio-material infrastructure that now sustains the nested market (for the concept of infrastructure see Ploeg *et al.*, 2012).

Figure 10.8 Preparing a delivery to Beijing

So far this particular nested market is producing interesting results. It is making authentic products available to urban residents at reasonable prices and with relatively low transaction costs. It has also created accessibility: the urban dwellers can easily visit the village and even stay overnight. In the village it has generated multiple effects. First, the production of authentic produce has been considerably enlarged. Without the nested market this would not have been possible.[20] The new market has also helped to raise incomes in the village. And for several villagers it has proved to be an important learning ground.

It should be noted that it is not just peasants (or peasants and consumers' groups) who are initiating and organizing new, nested markets (if needed, in co-operation with state agencies). Sometimes the state takes the initiative directly, retaining responsibility for the subsequent functioning of the nested market. One very interesting example is the new market for 'land development rights' that is being constructed in Chengdu (the capital of Sichuan province). To date, Chinese peasants are enabled to dispose of their land-use rights. They might lease this right to others, or sell it (sometimes in disguised ways). What often happens is that village committees reobtain the use-rights and then sell them for much higher prices to private companies in order to 'develop' the land (or they develop it directly themselves in order to obtain new income flows from renting or selling houses or industrial buildings). This land development can provide huge windfall profits. What complicates the picture is that the possibilities (or rights) to develop the land are severely limited (in China as a whole there is a so-called red line – no agricultural land may be converted into urban land without equivalent compensation elsewhere). Thus a new market for 'land *development* rights' has been defined (see Ye and Legates, 2013:248–64) that makes it possible to shift these possibilities from one place to another[21] and also ensures that these development-rights stay with the farmers and peasants (until they sell them). Thus farmers can shift the use-right to others while maintaining the development-right. When an opportunity emerges they can sell this latter right and thus benefit from the value it represents. This mechanism evidently smoothens the land market, greatly enlarges the mobility of land (i.e. the possibility to develop it) and makes the owners of the use-rights (farmers and peasants) into beneficiaries who participate in the distribution of the windfall land development profits. Evidently, this removes many frictions that otherwise hinder development.[22]

Markets as a point of reference

Market systems as they are known in the West increasingly govern far wider segments of society than solely the production and distribution of commodities. Education, security, health care, etc. are increasingly perceived, and treated, as commodities. This is intrinsic to the neo-liberal project that is seemingly being implemented everywhere. Under this regime the delivery and use of these services are increasingly thought of as needing to follow the rules of the market. More specifically, the search for profitability becomes a, if not the, guiding principle.

This is occurring in universities, in hospitals and many other realms where it was previously unthinkable.

Agricultural research and development (R&D) is a crucially important field. For many decades it has, nearly everywhere, been a public task – distanced from the logic of the market. Overarching interests such as food safety and security, self-reliance and/or the needs of agricultural producers operated as the guiding principles. Increasingly though, this situation has changed. First, agricultural R&D was oriented (through cost/benefit calculations) to achieving the largest possible macro-economic gains. This led R&D to increasingly focus on areas with optimal conditions for production (flat, fertile and irrigated land, with good infrastructure) and on crops and production methods that could yield rapid gains in productivity.[23] The Green Revolution and the technologies it spawned is exemplary of the innovations that this approach produced. In more recent times agricultural R&D has been increasingly privatized and the remnants of public R&D was carried out in accordance with the interests of the 'top sectors' of the economy. As such agricultural R&D has become increasingly aligned with the interests of large food empires and their focus on quick gains, high returns and greatly extended control over the food chain. Genetically Modified Organisms are exemplary here.[24]

While extended control and the wish to extend it further are diseases that are far from alien motivations in China, agricultural R&D has taken a route that differs fundamentally from the Western one. On the whole, the Chinese agricultural R&D system has produced outstanding results. Two symbolic examples of its success are the 'super rice' (yielding more than 10,000 kg/ha), developed by Dr Yuan Longping's research group and the Nong Da 108 Maize Variety[25] developed by the research group of Professor Xu Qifeng at China Agricultural University (College of Agronomy). This latter research group worked very closely with peasants.[26] One could argue that these two examples resemble the Green Revolution approach – at least to a degree. However, another groundbreaking example shows a marked difference with the Green Revolution paradigm. This example centres on the 'tea oil plant'. It took 20 to 25 years of skilful plant breeding to increase the oil content of Tea Oil Camellia (*Camellia oleifera Abel*), which is now a major source of vegetable oil for peasant households (see Figure 10.9). This length of time and dedication – but especially the choice of this plant for further breeding – are very unusual for a Green Revolution R&D project. Tea Oil Camellia is not a commercial crop. It provides oil that is used in rural households (basically in the kitchen for cooking). Moreover, the selected and improved variety is appropriate for cultivation under marginal conditions (areas with low soil fertility, water scarcity or hilly conditions, etc.). Thus, the project was not oriented towards favoured areas, as is much Green Revolution R&D, nor does it require or assume the application of large amounts of agro-chemicals. It is a 'simple' plant that can be cultivated in small, marginal fields belonging to small peasant households. All this implies that the improved *Camellia oleifera* is very different from the typical outputs of Western agricultural R&D systems. The latter design and organize R&D as part of the market system; in China (at least in this case) non-market

Figure 10.9 Tea Oil Camellia in one of the Chinese research stations

considerations (national self-provisioning of oil, support for marginal areas, strengthening the economy of peasant households, etc.) also influence agricultural R&D. In peasant areas, such as those of Hunan and other central Chinese provinces, this makes a considerable difference.

Multifunctionality: further distantiation

Multifunctionality means that the resource base of the farm is not used to produce just one single product for one specific market (Renting *et al.*, 2009). Instead, one and the same set of resources is used to produce a wider range of products and services. The resource base has, as it were, more 'functions'. Multifunctionality[27] carries the promise that the costs of producing two products or services (A and B) together, using the same resource base, are lower than when A and B are produced independently of each other (Saccomandi, 1998). In addition, multifunctional farms are not linked to just one market but to a wider range of markets. This reduces the risks and provides the farm with considerably more flexibility. If one market is unpromising, the farm can produce more for the other markets. Multifunctionality is thus another mechanism for distantiating the farm operations from the power of market forces, which often act as ordering principles. Instead of being, more or less, constrained to 'follow the logic of the market', the producer has more room for manoeuvre. S/he can choose, on the basis of his/her *own* interests, prospects and experiences, the best path to follow (or which 'portfolio' to develop) and which 'mix of markets' is best suited to his/her *own* views and preferences.

Chinese agriculture has been multifunctional since ancient times (King, 1911). The forms it has taken vary from combining rice growing and fish breeding, farming, forestry and fishponds, or horticulture and trading – evidently the combination(s) that could be chosen and developed depended on agroecological conditions, socio-economic relations and family circumstances.

Currently, multifunctionality is being further developed into a sometimes dazzling array of new activities, involving new products, new services and relations with new markets. It embraces different forms of agro-tourism, such as *nongjiale* discussed above, but also the many small restaurants emerging along secondary roads where *peasant food* is served to passing drivers and tourists (see Figure 10.10). Multifunctionality can also include:

- drying and peeling wooden trunks (from the farm's own forestry lands) in order to sell them for a higher price to the paper industry (see Figure 10.11);
- the use of a three-wheeled tractor (an essential part of the farm's own resource base) to provide transport services to mining industries, brick factories, road construction sites, etc. (see Figure 10.12 and Chapter 2); and
- harvesting the remnants of glass noodle production to convert into cattle feed (Figure 10.13) – a spin off from another multifunctional activity, the manufacture of glass noodles.

Figure 10.10 Agro-tourism

Figure 10.11 Processing wood for the paper industry

Figure 10.12 Transporting rocks

Figure 10.13 Drying waste from glass noodle production to obtain cattle feed

These are all new multifunctional activities that help, just as the more classical forms of multifunctionality did, to increase farm incomes and which, simultaneously, help distantiate the organization and development of the farm from the logic of the market.

By way of conclusion

All in all, the analysis of food markets in China – and especially the comparison with the West – flows into a, clear, albeit somewhat uncomfortable, conclusion. In the West the production, processing, distribution and consumption of food are represented as being governed by the market. This market discourse is dominant and is generally understood as explaining why the food system functions as it does. Farmers are told to structure and develop their production according to the logic of the market. Equally, agrarian policies are oriented towards, and reflect, market dynamics. Yet in reality there are fewer and fewer *real* markets. Market places have become displaced by value chains which, de facto, relocate competition to the inter-chain level. Transactions within the chain are governed by extra-economic forces orchestrated from the centres of the different chains.

If we look at Africa (see Chapter 1), where this neo-liberal and chain-centred approach was introduced in a roughshod way during the Structural Adjustment Programmes imposed by the World Bank and International Monetary Fund, we can only infer that this contradictory constellation is, to put it mildly, not always a recipe for success and good performance. This applies especially, though not exclusively, to the Global South.

In China one can find, as it were, the exact opposite of the Western constellation. There are markets everywhere and there are multitudes of small, medium, large and very large traders. However, none of them is able to control the markets (as occurs within Western value chains). While there is much government rhetoric about markets (markets embedded in a 'socialism with Chinese characters'), in practice the state is also implementing and developing a comprehensive agricultural policy (of the *Sannong* type). This is in stark contrast to the West which (with few exceptions) has been systematically deactivating agricultural policies. For example, China has started to subsidize grain production while the European Union has eliminated such subsidies. Agricultural policies in China do not reflect and strengthen the dynamics of the markets (as is increasingly the case in large parts of the West). Rather they function as mechanisms that *countervail* the markets. This also occurs in agricultural R&D. In the West R&D follows promising market opportunities (and increasingly the interests of the large value chains). In China R&D focuses on social and ecological needs and explicitly includes marginal conditions.[28] Meanwhile, although peasants produce for the markets (they are responsible for more than 90 per cent of all the food consumed in China), they actively *distantiate* the organization and development of their farms from the markets. This occurs through mechanisms such as circularity, multifunctionality, social networks, self-provisioning and cultural repertoire, all of which have been discussed at some length in the preceding chapters.

Thus, we are facing here two *chiastic* structures[29] – one at the level of the agricultural units of production (the farms), the other at the level of the food system as a whole.

First, farms in the west generally speaking no longer produce for a market but for specific value chains. At the same time, the organization and development of production is strongly ordered by markets relations (especially those between downstream and upstream markets). In China, by contrast, farms produce for the *market* (embodied in and represented by market places), but their organization and development is actively distantiated from these markets.[30]

The second *chiasm* regards the food system as a whole. In the West the food system is said to be governed by markets, but there are hardly any market places any more. The markets here tend to be merely a shield that allows for the undercover operations of value chains or food empires. Here again China represents, at least to some degree, the opposite. Market places play an important, if not decisive, role and government policies effectively influence the food system as a whole (although they are organized under the veil of market support and adherence to World Trade Organization requirements).[31]

Together these two chiasms bring about, at a higher level of aggregation, another probably even more astonishing chiastic structure which makes China look like Alice's Wonderland – at least when seen from a Western viewpoint. Left is right and the top is the bottom. And visa versa. Everything changes and together things are linked in what appears to be inverted ways. Production *distanced* from the market as an ordering principle makes for dynamic and well-functioning food markets that effectively link the urban and the rural. This seemingly astonishing relationship also operates the other way around. The current market places, where food is central, are strongly driving forward agricultural production and farms.[32] This applies a fortiori to the newly emerging nested markets: they trigger production not realized so far and involve producers who previously functioned at the margins.

All this is in strong contrast to the way that relations have evolved in the West.[33] Here large market systems (as value chains) which in practice often exclude the interactions and mutual adjustment of demand and supply, tend to favour sclerotized farming structures (that have hardly any flexibility or resilience left). Simultaneously the latter are fuelling the former, not only with cheap raw materials, but also and especially with value that is appropriated.

This contradictory panorama also has a symbolic level. In the West markets are presented (and apparently accepted) as given. Markets are just there, they are self-evident realities that go beyond (and should not be distorted by) any intervention. Their presence and hegemony is indisputable.[34] In China, on the other hand, markets are understood as being actively constructed, adapted, negotiated, contested and (eventually) recontested. This is not only the case in intellectual debates. It is remarkable that any setback in the markets is seen by Chinese people as being related to one or another mistake of the government – whether or not this is actually the case.[35] Ironically, this makes for the last *chiasm*: a West which defines itself as dynamic but which rigidifies its main institution (the market) versus an East that is often understood as a ponderous giant but is actually able to juggle with markets.

Notes

1 One of the sections of this chapter (on nested markets) is partly based on an article about rural development occurring through the construction of new markets by Ploeg *et al.* (2012) and a paper by Wu *et al.* (2015).

2 The irony is that this intelligent socio-material infrastructure (see Ploeg *et al.*, 2012 for an explanation of this concept) was not designed or created on purpose. It was, in fact, an unintended consequence of traffic policies. In order to avoid many heavy lorries and a multitude of three-wheeled tractors entering the centre of Beijing, a regulation was created prohibiting them from crossing the fourth ring road. This made transfer hubs necessary. One of these developed into Xin Fa Di. The traders now very much like this arrangement. As one of the managers of the market explains: 'You can't chase them [the traders] away, they only go when they lose money. Neither can you invite them, they will only come if they can make money and then they come by themselves.'

3 These peasants pay an entrance fee of 2 yuan per day.

4 There are large retail chains operating in Beijing, such as Walmart, Carrefour and Chao Shi Fa. They have their own storehouses and managers for purchasing food in Xin Fa Di. They are constantly buying and their transport trucks deliver twice a day to their shops in Beijing.

5 There is no direct regulation ('we can't, this is a market'), but if supply is low (due to heavy rains or blocked highways or whatever) entrance fees and rents for space are waived and Guanxi relations are used to make sure sufficient supplies arrive at the market. In addition, the market is now building storehouses with a capacity of 120,000 tons. This is being financed by the enterprise itself, which argues that building strategic

infrastructure for the city is in the public interest. Within Beijing there are other storehouses.

6 The enterprise rents the available spaces to the salesmen. Buyers have to pay a fee when leaving. Beyond this a range of trading facilities is supplied. These include the certification of food products and the provision of market information through television, the internet, bulletins and social media.

7 Here one has to take into account that the first half of 2009 was characterized by very low off-farm prices, especially in the dairy and meat sectors.

8 This image does not appear at all unusual until one comes to consider that the proprietary seed market (brand name seeds) accounts for 82 per cent of the global seed market, 67 per cent of which is controlled by ten companies (empires) of which Monsanto is the largest accounting for 27 per cent of the world's global seed market (Clayton, 2015).

9 Critical discourse in the West finds it difficult to elaborate on this kind of conclusion, due to it equating the market to an 'evil'. If markets are equated to capitalism, and only to capitalism, any anti-capitalist stance is automatically translated into an anti-market position. In today's circumstances such a point of view is not very helpful. It certainly does not contribute to understanding why peasants construct new, nested markets (as discussed later in this chapter).

10 The huge number of clients derives from the expansion of the Chinese middle class that is now estimated to embrace some 400 million people. Most of them have a rural background and they love to spend weekends or short holidays in the countryside.

11 These levels correspond with the three stages through which agro-tourism developed. The first is characterized as the *nongjiale* period (before 1994). Here local peasant initiatives were central; they were limited, however, to the level of the single farm and household. Next there followed a period mostly characterized as one of folklore tourism (1995–2000). In this period the scale of operation moved to village level. The architecture of a village as a whole, its rituals, festivals and/or culinary traditions were put centre stage. Finally (after 2001), attention moved to yet another scale: the landscape as a whole. This approach embraces several villages in the same package. Scenic beauty, attractive roads, walking paths, panoramic observation points, archaeological monuments and/or natural parks have become the new, overarching points of reference (see Wu *et al.*, 2015; for a discussion of the third stage, Ploeg *et al.*, 2012).

12 We refer to the market for organic produce, the 'Green Food' market, the market for eco-agriculture and the 'one-village-one-product' market. The latter corresponds to the markets for regional specialities in Europe.

13 Their production is presently mostly limited to household needs and social networks. This is associated with a typical division. Products intended for household consumption and surrounding social networks are often traditional varieties and no (or hardly any) chemicals are used. By contrast production for the market is grounded on new varieties (sometimes including genetically modified organisms) and a widespread use of chemical fertilizers, pesticides, herbicides, growth hormones and so on.

14 In the case we studied there was a total of around 300 consumers with a stable nucleus of forty. The products included dried sweet potatoes, glass noodles, pork, free-range chicken, free-range eggs, tofu, fresh vegetables, dried persimmon and herbs.

15 Around this producer committee (consisting of two members of the village committee and two other villagers) there is a stable group of thirty producers.

16 In a way we are dealing here with modest forms of economic democracy. More specifically, in this respect the nested market helps to introduce elements of food sovereignty: people exert the right to decide how the production, distribution and consumption of their food is organized.

17 Food sovereignty can loosely be defined as the right of people to determine their own food and agriculture policies; it is, in essence, about the democratization of food and agriculture. See Via Campesina 2007, IAASTD 2008 and Wittmann *et al.*, 2010.

18 These included typical socio-cultural lock-ins such as urban people imagining that pork meat comes without fat. When real pork meat (especially that raised at a small-scale level in the villages) is seen and tasted it is, consequently, perceived as being 'too fatty'.

19 Excursions to the village and seeing local activities, such as pigs being slaughtered, helps a lot in this respect.

20 This is especially due to the (relative) security that the products will be sold, to the somewhat higher prices and, especially, because there is no longer a need to go to the county town and try to find interested traders. In more general terms, the price they receive is somewhat (but not excessively) higher and the transaction costs are considerably lower.

21 This is made possible by, for example, concentrating housing in villages more closely together and then selling the space created to cities in need of expansion.

22 'The basis of the policy is that rural development rights clearly belong to the farmers.' (Ye and LeGates, 2013:251). The authors also observe that this construction as a whole is potentially open to corruption.

23 One consequence of this in many parts of the world is that less endowed areas suffer marginalization. In China this tends to be different – and this considerably helps peasant agriculture.

24 It is well known that the innovations stemming from such an approach are not necessarily in line with the interests and prospects of peasant producers. When such a mismatch occurs this very seriously hampers the processes of agricultural growth and development.

25 *Nong Da* means agricultural university.

26 This is a typical example of what, in international debates, is referred to as 'Mode 2' research. According to Hessels and van Lent, 'Mode 2 knowledge is produced in the context of application by so-called transdisciplinary collaborations. Moreover, scientists are more reflexive and they operate according to different quality criteria when compared with the traditional disciplinary mode. The new mode of knowledge production has been coined 'Mode 2', and it is not believed to replace Mode 1, but to supplement it.' (2008:740). Mode 2 of technology development is far more focused on the 'context of application' than a traditional academic approach which is often relatively isolated from real life problems. As such it tends to be far better in dealing with novelties and heterogeneity. Ye (2000) provides six ways in which the Mode 2 features of the Chinese agricultural R&D system might be strengthened.

27 In the literature these are also referred to as 'multi product firms'.

28 In the West it was assumed that 'optimal solutions' designed to function under 'optimal conditions' would also render the best solution (albeit less productive) under 'sub-optimal conditions'. Any serious inquiry into the history of agronomy would have shown that this is not true. But such an inquiry was not undertaken. The assumption functions well as an ideological veil that permits highly selective R&D.

29 A chiastic structure relates two or more clauses to each other through a reversal. Chiastic structures represent a twist in the normal sequences of events. The concept of *chiasm* is derived from the Greek capital X (*chi*). The X suggests an inversion of the interrelations between different elements. A chiasm assumes two pairs of elements, say A1 and A2 and B1 and B2. Normally there is parallelism: A1 translates into A2 (A1 –> A2) and the same applies for B1 and B2. In a chiasm this is inverted: A1 goes together with B2 and B1 is associated with A2. Thus there is an unexpected crosswise arrangement of concepts (and/or empirical elements). A chiasm represents a rupture in, and critique upon, unilinearity. The particular chiasms discussed here unveil the *complementarity* of market and non-market (as opposed to the often assumed *dichotomy*) and the *complexity* (as opposed to the *straightforwardness*) of the relations between market discourse and the empirical realities of markets as well as those between production *for* the market and distantiation *from* the market.

30 See also Ploeg, 2008.
31 There are more countries doing so. Canada, for example, is one.
32 It is very tempting, of course, to ask how this typical Chinese constellation was created and has developed since the beginning of the 1980s. A historical reconstruction of the making of this particular market structure is beyond the scope of this book. But one could hypothesize that the absence of large corporations in the first, say, 10 to 15 years played an important role: it allowed the myriads of Chinese traders to take and develop their positions and having secured these they were not very eager to relinquish them.
33 Other authors who unveiled chiastic structures (without naming them as such) include Zhang and Donaldson, 2013 (on the unexpected consequences of the privatization of land) and Donaldson, 2011.
34 Underneath this veil, though, the markets are hijacked by large empires.
35 If the price of, say, fruit is high the government is blamed. When the prices are low the government is also blamed.

11 The relevance of the Chinese experience[1]

Jan Douwe van der Ploeg

The large number of peasant holdings in China and the associated peasant nature of its agriculture are not simply matters of historical legacy. Rather they are a result of a major reversal that occurred at the end of the 1970s when there was a widespread peasant rebellion that shaped China's agriculture into what it is today. The most visible and best documented part of this rebellion was the Anhui uprising (Du, 2006; Gulati and Fan, 2007). Here, in the small village of Xiaogang eighteen peasants concluded that they would be better off begging in the streets than continuing to work in the collective system that was constructed from the mid-1950s onwards. The eighteen secretly divided the land of their production team into individual plots and signed a document promising each other to take care of the children of any of them who might lose their lives in what evidently was a full frontal confrontation with the state and the Party.

The rebellion was not just limited to Anhui province. In Guizhou province the peasants redistributed a far larger proportion of the land than those in Anhui (White, 1998). Even during the Cultural Revolution, officials in Guizhou recognised the existence of these private plots and established production quotas for households rather than for production brigades (Donaldson, 2011:55). Thus, a widespread but subterranean[2] movement undermined and brought down the institutional structure of collectivized agriculture. Enlightened leaders of the Party (first at municipality and provincial levels and then later at the national level) tolerated the changes that seemed impossible to rein in and then started to actively support them (as explained by Du, 2006). This then resulted in a radical, massive and accelerated land reform which was essentially redistributive. This reform introduced the Household Responsibility System (HRS): each peasant household had access to some land and became (in a step-by-step process)[3] responsible for the production to be realized on the land which was now theirs.[4] It was the greatest land reform the world has ever seen and one which, unlike many others,[5] was highly successful.

The really exceptional feature of this land reform, though, lies elsewhere: it was the first redistributive land reform in which the newly introduced man–land relation *equalled* the average man–land relation.[6] This may sound enigmatic to outsiders but the political, economic and historical significance of it was, and still is, exceptional.

A truly exceptional land-reform

All land reform processes (be they redistributive or aiming at the creation of large state controlled cooperatives) operate from the concept of an *economic holding.* This is the carefully calculated amount of land to be transferred (directly or indirectly) to a household so that this household can make a reasonable income – assuming they work the land properly. Central to this calculation are the given land use types, the reigning (and/or expected) market and price relations, a so-called 'saving quota'[7] and the available (and/or expected) technological profiles. Thus, the 'economic holding' critically reflects the evolution of markets and the process of technological change. More generally speaking, it reflects the conditions created in and through the process of capital accumulation. As a consequence, the calculated (and implemented) economic holding tends to differ significantly from the *average* man–land ratio. More and more land is needed for profitable farming (and the associated shift of social wealth from the countryside to the cities). As exchange relations between town and countryside (between industry and agriculture) change to the detriment of the rural, the economic holding moves further away from the average, historically delivered, man–land ratio.

In 1952 a land reform was implemented in Egypt. The calculated economic holding amounted to 2.6 acres. That was 44 per cent above the average land–man ratio (which then equalled 1.8 acres). Thus, as Doreen Warriner euphemistically concluded, 'some displacement was needed' (Warriner, 1969:413).

Other countries followed the same pattern.[8] In Iran (1963) the calculated and then subsequently implemented economic holding was 80 per cent beyond the average ratio. In Tunisia (in 1964) it was 160 per cent. In Chile (during the Frei reform implemented between 1965 and 1970) the difference was 260 per cent and in Peru (1969–1974) it reached 400 per cent.[9] Twenty-five years later Peru's newly developed Majes irrigation scheme established an economic holding that was ten times higher than the average (Vera Delgado, 2011). Thus, land reform carried out under the aegis of 'giving land to the tiller' has almost everywhere effectively become a politico-economic instrument that subordinated agriculture to the requirements of processes of capital accumulation: it excluded large portions of the (usually growing) rural population from any access to the land. This paradox explains why land reforms have been so strongly and enduringly contested and opposed in many parts of the world (Ploeg, 1974; 1990).

While there have been temporary exceptions (Bolivia in the 1950s is a notable case) as well as many local deviations triggered by peasant struggles (Catacaos in the North of Peru from 1969–1989 is a notable case here),[10] in national terms China represents the main radical and enduring rupture to this worldwide trend towards marginalization and exclusion.[11]. The land was divided in a strictly egalitarian way (according to the number of household members and taking into account different soil qualities) and all households were included. Living in the countryside meant (and means) that one has the right to land. This is an inalienable aspect of *Hukou*, the set of rules that also regulates access to educational facilities, medical care, rural pensions, residence (at least in the past) and (in recent years) the right to be exempted from taxes.

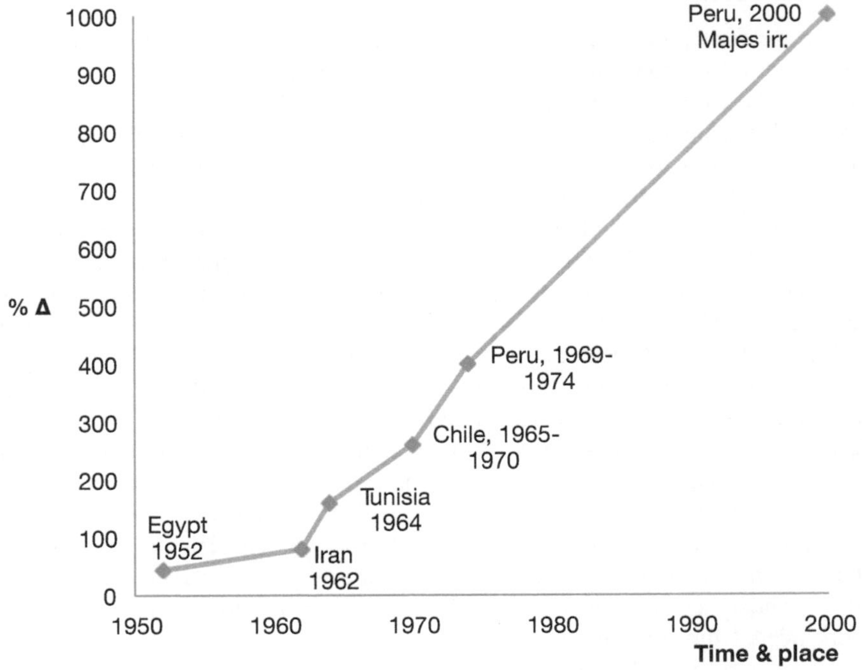

Figure 11.1 The development of the economic holding compared to the average man–land ratio

This egalitarian redistribution of land (within which all rural households gained access to it) clearly reflected the power relations between state and peasantry, the history of the Communist Party, the egalitarian tradition of the Chinese peasantry and the grassroots nature of the land reform itself. It started as a rebellion;[12] peasants implemented their own reforms and the state could do little else than simply follow and legitimize the subterranean shifts that were going on across the country.

This radical rupture with the main trend had a range of repercussions. In the first place it made the combination of the peasant household and the village community into a strong land–labour institution that would put its imprint on many of the developments that occurred in the following decades (part of this is discussed in Chapters 2, 4 and 7). Land again became a crucial medium that tied different generations together (see Chapter 3) and which turned the workers in the communes into peasants. In the second place, the institutionalization of the HRS translated into a strong increase in production and productivity (in the 1978–1984 period total agricultural output increased by 42 per cent),[13] an increase that later continued through the ongoing labour-driven intensification process that we discussed in Chapter 4. As a matter of fact, the possibility of increasing yields was one of the concrete triggers of the peasant rebellion. To quote a leader of the Anhui rebellion: 'When we started working as peasants again, we were able to realize high yields [. . .] When you have your own land you care better for the plants.' (Ploeg, 2013:126–7). Third, it provided (especially from the beginning of the 1990s onwards) a growing labour force for the quickly expanding industry – a labour force that could fall back on farming in case of need. This also prevented the emergence of extended slums (see Chapter 2) and the newly established town–countryside relations (which involved a flow of resources from the urban to the rural) allowed, especially from the 2000s onwards, for what we previously described as a 'reinvention' of farming (see Chapter 7). Finally, the combination of the egalitarian distribution of land and the urban–rural flow of resources that later followed has created, especially since 2008, a huge domestic market that has helped to reduce the effects of the economic crisis. This last element (peasant households becoming a cornerstone of the internal market) is an ingredient that is lacking in many other rural countries in the Global South – which (ironically) translates into a major limitation on the accumulation of capital (Mazoyer and Roudart, 2006:483 onwards).

The relevance of an unchained peasant agriculture

The reshaping of China's agriculture into a *peasant* agriculture at the end of the 1970s had a multifaceted and far reaching impact on subsequent developments. In the process that followed (regaining partial control over the marketing of produce, being relieved from tax obligations and obtaining ever more control over the main decisions concerning the organization and development of production) the Chinese peasantry was literally 'unchained'. The freedom from imposed limitations and restrictions and the freedom to organize production according to their own insights, interests and prospects were considerably enlarged. The relevance of this unchained peasant agriculture can best be illustrated through a comparison. I opt here to compare China's peasant agriculture with the agricultural sector of the Netherlands which is strongly, albeit far from exclusively, entrepreneurial in character. Peasant agriculture tends towards the construction (and maintenance) of a self-owned resource base that allows for farming that is not firmly connected with upstream markets (for inputs, machinery, credit, seeds and so on). By contrast, entrepreneurial agriculture is firmly based on resources mobilized from different markets. Thus, it is heavily engaged with (and highly dependent upon) upstream markets.

The key data in this comparison are summarized in Figure 11.2. This shows that in the Netherlands in 2010 the total Gross Value of Production (GVP) of agriculture was equal to €24,850 million. Total costs (associated with buying, leasing and/or lending of the required resources) ran to €19,212 million. Thus, the Net Value Added (NVA) was €5,638 million, which is 22.7 per cent of agricultural GVP.

For China, in the same year, total agricultural production generated a GVP of 6,931,980 million RMB ('people's money') – roughly equal at that time to €693,198 million. The costs equalled 2,878,620 million RMB. Thus, the remaining

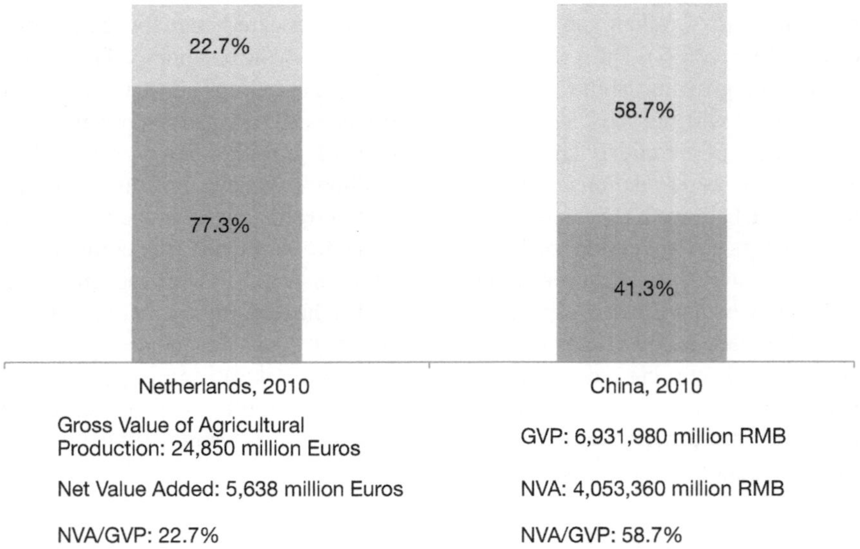

Figure 11.2 Comparing Chinese and Dutch agriculture

NVA reached a level of 4,053,360 million RMB. This is equivalent to 58.7 per cent of the total agricultural GVP.

Consider the difference: 58.7 versus 22.7 per cent. Chinese farmers retain almost three times as much of the GVP of agriculture as their Dutch counterparts. This huge difference points to the *current* relevance of peasant-like agriculture.[14] If China's agriculture were structured like Dutch agriculture, the total NVA from China's agriculture (available to pay the incomes that peasants derive from farming) would not equal 4,053,360 million RMB but just 1,573,564 million RMB. If such a hypothetical shift did occur, this would have dramatic consequences for poverty levels, lead to a contraction in the domestic market, affect the quality of life in the countryside, increase pressure to migrate to the towns and also stymie the process of agricultural growth itself.

In reality the difference is even bigger than the data on the NVA as percentage of GVP suggest. This is because in China most of the NVA can be used as income by peasant households, while in the Netherlands interest payments on credit will take a major slice of the NVA that the farm household actually sees.

The wider significance of China's peasant agriculture

The relevance of China's peasant agriculture is not only located in its significance for Chinese society and the way it contributes to the welfare and well-being of its rural people: it also resides in its contribution to the way we think and theorize about farming generally. The Chinese experience clearly shows that it is possible to involve the whole of the rural population in land reform processes (as opposed to including only a minority and excluding a majority of rural people). The democratization of the access to land does not necessarily provoke stagnation and widespread poverty. On the contrary, the Chinese experience convincingly shows that it is highly possible to realize ongoing growth in both production and productivity (and ensure high levels of technical efficiency) through labour-driven intensification. This is in contrast to the notion that modern agriculture necessarily requires technology-driven intensification which excludes many people. Of equal theoretical and political importance is the finding that the development of peasant agriculture can lead to both agricultural growth and poverty alleviation. The Chinese experience has convincingly proven (over a period that now covers 35 years) that agricultural growth need not necessarily be a selective process that only benefits a few while marginalizing many (and pushing them, in the end, towards the cities).

More specifically, the Chinese experience shows that rural women may very well be a major social force in agriculture and its development. It also proves that labour migration does not necessarily imply a weakening of farming – provided that this outmigration is circular. Temporary outmigration can actually substantially contribute to renewal and/or a 'reinvention' of farming through flows of new experiences, ideas, contacts and resources.

The Chinese experience also indicates that the sense of belonging – and of land tying people together – cannot be considered as a thing from the past that dwindles away with progress. It can be an important cultural value even (or maybe especially) in societies undergoing far- reaching transitions. The same experience seems to also suggest that metropolises may, in the end, generate new processes of re-peasantization that occur both within and outside these metropolises.

Resetting the agenda of agrarian sciences

While post World War II agrarian sciences have done little to help develop a better understanding of the reality, dynamics and requirements of peasant agriculture (to put it euphemistically), the Chinese experience clearly demonstrates that a proper understanding of peasant life and peasant production is a task that cannot be ignored. This applies especially when the performance of Chinese agriculture is compared to that in, say, Africa, the Andes or Europe. Developing such an understanding can never be a one-off exercise. The constant reinvention of farming by peasants themselves requires ongoing elaboration of insights and theories that reflect new realities and which, when needed, helps to strengthen them. This also means that research agendas should be related to the real concerns of peasant producers.

Agrarian sciences have to reset their horizon of relevance and explicitly include the realities, dynamics, needs and requirements of peasants and peasant agriculture. Peasant agriculture should be recognized as an immensely important land–labour institution and as a potentially progressive force that might help to transform our societies. This resetting of agendas needs to explicitly include the hitherto neglected peasantries of the rest of the world.

More specifically it needs to be recognized (following the Chinese, but also other experiences) that societies in which capitalist relations of production are clearly hegemonic can also harbour large economic sectors that are non-capitalist. The latter may be indispensable for society as a whole and may also develop into sources of rebellion and change.

It is argued here that one of the recurrent debates in agrarian sciences – small farms versus big ones – is not such a decisive issue. The question needs to be located within the prevailing societal context. What matters more is *what is done* with the land and, more generally, with the available social and material resources and how agricultural practices relate to and harness living nature. In turn this implies, first, that the political economy of agriculture should explicitly include *material* questions. What are the main resources? How are they used? How are they developed? And, what is the inbuilt time horizon? Second, we have to leave behind the notion of the market as an abstract and self-regulating system that assesses the best possible equilibria in the realms of production, distribution and consumption. Markets need to be seen as concrete places where transactions occur and where products and services are channelled in specific (and mutually contrasting) ways. That is to say, some markets function well, others don't. This means that, in policy, the art of balancing is crucially important. All markets are governed, but the modes of governance differ considerably. In consequence, the outcomes can differ greatly.

A more pleasant world?

The image in Figure 11.3 is a painting by the Chinese artist Ao Song. It illustrates the big dilemma China's agriculture faces: peasant farming or corporate farming? This is an issue that is often framed in highly distorted ways. One of these is the assumption that the two forms might more or less happily coexist, each contributing to the task of 'feeding the world'. But in this painting Ao Song shows very clearly where the water is being pumped. The happy coexistence is an illusion partly because capitalist agriculture as it exists today is built upon, and intrinsically requires, ongoing expansion.

Throughout this book we have argued that peasant agriculture carries an important promise for the world. Ao Song's painting illustrates this perfectly. It shows that peasant agriculture contains a fundamental balance: one that links and unites people and the land. The peasant side of the painting shows richly chequered fields and many farmers. It reminds us of Henri Mendras' beautiful words: 'If the world becomes one huge garden, where we need to cultivate every corner to produce enough food [. . .] then it is time that people become again gardeners.' (Mendras, 1976:213; our translation). That is indeed what Chinese peasants are: gardeners who care very well for the land and are able to realize very high levels of production in tiny plots – because that is what gardeners do.

The notion of gardening and gardeners is not alien to the debates about China's agriculture. As early as 1974, Sterling Wortman, who was heading a delegation of USA plant scientists that visited five different communes, observed that the crops

Figure 11.3 Where to go?
Source: reproduced with permission of Ao Song

looked very healthy and the productivity was uniformly good: 'in a sense there is no farming in China – even the grain crops are "*gardened*".' (Wortman, 1974; italic added). He concluded: '(o)ne does not sense the hopelessness evident in some other countries. Rather, there is a hope for a better future.' (ibid.).

On the other side of the river that separates these two worlds in the painting the fields are empty (apart from the machines). The contrast demonstrates that a world with peasants is certainly a more pleasant place than a world without them. For peasants do not only produce food, they produce a pleasant world as well. They make a 'green and pleasant world' (Newby, 1980).

Of course, this side of the river (i.e. peasant agriculture) is also affected and conditioned by the capital relations that affect society as a whole. It is not located outside this world – it is not the 'pure place' without distortions or problems that some people dream of (see e.g. Goodman, 2004; Guthman, 2008). As part of this world, 'this side' is as much affected by capitalism as it is a response to it. As such it is a partial alternative and a cradle of resistance. And that makes it significantly different from the 'other side'.

Exceptional, but not from another planet

The phenomena of smallholdings, peasant strategies, multiple job holding, multifunctionality and looking for new, interesting markets (if not actively constructing them) are not exclusive to China – they can be encountered and have significance in many parts of the world, as illustrated below.

Smallholdings

Take the notion of a smallholding. According to the accepted definitions there are extensive smallholding sectors everywhere. In the USA, Poland and the Netherlands. In Russia and Ukraine. In Peru, Colombia, Bolivia, Venezuela, Brazil and Argentina. There are also large smallholding constellations in South Africa and Morocco (and in all the countries in-between). Smallholdings are, in the first place, small when compared to larger holdings. Second, they are usually too small to render an adequate income. It is assumed that these two defining characteristics overlap or are even identical. Having a relatively small acreage, or economic size, implies that the income will be too low, and deriving a low income from the farm means that it is de facto impossible to enlarge it. Thus smallholders are entrapped – at least that seems to be the case within this particular view. In practice this view is strongly related to agricultural policies that are supposed to set smallholders free from the assumed misery they are caught up in, by moving them out of agriculture. If not voluntarily, then with a few well-meant nudges.

A smallholding is a relational concept. The size of the average farm (and thus the ceiling that defines the smallholding) depends on demography, the amount of available land, ecology, history and town–countryside relations. The existing land–labour institutions, technological development and markets also make their imprint on the size of the average farm and on the incomes that can be derived from it. But what is done on the smallholding is also crucially important: how are the available resources combined and used? Having a one hectare holding dedicated to dairy farming implies a smallholding (unless feed and fodder are bought or the commons are actively used). But one hectare with vegetables (or herbs) can often represent a respectable economic operation capable of generating an adequate income.

Whether or not a farm is a smallholding depends on a range of contextual factors (many of which have been referred to above). It is curious, though, that schemes of classification rarely put such contextual factors centre stage (although constructing such a scheme is far from impossible, as demonstrated in HLPE, 2013:41–4). In the existing schemes of classification it is the *intrinsic properties of the farm* that define whether or not it is a smallholding. Acreage, economic size, number of animals and/or the value of capital goods are the most commonly used yardsticks. This makes the concept of smallholding into an ideological construct par excellence. Acreage and economic size (and for that matter other forms of operationalization as well) do little to inform us about the position of the farm vis-a-vis other (larger) farms, nor about the level of income. This is because they do not take into account farmers' responses. As suggested above, a 'small' one hectare farm might be changed into a 'large' one hectare farm by changing the cultivation scheme. Similarly, a small economic size might be countered by reducing costs and/or obtaining better prices for the produce.[15] It is important to note that we are not talking about random deviations. These are systematic differences that arise because peasant strategies always centre on how to make the most out of a smallholding.

If a smallholder runs up against the limits of making the best out of a small farm, s/he can also shift to employing additional repertoires, such as pluriactivity or multifunctionality. Many people actively choose such a combination, partly because it seems to combine the best of two different worlds. Then new balances start to govern the development of the farm.

In terms of the dynamism of the agricultural sector smallholdings cannot be considered as inevitably doomed to disappear (thus making the previously 'medium' farms the next category of 'small' ones). There definitely is no single logic, no unique pattern, that implies that large farms will get larger and small farms will disappear. Of course, some smallholdings are being closed down (for whatever reason). But at the same time new smallholdings are emerging. Equally some large holdings might be discontinued (again for different reasons) and give rise to the birth of new smallholdings. Many farms now classified as large farms grew out of smallholdings.

There are also significant periods when major reversals occur. This is especially the case during and after economic crises when larger farms are more likely to collapse while smaller ones are more able to survive. This happened in the aftermath of the 2008 crisis: many large-scale, intensive, high-tech farms had a negative cash flow (especially when involved in ongoing expansion), while smallholdings were able to continue and sometimes hardly felt the cold winds of the crisis (see Oostindie *et al.*, 2013; Broekhuizen *et al.*, 2013).

In the USA, farm size is defined by an economic criterion: gross product. Small farms are those with a gross product (or total sales) of less than 250,000 US dollars.[16] According to the 2007 Agricultural Census 1,995,133 farms fell into this category. This equalled 91 per cent of all American farms. Between 2002 and 2008 the number of small farms rose by 118,000.

In the Netherlands there were 67,481 farms (discounting small, so-called 'hobby farms') in 2013. Using standardized output (SO) as the criterion, 10,410 of these could be classified as large (17 per cent). They had a SO of more than €500,000. Farms with an SO of between 3,000 and 25,000 €/year were considered as small. There were 17,845 of these (or 26 per cent of the total).

In the European Union (EU) as a whole there are nearly 12 million farms. Sixty-seven per cent of these have less than 5 hectares. Policy makers are giving increasing attention to small farms (Davidova *et al.*, 2013).

In Brazil the notion of a smallholding is strongly linked with that of family farming. Family farming is legally defined: being less than four times the size of a '*módulo fiscal*' is one of the criteria (this refers to economic size). The other criteria are mostly using family labour, living mainly on the farm and the farm being managed by the owner and his/her family. Together this definition delineates small family farms from the large, corporate holdings. Brazil, like many other Latin American countries, has a dual farm structure: large *latifundia* and small *minifundia*. Between 1996 and 2006 the number of small family farms grew by some 400,000 units, due both to land occupations headed by the *Movimento dos Sem Terra* (MST) and parallel land reform processes.

For the world as a whole it can be argued that at least 72.6 per cent of all farms belong to the category of smallholding (with less than 1 hectare).[17] In synthesis: smallholdings are everywhere. But are they relevant? Do they matter? What do they contribute to the economy and society? Is smallholder agriculture a relevant phenomenon at all?

The answer to all these questions is affirmative. If smallholders have access to the space they need (Halamska, 2004), that is if conditions are not too adverse, smallholder agriculture is productive, sustainable and efficient. It generates considerable employment, strengthens the wider rural economy,[18] helps to sustain and enlarge an internal market for wage goods[19] and is an important pillar for food and nutrition security. We previously referred to small farms in the USA. According to the US Commission on Small Farms 'the small farm [is] the cornerstone of our agricultural and rural economy'. The Commission also felt 'that a sustainable rural renaissance can be anchored in a vibrant, dynamic, small farm sector' (USDA, 1998). The experience of Brazil is equally revealing. While small family farms cover only 24.3 per cent of the total agricultural area, they generate 38 per cent of the total value of production and 74 per cent of all agricultural employment. They also supply a high percentage of the produce domestically consumed in Brazil: 58 per cent of all milk, 50 per cent and 59 per cent respectively of chicken and pork, 46 per cent of maize, 70 per cent of beans and 87 per cent of cassava. These high shares emerge also because the *latifundia* have a strong export focus – but even so they underline the strategic contribution of the smallholder sector to food and nutrition security.

Peasant strategies

While the notion of a smallholding refers to the unit of production and its relative size, the concept of peasant strategy refers to the way in which this unit is managed and developed. Most smallholdings are organized and developed in a peasant-like way, but peasant agriculture is not just limited to smallholdings. And just as smallholdings are far from limited to China (albeit a smallholding nation par excellence), peasant strategies are also not just limited to China. All over the world, in the Global North as much as in the Global South, peasant strategies applied to agricultural units of production – especially, but not exclusively, the smaller ones – mean that these holdings are able to face, and to respond to, global markets and the squeeze and volatility they bring.

A decisive feature of peasant strategies is that the logic of the market is held at considerable distance from the decisions that govern the organization and development of the farm. Of course, if there are attractive opportunities (new profitable crops or other worms to be picked by early birds), the peasants will be there. They are not stupid. Operating as a peasant is about building an autonomous resource base and improving technical efficiency (i.e. producing more with the given resources) in order to improve the well-being of the household and to lay, eventually, a foundation for the next generation. Peasant farming is about the aspiration for emancipation – it is the materialization of this aspiration.

We are not talking about the aspirations of those who already have access to (a lot of) capital and who are engaged in capitalist relations of production (and exploiting the work of others). We are talking about the aspirations of the have-nots of this world. Even seemingly well-established farmers face the threat of becoming have-nots, as markets deteriorate, credit needs to be suddenly repaid or new quality standards for produce (perhaps involving heavy investment) are formulated. Peasants engage in co-production (in the conversion of living nature into food) in order to fulfil their aspirations and they do this by governing the process of production with a precise set of Chayanovian balances (Ploeg, 2013). In this respect they fundamentally differ from agricultural entrepreneurs who tend to operate as brokers: they contract land, machine services, credit, seeds, inputs, etc. and then sell the produce. Such operations are governed by the market, production is organized, developed and, if need be, terminated according to the logic of the market. Here agricultural production is a *financial operation*. Ironically, in today's markets, agricultural production organized as a financial operation is increasingly difficult to sustain. It is precisely here that peasant farming has so much relevance: it secures the continuation of food production in a hostile environment. And that is also why peasant strategies are not limited to particular corners of the world but are widespread. They can be found everywhere – simply because, from an institutional point of view, it is the best way to organize agricultural production. It is the land–labour institution that can even withstand 'modern times'.

Pluriactivity, circularity and multifunctionality

Re-reading the accounts that describe the massive processes of migration in Europe that occurred in the post World War II period (from the Mezzogiorno in Italy to the north of Italy and later to Germany, Belgium and the Netherlands; from Tras-os-Montes in northern Portugal to Switzerland and France; from Krakow in Southern Poland to Chicago in the USA; etc., etc.), one can be struck by an aspect hardly noticed initially. This is that many of the male labour migrants (women often stayed home and took care of the farm) returned (sooner or later, but mostly later) to their villages and invested their savings in (re)constructing their houses, establishing new rural enterprises and/or in their former farm enterprises. There was a circularity – not as much as in China (see Chapter 2) but even this limited circularity had notable and positive effects on agriculture and the rural economy. Nonetheless, the dominant storyline neglects this circularity and its effects. Progress was said to emerge from the state-controlled process of modernization.

A similar obfuscation occurred, and still occurs, when it comes to pluriactivity and multifunctionality. Take the Netherlands: on some 80 per cent of all farms the man or the wife (or both) has an extra-agricultural job. Depending on the economic circumstances this provides, on average, 40–60 per cent of the total family income. Without pluriactivity much agriculture would be impossible in the Netherlands. The same applies to multifunctionality which is now an important source of additional income in many households (depending on the definition up to 50 per cent of all farming families are involved in multifunctional activities). Because of these 'additional' activities they can continue with agricultural production. Nonetheless, in the dominant discourse it is the markets, investments in scale enlargement, and new technologies that 'move' farming forward.

There is in the Netherlands, as elsewhere, a small segment of large-scale, specialized farms that do not depend on pluriactivity and multifunctionality. Whether they fare well is another question – especially when the volatility of current markets is taken into account. The point, however, is that the dominant vision and views define a horizon of relevance that only includes this segment and which therefore misses a range of other, highly relevant mechanisms. 'Modern times' do not want to acknowledge such mechanisms and this is also true of policy makers.

Rural and agricultural policies

Markets and policies are increasingly represented as presenting a dichotomy. This applies a fortiori to markets for food and agricultural products and to rural and agricultural policies. More market is assumed to imply less policy and vice versa. If one is more efficient, it is assumed that the other is less efficient. And if one side of the equation represents order, the other side is viewed as a distortion. In short, markets and policies increasingly figure as polar opposites.

In this respect China seems to represent, at least at first sight, an exception (or, as many economists see it, the deviation from the rule). Markets are politically shaped and embedded in often complex sets of rules, many of which are defined and actively implemented by the state. Markets are connected or separated from each other through institutional arrangements in which the state has an important say. This applies to the interrelations between local markets, regional markets, newly emerging nested markets, the national market, export markets and the world market. There is an explicit policy of *sannong* because it is believed that food security, peasant incomes and the quality of rural life cannot be left to the vagaries and dull compulsion of the markets. Simultaneously there is reflection on the evolution of markets, the impact of policies and the intertwinement and interactions between the two. Such reflection brings, and is informed by, widespread experimentation. Lower levels of government are allowed to implement and test alternative policy designs. Then new policy approaches result in the construction of different markets, with different actors operating in them, with different results. This is convincingly shown by Donaldson in *Small Works* (2011). In consequence rural and agricultural policies are continuously evolving. They are moved forward by and through a complex and richly chequered whole of frictions, contradictions, objectives, experiences, lessons learned, mistakes, misinformation and, sometimes, completely biased images. It is indeed 'crossing the river by feeling the stones'.[20]

As argued above, China increasingly looks like an anomaly. While the neo-liberal project seems likely to eliminate rural and agricultural policies (the demolition of the EU's Common Agricultural Policy is a point in case), China's case represents the opposite. Subsidies on grain and other important commodities and market regulation through quotas and the like are disappearing in the EU; they are, however, being actively introduced in China – often in new and creative ways.

In reality, of course, China is far from being an exception. Intelligent and goal driven agricultural policies have been at the roots of the post World War II take-off in Japan, South Korea and Taiwan. They also helped many European countries to go beyond the disastrous crisis of the 1930s and build prosperous and booming agricultural sectors in the post-war era that turned Europe from a net importer of food into a continent that is more than self-provisioning (with one major exception which also applies in China: soy). These historically delivered agricultural and rural policies are now being demolished – but two observations are needed here to put things in context. First, the governance of food and agricultural markets is not disappearing. It is *shifting* from states and supra-national bodies towards food empires – the extended, corporate networks that increasingly control the food and agricultural markets. They reorder markets to bring them in line with their interests and prospects. Second, many states and supra-national bodies such as the European Commission are disguising their agricultural and rural policies. Energy policy, spatial policy, food safety policy, health policy, environmental policies, policies for protecting landscapes, natural values and biodiversity, cohesion policies and more are now becoming the main levers to gain (or regain) at least some grip on agricultural production and the countryside and to resist dominant corporate interests. The balance of contrasting socio-political blocks is often decisive here.

Perhaps China is better off because policies are still called policies. And they are *structured* as policies – not disguised but explicit. Who knows? But one thing is certain: it is ludicrous to refer to China as the last Moloch, practising something as terribly outdated as rural and agricultural policies.

By way of conclusion

Features that appear at first sight very much Chinese (if not almost exclusive to China) are more widespread than is normally assumed. Smallholdings, for instance, can be encountered almost everywhere in the world and are far from irrelevant. Peasant strategies are omnipresent and probably far more important than most people (and, ironically, especially the experts) believe. Pluriactivity and multifunctionality can increasingly be encountered in agrarian landscapes everywhere – not as exit strategies (as the same experts argue) but as mechanisms that clearly promote resilience and continuity. Circularity in rural–urban relations is not limited to China but has been important in many regions. The same applies to rural and agricultural policies: governance is needed everywhere, for better or for worse. What matters is *who* governs, in which *way*, with what *objectives* and with what *effects*.

The conundrum of China's agriculture can probably be explained by the fact that these features systematically coincide, intertwine and mutually reinforce each other. What is kept separated elsewhere and treated as something that is to be surmounted as soon as possible is – in China – an undeniable fact of life, tied together into a knot that nobody can untie. There is no master plan underlying

China's agriculture – it is the outcome of the particular way that history has evolved. Yet this does not exclude the possibility of searching for an implicit paradigm that lies within it. This will be the aim of the concluding chapter.

Notes

1 This chapter is partly based on a seminar text entitled 'Strengths and Weaknesses of the Chinese Peasantry: A Comparative Perspective', presented on 24 October 2013 at the College of Humanities and Development Studies, China Agricultural University, Beijing. The chapter also uses fragments of *Peasants and the Art of Farming* (Ploeg, 2013) and a report that the author co-authored (HLPE, 2013).
2 The 'subterranean' nature was even reflected in the terms used to describe the newly emerging patterns. Initially the redistribution of the land and the restoration of an individual share in total production was referred to as *da baogan* which means something like 'big contracting to you'.
3 Production quotas were dismantled, while markets and the required new support structure were constructed (see Ye *et al.*, 2010 for a succinct description of this process).
4 In fact the *ownership* of the land stayed with the production teams; the peasant households obtained the usufruct rights.
5 The title of Thiesenhuisen's book (1995) about recent land reforms is revealing – *Broken Promises.*
6 The foundation for this had already been laid before, during and immediately after the foundation of the People's Republic of China in 1948.
7 This 'saving quota' represents a certain proportion of value that was supposed to be shifted, through the banking circuit, to the urban economy. In the Peruvian land reform of 1969 onwards, for instance, this was set at 25 per cent of the calculated income of a farming household.
8 It would be pointless, of course, the compare the absolute size of economic holdings between different countries or different epochs. However, the economic holding expressed as percentage of the average man–land ratio is relational and therefore can be compared. In countries where land reform has gone on for several decades (as in Mexico and India, for example) an ongoing increase in the economic holding can be observed as time passes.
9 An extensive discussion of this data plus all the relevant references can be found in Ploeg, 1974 and Ploeg, 2006.
10 This case is synthesized in Ploeg, 2008, Chapter 3.
11 Ironically, the other radical exceptions are Japan, Taiwan and Korea where, after World War II, the Americans induced and implemented radical reforms that established economic holdings that were identical to the average man–land ratio. This had far reaching and positive effects for the development processes that followed (see Henley and van Donge, 2013).
12 It was not the outcome of a grand design (of 'seeing like a state') as was the case in many of the other land reform processes referred to above.
13 Per capita incomes grew even more (Netting, 1993:252)
14 I am well aware that many other experts will interpret this data in a different way arguing that the relatively higher cost levels of Dutch agriculture (and consequently, the lower NVA–GVP ratio) simply reflects the high technological development and associated intensity of Dutch agriculture. In comparison, China's data would be a reflection of a low level of technological development (less use of embodied capital goods, such as mechanization and automation and a lower level of input use). With further development, they would argue, China will converge with the Netherlands: NVA as percentage of GVP will 'inevitably' be reduced. Evidently, this reasoning is flawed.

The yields realized in China's agriculture are (all circumstances taken into account) as high if not higher than in Dutch agriculture. China's agriculture cannot be defined as underdeveloped or backward. The point is that it has taken (partly due to the introduction of the HRS) a *different* developmental trajectory. Its technology differs from that in use in north-west Europe –which logically follows from the relative factor prices as Hayami and Ruttan (1985) would argue. China and the Netherlands cannot be ordered in terms of inferiority and superiority, nor in terms of vanguard and backward positions (as if development were a unilinear process). Both China and the Netherlands have very well developed agricultural systems, but they are different. They have followed different routes and continue to do so. A typical difference is that in China agricultural intensification is basically labour-driven, while in the Netherlands it is technology-driven.

15 Economic size is calculated on the basis of the average gross value added that might be realized with specific animals and crops (thus different animals and crops can be added together to give an aggregate economic size of the farm). However, smallholders typically have lower costs than those of the average farm, let alone those of the larger farms. Equally, a smallholder might be engaged in direct marketing, thus getting considerably better prices than that achieved for produce that is sold through conventional channels. In short, measures of economic size are systematically biased against smallholdings.

16 The vast majority of these had a gross annual product of less than 10,000 US dollars.

17 This is based on an eighty-one country subset of Food and Agricultural Organization World Programme for the Census of Agriculture (FAO-WCA) that covers two-thirds of the world's total population. This admittedly is an inadequate data set, but the only one currently available at the world level.

18 It is 'the engine of rural non-farm growth' (Haggblade *et al.*, 2007; see also J. Zhang *et al.*, 2006 and Mohapatra *et al.*, 2006).

19 So-called 'consumption linkages' are weak in estate-led agricultural growth but stronger in smallholder agriculture (see Haggblade and Hazell, 1989; Janvry and Sadoulet, 1993).

20 This expression is from Deng Xiao Ping. It was a call for pragmatism. Later on Ye *et al.* (2010) used this expression to title their synthesis of 60 years of rural and agricultural policies in China.

12 Putting the jigsaw together

A return to paradigmatic issues[1]

Jan Douwe van der Ploeg and Ye Jingzhong

Debates about the peasantry easily, and far too frequently, degrade into dichotomies. The existence of a peasantry is either recognized or bluntly denied: one is either for, or against, peasants.[2] The same is true of the notion of modernization – often understood as the universal process that necessarily heralds the 'death of the peasantry' (Hobsbawn, 1994:289). According to different ideologies, modernization either turns peasants into capitalists and proletarians or into agricultural entrepreneurs and urban workers.[3] Here again, people, institutions, and social movements take different stances: they are either in favour of, or against, modernization.[4]

Similar positions can be encountered in the discourses concerning China. Depending on the epoch and political orientation some discourses centre on the peasantry as a progressive force that positively contributes to the development of the nation, while other discourses view it as the main hindrance to change (Schneider, 2014). Equally, people are in favour of or against the main elements of modernization that are being promulgated and implemented in China today. This has led to some raising the question of whether there was (or could be) a 'Chinese' modernization that differed fundamentally from the 'Western' one (Day, 2013:81).

Ironically, these firm and dichotomized positions mostly lead us away from the main lesson to be drawn from contemporary history: if there is a strong peasantry modernization is possible but not needed, but if the peasantry is weak, modernization is impossible. And when modernization is imposed from outside, it creates a disaster. This becomes clear when one examines the rural exodus (a key element of modernization).[5]

Redundancy or tragedy

When demographic growth slows down, total social wealth increases, its distribution becomes more equal, urban economies generate more employment and there are more possibilities to travel; people will surely then move from the countryside to the cities, the number of farms may decrease and an enlargement of the remaining ones may occur. This is one way that modern history can go.[6] And whenever and wherever it goes this way there is no need to artificially stimulate this process. People themselves will decide to move or else the next generation (or at least part of it) will do so. Trying to hasten this process will only generate frictions. The only useful thing that can be done is to foster *additional* conditions for this process to occur smoothly.[7]

If, however, one or more of the conditions outlined above is lacking (for example when there is generalized poverty, a lack of urban employment, high demographic growth, large spatial and social inequality, and/or an overcrowding of the cities) it is downright counterproductive to try to induce and/or accelerate a rural exodus. It will not only generate all kinds of frictions, but will sooner or later bite its own tail, turning development into stagnation and regression.

The main problems occur when the state adopts the modernization of agriculture as a politico-economic project – or what Scott (1998) calls a megaproject. Such a project has often been adopted in order to drastically reorganize the agricultural processes of production through the introduction of new technologies, new organizational models and a far-reaching repatterning of the relations between agriculture and the markets. Thus, the project implies a drastic reduction of employment levels in agriculture, the enlargement of farms and, as a consequence, an outflow of people from agriculture and rural areas. As Ye and LeGates put it: 'labour is to move out of agriculture and capital (in the form of increased investments, new technologies, etc.) is to move in.' (Ye and LeGates, 2013:230). This modernization is reflected in what has come to be known as the 'Western paradigm' and modernization can now, in hindsight, be seen and understood as the operationalization of this paradigm. In those places where modernization occurred for the first time, it has turned out, at least initially, to be relatively successful – according, that is, to its own standards. However, in precisely those places where it apparently succeeded (such as Europe and the USA) it was not really needed.[8] Since the required conditions (as outlined above) were present, a shift of people from the countryside to the towns, an enlargement of the remaining farms and a substantial increase in agricultural production would have occurred anyway (albeit probably with a different rhythm and contrasting results). As a matter of fact such a shift was already taking place – at least partially. But wherever these conditions are only present to some extent, or completely absent (as is the case in large parts of the Global South) such modernization is doomed to fail. It only contributes to an increase in poverty, a stagnation of agricultural production, a contraction of internal markets and increased dependency.

And that is precisely what has occurred: in the Global North the modernization of agriculture was largely redundant. It was not really necessary. To a large degree modernization was a farce.[9] In the Global South, though, it became a tragedy. This has been noted and widely discussed in China (Wen Tiejun, 2004) and contributed to the formulation of the, alternative, *sannong* policy (Day, 2013:103–4).

In a way, modernization in the Global North came down to a reorganization of time. Trends that would have occurred anyway were brought forward. Figure 12.1 helps to clarify this point. It shows, in the first place, the *tendence lourde* (the long-term trend): the slow but persistent trend of increasing yields (shown here with levels of milk production per milking cow per year for the Netherlands from 1950 onwards). We can take these steady but persistent increases in milk yields as emblematic of agricultural development as a whole.[10] Breeding a more productive herd used to be, as was agricultural growth in general, a labour-driven process. The quantity, and especially the quality, of peasant labour were decisive in this respect. Thus, increases in milk yields are a perfect *pars pro toto*: they are due to the improved craftsmanship of dairy farmers, their growing knowledge of breeding, better care of the animals (including improved grassland production and cattle feeding), more effective organizations for artificial insemination and so on. The dotted line in Figure 12.1 gives a graphic synthesis of how cattle breeding was modernized: *as a temporary jump*. This occurred through the introduction of Holstein cattle, substantial increases in the use of concentrates for cattle feeding, new stable layouts, etc. This translated into an initially rapid and seemingly impressive acceleration of increases in milk yields. However, as suggested by the dotted line in Figure 12.1, these increases later slowed down and eventually even stagnated.[11] After a few decades the line of classical breeding once again overtook the newly set levels that resulted from modernization.

It could be argued that if (but only if) there had been a scarcity of milk and dairy products at the end of the 1960s, the organization of such a 'leap forward' would have been justified – the more so if such scarcity were accompanied by human malnutrition. Ironically, the opposite was the case. At that time Europe

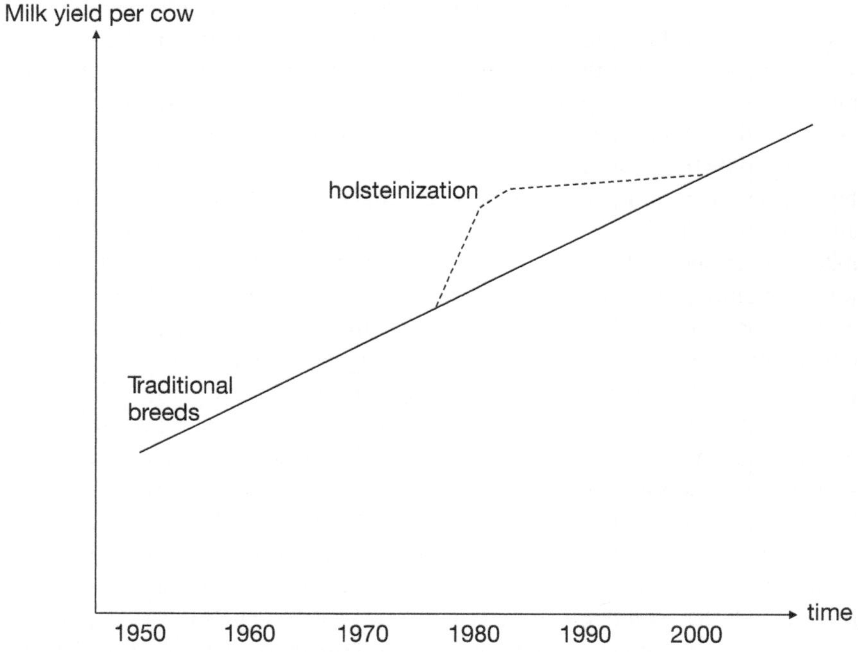

Figure 12.1 The modernization of dairy farming

Source: our own elaboration

wrestled with 'butter mountains' and other expressions of overproduction. But other interests were also involved. Having an overproduction of milk helped European dairy industries to conquer new export markets (especially in the Global South where local dairy producers were outcompeted en masse). Alongside this, modernization also represented an adieu to the way in which cattle breeding was organized in peasant communities. Breeding shifted to specialized expert-systems and associated enterprises. Much peasant labour and knowledge became superfluous and was lost.

When Holstein cattle were first introduced from the USA into Europe (from the late 1960s onwards) this represented the possibility of an impressive leap in milk yields. Additional changes, such as the shortening of time horizons in selection, a massive increase in the use of concentrates and other elements further strengthened and sustained this leap (see Ploeg, 1987 for a detailed analysis). As a consequence, local breeds such as Friesian, Brown Swiss, *Rossa Reggiana* and many others seemed to lag behind, leading many dairy farmers to change to the new Holstein breed. Nevertheless, after two to three decades a slowdown followed and the sturdy continuation of 'old fashioned breeding' (done by farmers with their traditional breeds) generated yields comparable to those of the Holstein breed.[12] In this respect the modernization of agriculture essentially comes down to the anticipation and acceleration of future developments. It *brings forward* results that probably would have occurred anyway. The results were impressive (milk yields at the time skyrocketed). However, it was also redundant. Nobody needed more milk, there was already a surplus in Europe.[13] And these accelerated changes came at a price, both at the macro and the micro levels.[14]

That said, the proven capacity to shift future potential to the present greatly enhanced the fetishism that currently characterizes mainstream agrarian sciences. Improved varieties, new breeds, new technologies, new organizational models are now seen as the panacea for each and every problem. This fetishism typically excludes any critical inquiry into the contextual situation in which these new solutions will be applied. New technologies are seen as by definition as being superior and therefore as representing progress per se.

The technological fetishism of today's agrarian sciences possibly helps to explain the authoritarian character of most programmes for the modernization of agriculture. If particular *things* are superior – such as high yielding varieties (HYVs), Holstein cattle, cubicle sheds, large-scale farms, etc. – they simply must be applied. If there is opposition, it is written off as 'resistance to progress'. Thus, modernization frequently translates into highly coercive systems that aim to impose supposedly needed changes, irrespective of context. Many ministries of agriculture indeed became a state-within-the-state (Visser, 2010).

In the Global South the modernization of agriculture has so far been mainly a tragedy. Over the decades, the packages of new HYVs, increased use of inputs (notably fertilizers, pesticides and herbicides), mechanization and credit were introduced with a special focus on the areas and farms offering the best conditions. However, in contrast to Europe and the USA, most developing countries could not afford the expensive schemes to regulate markets and to sustain off-farm prices

(Abramovay, 1992). Thus, at the farm level it proved to be difficult to sustain the transformation of production, especially given the higher costs involved. A few bad harvests or some years with low prices undermined the continuity of the newly introduced practices and indebtedness triggered a downward spiral. Later on, Structural Adjustment Programmes (meant to reduce public spending) further eroded the basis for a seamless application and continuation of 'modernized farming'. As a consequence, peasant households started to look elsewhere for (parts of their) income, labour input into farming was reduced and as a result yields started to decline. Credit schemes added another set of such downward pressures. In order to avoid risks, larger areas were sown and yields per hectare went down. Thus the potential benefits of the new packages were undermined. In synthesis: after a certain point in time, growth rates declined and the rural exodus gained momentum. The effects of the latter have been dramatic. The cities and metropolises of developing countries did not offer sufficient employment opportunities to those who left the countryside and a 'planet of slums' was created (Davis, 2006).[15]

This tragedy is still with us. While the urban sectors cannot possibly absorb the youth from the countryside (see Losch, 2013 on Africa), patterns of agricultural development still hinge around the assumption of the need for, and desirability of, greatly reducing labour input.

Within this tragedy another one is unfolding. This relates to the role that agrarian sciences play. Poverty in the countryside and stagnation of agricultural production are understood as a sign of traditional agriculture having reached its technical ceiling (Schultz, 1964), just as malnutrition and hunger in the cities are understood, in the dominant view, as a result of inadequate food production. Wherever, and whenever, these phenomena emerge, the proposed solution is the same: productivity needs to be increased through yield increases. Thus, with the best of all possible intentions new technologies (e.g. new seeds with the associated packages of inputs, machines and equipment) that carry the possibility of increasing the productivity of both land and labour are designed. This search centres on finding the so-called 'optimal solutions'; the ones that render the highest yields under what are called 'optimal conditions' (flat fertile lands, no stones, sufficient availability of water, energy, credit facilities, etc.). It is assumed that even under 'sub-optimal conditions' these optimal solutions will still give the best results. However, once released, the new technological packages will be applied only, or especially, by the better-off producers operating under relatively good conditions. They gain a relative advantage while producers unable or hesitant to apply the new technologies suffer a reduction in their competitiveness and are marginalized. As a result, production increases on some farms, but on others it decreases. The aggregate effect might well be insignificant.

The regional effects can vary significantly: some areas (especially deltas and irrigation schemes) become growth poles, others become declining *hinterlands*. This is the tragedy of agrarian sciences: what is promised on the one hand systematically slips away on the other. This process explains why at least five decades of massive development interventions in Africa that have tried – again and again – to develop agriculture have only resulted in reproducing and relocating stagnation. But things can be different as shown in Chapter 10 where we described a long term Chinese agricultural research and development (R&D) project that improved *Camellia olifeira*: a marginal plant for marginal areas.[16] In the Western paradigm this would have been an irrational choice.[17] In the Chinese approach to agriculture, though, it was a strategic and self-evident option.

The transfer of the modernization project from the Global North to the Global South has been legitimized in a twofold way. In many parts of the Global North it has been apparently successful – and in most places of the Global South there seems to be no alternative. It is precisely at this point where the need for an alternative paradigm emerges. Could modern agrarian history in the Global North have proceeded in a way that differed from the path implied by the modernization project? And, more significantly, are there alternatives to the repetition of the modernization trajectory in the Global South?

These two questions are not purely speculative. Nor are they disconnected. It should be noted that agricultural development in the Global North was never a singular and uniform process – instead, it has been, and is still, a highly differentiated process. One sub-process has been fuelled and structured by the mega-project of the state – and, in the case of the European Union (EU), the supranational state – but several other highly significant sub-processes occurred as actively constructed *responses* to modernization. Examples of these include low external input agriculture and multifunctional farming – both of which are blossoming throughout Europe and both of which present real life alternatives to modernization driven by agrarian science and policy. The problem, though, is that those who view the world through the prism of the Western paradigm are barely able to perceive these contrasting sub-processes and, even if they can, are always ready to write these off as 'exceptions to the rule': as anomalies, devoid of any theoretical or socio-political relevance.

Are there alternatives in the Global South? Are there alternatives to a robotic repetition of the modernization of agriculture that occurred in the North? Here again the problematic side of the Western paradigm comes to the fore. It provokes, in an a priori and seemingly convincing way, a negative answer – a definitive no – to the question. It excludes curiosity, it blocks experiments, it prevents possibilities of endogenous growth and development and it imposes harmful forms of institutionalized ignorance.

This is why it is so important to critically reflect on the Western paradigm and to try to elaborate – from the grassroots level upwards – paradigmatic alternatives.

Lessons to be learned from Chinese agriculture

China's agriculture is not perfect nor should it be held up as sacrosanct. There are problematic sides to it and it is not, and cannot be, a model to be followed elsewhere (if only for the exceptionality of its man–land ratio – see Chapter 11). Nonetheless it contains features of particular importance.

Above all, China's agriculture and its recent development shows that food production and provisioning can be patterned in ways that differ radically from the iron logic suggested by the Western paradigm. It shows that there is not one single and linear way to develop farming and food provisioning – there is, instead, a *plurality* of alternative possibilities.

As argued throughout this book, the Chinese experience convincingly highlights a number of such possibilities and, at the same time, shows their feasibility. These possibilities are summarized in the following list.

1) It is possible to combine agricultural growth and poverty alleviation (as opposed to making agricultural growth a selective process that impoverishes many and benefits just a few).
2) It is possible to involve the whole of the rural population in transitional processes (land reforms included) that result in increased production, higher incomes and a more attractive countryside (as opposed to moving large segments of the population out of agriculture and turning the countryside into a featureless monoculture).
3) Labour-driven intensification can achieve productive results that are comparable, if not superior, to those resulting from technology-driven intensification.
4) Labour-driven intensification also results in less waste and is far more sustainable in terms of resource use.
5) Women can play a major role in continuing and developing agricultural production.
6) Rural outmigration does not necessarily imply a weakening of farming, provided that it is embedded in circular patterns that link rural and urban economies.
7) Multifunctionality is not just a 'luxury' that only rich countries can afford. In developing countries it can be a powerful vehicle for strengthening peasant and smallholder agriculture.
8) Societies in which capitalist relations of production are hegemonic can also have significant sectors of the economy that are non-capitalist. These sectors can be more productive and more attractive than they would be if capitalist relations prevailed.
9) Although the sense of belonging might be at odds with the notion of globalization, it can also represent a powerful productive force. This applies especially to the relation between 'man' and land.
10) When metropolises become too large they generate processes of re-peasantization, both within and outside the metropolis.
11) Big metropolises do not need large-scale farms for their food provisioning. They can also be fed by peasants and smallholders, provided there is an adequate marketing structure in place.
12) In this respect, the presence of market places embedded in an appropriate socio-material infrastructure is preferable to market dominance by value chains.
13) The emancipatory aspirations of peasant families are the main driver of sustained agricultural growth and of food security.

14) Peasant agriculture is not stagnant: it is constantly renewing itself.
15) More specifically, peasant agriculture does not lose its momentum when it meets 'technical ceilings'. On the contrary, it continuously moves the 'frontier function' forward.
16) The question of small or big farms is not the decisive issue. The overriding questions are what is done with the land, by whom and for whom, and who or what is in control. Equally important is the question of what is done with the social wealth produced by agriculture.
17) Through labour intensive and sustainable methods smallholdings can realize high levels of production and productivity and, if the town–countryside relations are favourable, they can also generate acceptable incomes.
18) Hence, insisting that particular farms are 'too small' is nothing but an ideological statement – it critically ignores the malleability of both agriculture and the relations between agriculture and the rest of the economy.
19) Recent Chinese history shows that an initial 'urban bias' (heavy taxes on farmers, low farm gate prices, etc. that resulted in an overall flow of capital from the countryside to the cities) can be converted into a policy that supports farming.
20) Modernization does not necessarily imply a rigid spatial and institutional separation between agriculture, industry and the provision of services. The three can go together as demonstrated, first, by the Township and Village Enterprises (TVEs) created in the 1980s; second, (albeit in a different way), by the massive process of labour migration that started in the 1990s; and third, by the ongoing reinvention of peasant farming (see Chapter 7). One of the consequences is that the issue of farm size cannot be framed in the same way as in the West.
21) When there is too much surplus labour in agriculture, this problem can be resolved at the level of the *rural* economy (i.e. the shifting of labour to the *urban* economy, which is presented in the Western paradigm as a structural necessity, can very well be avoided).

There are further specific arguments that relate to the agricultural sciences:

1) Trying to understand the logic of peasant life and peasant production is a task that cannot be neglected or ignored (even if there are strong pressures, especially in academia, to do so).
2) Peasant farming is continuously reinventing itself – this gives even more grounds for ongoing research into it.
3) Since World War II science has done little to help understand or strengthen peasant agriculture. It has mainly generated ignorance.
4) The research agendas that underlie the development of new agricultural technologies need to reflect the real situation, needs, possibilities and limitations of peasant agriculture.
5) While markets can be conceptualized as abstract systems that create, through an invisible hand, equilibria between demand and supply (and thus set prices)

it is more useful to understand them as specific places where concrete transactions take place. Equally, markets are part of wider networks that link producers and consumers. These places and networks are governed in specific ways, which can be favourable or detrimental to peasant agriculture, depending on the way the markets and networks are structured and governed.

6) It needs to be recognized that peasant agriculture is an immensely important land–labour institution and is also capable of playing a progressive role in societal transformation.
7) Agricultural problems cannot be studied in isolation. They need to be contextualised in terms of contributing to or hindering the construction of food sovereignty, the attractiveness of the countryside and the well-being of farmers (i.e. *sannong*).

These points are not random nor are they disconnected – they are neatly interwoven. There is, we believe, a *paradigm* underlying these points, one that lies hidden in the structure and evolution of China's agriculture (for an agrarian history of China see King, 1911 and Netting, 1993). For reasons of simplicity we will refer to it as the 'Chinese paradigm'. This paradigm does not primarily deal with issues of class, peasants' role in development and/or the construction of socialism (these issues have been central in Chinese debates over the last decades – see e.g. Day, 2013). Instead it focuses on *how the agricultural process of production is (to be) structured and developed.*[18] It is precisely this respect that significantly, and in many ways, sets the Chinese paradigm apart from the agricultural paradigm of the West. What is impossible within and according to one, is quite feasible and valid in the other.[19]

Of course, the Chinese paradigm which we present here is not the only possible way to understand China's agriculture. Part of today's confusion (see Chapter 5) is that different expert-systems and many scholars within China increasingly use the Western paradigm to analyse Chinese realities and use this as a basis for proposing new agricultural policies and development trajectories. Ye (2015) gives an extensive overview of the way modernization is increasingly engrained in Chinese politics. However, there is no hard and fixed correlation between paradigms and empirical realities. It may well be argued that the outstanding features of China's agriculture, as outlined above, are due to an underlying and implicit paradigm that differs considerably from the rhetoric of current modernization policies. Modernization, as articulated by the Chinese state apparatuses, is not the best possible lens for gaining an understanding of China's agriculture. Far from it. This also applies in the West: the Western paradigm is not the only possible tool for understanding the historical evolution of European agriculture and/or the contemporary realities of low external input agriculture or multifunctional farming in Europe. Ironically, the 'Chinese' paradigm might considerably help us to improve the understanding of some specific aspects of European realities.

The Chinese agricultural paradigm

Having arrived at this point we can be brief. In substantial terms the Chinese agricultural paradigm can currently be summarized as follows (currently, because this paradigm is historically evolving).

It is perfectly feasible, even in the twenty-first century, to organize agriculture in a peasant-like way. This expresses a form of farming that is essentially grounded on ecological capital which, through skilful and industrious labour, is converted into food and other products and services. A central aspect of this peasant agriculture is a dedicated and knowledgeable workforce for whom farming is an important and valued part of their lifestyle (that it is to say that farming is not just an exchangeable economic activity). It offers young people the possibility to go to the cities and to take up farming again when they return. Under favourable circumstances, peasant farming might generate prosperity, resilience and autonomy.

Peasant agriculture is *linked to* the urban economy through three strategic mechanisms: through circular labour migration; through multifunctionality, which allows the peasant farms to relate in multifaceted ways with a wide range of urban markets; and through supportive policies (at least at present) that carry a 'rural bias' in that they are pro-rural, pro-peasant and, if necessary, pro-poor. Peasant agriculture in turn *contributes to* the urban economy by providing abundant, diversified and high quality food (both fresh and processed) – a supply that is both stable and relatively cheap. It also generates high levels of productive employment, which grows an internal market that, in turn, sustains industrial development. This peasant agriculture is *developed* through labour-driven intensification which is supported by resources that stem from labour migration and pro-rural policies. Novelty production and the exchange of knowledge among peasants are crucial aspects as is the additional support provided by agrarian sciences. Peasant agriculture *sustains* industrial development (and more generally the process of capital accumulation) through three mechanisms:

1) providing a supply of relatively cheap labour (and a 'safety net');
2) providing relatively cheap but good food; and
3) creating an extended internal market.

But peasants are not totally subordinated to food empires that squeeze farming to the bones in order to directly feed processes of accumulation.

The many indicated interrelations between these features require multi-level balancing: within households and villages, but also between economic sectors and between town and countryside. The state plays a central (and evolving) role in this balancing. It is needed because no simple imposition is possible and because there is not just one goal or objective. On the contrary, different interests, incommensurable objectives and contrasting perspectives need to constantly be equilibrated into a well-functioning whole. The role of the state in all this is sometimes positive; at other times it is negative. Above all the state's role is constantly being debated, negotiated, contested and renegotiated. In short: multi-level balancing is an intrinsic part of the Chinese paradigm.

As suggested above, this paradigm is in constant flux; it evolves. This implies that it often comes to the fore as a *critique* of practice. Here we might refer to the position of rural women (discussed in Chapter 8): they are still very far away from being autonomous or prosperous. In several respects the rural work they do is far from attractive – they do not positively value it nor do they feel positively valued. In yet other situations peasants are, de facto, subordinated to newly emerging food empires (under whatever guise). And nearly everywhere there are threats to access to, and control over, the land – threats that pose a grave danger to the autonomy that is central to peasant agriculture.

The first chapter of this book presented a range of specific features that characterize Chinese agriculture (see Table 1.1, page 7 for a summary). Several of these features were outlined in more detail in subsequent chapters and have been synthesized earlier in this chapter. Taken together, these features are specific aspects of the Chinese agricultural paradigm. Of course, this is only the case when the different features are shaped into a coherent, and internally more or less consistent, whole (as specified above). For example, it would make no sense to seek to link a multitude of small peasant farms with large food empires in order to provide the metropolis with convenience food. Such a constellation would not work: there would be too many frictions and the transaction costs would be far too high. Such an arrangement would be completely out of balance. This is precisely the reason why the growing hegemony of value chains (as food empires are called in hegemonic language) results in the increased marginalization of peasants and peasant agriculture, which are 'squeezed out' by large-scale entrepreneurial and/or corporate agriculture. However, food production by peasants linked to urban domains through a range of decentralized (and nested) markets may very well constitute a working whole, especially when consumers prefer fresh, unprocessed, foods. The same is true for many other combinations: the key is that the elements that make up the food system fit well together and form a well-functioning whole.

The Western paradigm has been made explicit in many handbooks. It underlies many development programmes and informs the objectives and mechanisms of R&D activities that aim to develop new technologies. It is omnipresent in the curricula of agricultural universities and in the international research institutes that belong to the Consultative Group on International Agricultural Research (CGIAR). It is the intellectual luggage of those who run ministries of agriculture, extension services and other expert-systems. It is incorporated within farm accountancy techniques and is strongly intertwined with the interests and modus operandi of food empires, large retailers and banks. In short, it is embodied in a socio-technical regime (Rip and Kemp, 1998), and is a reflection (and justification) of the way this regime operates.

The alternative, a Chinese paradigm, is far from explicit. Only parts of it have been explicated – most of it is implicit. That is, it remains hidden in practices, relations, potentials and achievements that mostly go unnoticed or are neglected.

The task of distilling the contours of the Chinese paradigm out of these practices is, however, an urgent one because it promises to be one of the important building blocks (see 'lessons' discussed above) of a new, and much needed, paradigm for farming, food and the countryside that provides an alternative to the Western one. Although it is abundantly clear that the construction of such an alternative paradigm should draw on experiences from all over the world, we limit ourselves here solely to Chinese experience and to comparing the Western and Chinese paradigms. Along these lines, some additional observations should be made. These fall into three categories. The first one concerns *relational* issues: how do the contrasting paradigms relate to the realities they describe and inform? Second, there are some *methodological* considerations, regarding the concepts that are central to each paradigm. The first and second categories deal with ontological questions. What, in the end, is farming? And, how should scientific knowledge relate to farming? The third category raises thematic issues about the *representation* of the many elements that, together, constitute the worlds of food, farming and the countryside.

Relational aspects of contrasting paradigms

As argued above, the Western paradigm is explicit. It has been made explicit during a long and complicated process that started in the 1930s in the realm of agricultural sciences.[20] Since then it has informed the construction of relatively new practices, interrelations and expectations which, in turn, have helped to further unfold this newly emerging paradigm and to legitimize it. The Chinese paradigm is largely implicit; it is still lying dormant in richly chequered realities that display many complexities and enigmas. Some people will even wonder whether there really is an underlying paradigm. At first sight, Chinese scholars, politicians and everyone else who deals with agriculture simply take the countryside, food production and the peasantry as givens. Chinese farming has a rich morphology but no one seems to try to go beneath its outer expressions in search of an underlying paradigm. The debates are mostly centred on class analysis and state–peasantry relations but hardly (or not at all) on the exceptional structure of the processes of labour and production, nor the impressive outcomes that they provide.[21] Peasants, farming systems, etc. are simply seen as 'facts of life' one has to deal with. However, in dealing with farming and peasants as they are *one implicitly takes on board a range of paradigmatic cornerstones*. Dealing with Chinese agriculture *as it is*, means dealing with smallholdings and how these can be further developed. It means dealing with the centrality of labour in agricultural production just as it implies coming to grips with peasants, peasant logic and peasant farming. Such elements necessarily form the skeleton of a paradigm – even if those involved are hardly aware of it.[22]

The Western paradigm conceptualizes agriculture as being governed by markets and technology. Technology defines the so-called 'function of production', that is the conversion of inputs into outputs. This function is a discrete S-shaped line on which each input level translates into a precise output level. If output levels fall below the S-shaped line or function, this points to failures of the operators involved in converting inputs into outputs. They are inefficient. With technological development the function moves upwards and more output is possible with the same inputs – or a similar output might be realized with fewer inputs. The exact position of the farm enterprise on this S-shaped curve is, in turn, defined by the relations that govern the markets. More precisely, it is the price–cost relation between inputs and output that specifies the optimum. Changes in market relations will induce a new optimum. Entrepreneurial capacities are needed to ensure that the optimum is repeatedly realized. In this way, markets and technology determine farming. Thus agriculture is essentially understood as the unfolding of the laws that govern the economy and the conversion of inputs into outputs (these latter include biological, chemical and physical laws).

By contrast, in the Chinese vision agriculture is conceptualized and represented in a radically different way: *it is what farmers are doing*. And as simple as this may sound, it makes for radical differences. By focussing on farmers' practices (as part of an organized flow of activities that unfolds over time and is conditioned by the circumstances under which farmers operate), actors and endogenous perspectives are put centre stage. And this leads to solutions that differ radically from those brought forward by the market and technology oriented paradigm of the West.

One important consequence of this is that the Western paradigm disrupts rural realities. It *denies* existing practices, and it particularly denies the potentials and promises that exist on the ground. Agricultural practices (wherever and whenever they are located) are generally described as *not yet* sufficiently developed, *not yet* applying the newest technological possibilities, *not yet* sufficiently entrepreneurial. And some time after these new possibilities are applied they again become outdated due to changing market relations and/or the availability of even newer technologies. From this framework it is rather futile to try to understand the reasons why practices differ from the new technological and economic opportunities – what matters is to know how the new optima are to be reached. *Getting Agriculture Moving* as the title of a once influential book (Mosher, 1966) described it was – and still is – the issue. By contrast in alternative approaches such as the Chinese, theory is a *reflection* of reality. It is an attempt to understand agriculture as it is and why it is the way it is. It is also about exploring the possibilities for further unfolding what is currently entailed in agriculture as it is – and communicating these, where needed and if possible, to farmers and others. While the Western paradigm focuses on two key words – *not yet* – the contrasting views compose their narratives around a different key word: *already*. Farmers are already tackling this or that problem; they are already changing, already contributing to increases in production. And so forth. In the Western paradigm agricultural and rural development are essentially *exogenous*; in the Chinese paradigm they are *endogenous*.

This central difference between denial and reflection is associated with a range of other relational differences. First, in the Western paradigm institutionalized science is taken to be the main driver of agricultural development: it constructs new technologies, unravels the markets and specifies the right economic decisions.[23] It also informs policy making. In contrasting views, it is the farmers themselves, the resources they control and the many networks that locate them in wider society,[24] that are seen as the main drivers. That does not mean that agricultural policies, technologies and/or markets are irrelevant. They can be highly important – but they are important only in so far as they are actively *translated* by farmers themselves into new courses of action (which may range from alignment to active resistance and which embrace all possible forms of deconstruction and reconstruction). Sometimes new policies can even emerge as the, more or less, direct outcome of farmers' actions – as was the case with the major change from Chinese communes and production brigades to new peasant units of production (see Chapter 11). This leads to a second crucial point: markets, policies and R&D that aim to develop or disseminate new technologies are key arenas where important battles take place. This raises another question. What is the role of the peasantry in all of this? Are they passively following the script dictated by markets, technology and policies? Or are they actively struggling in order to enlarge their autonomy and well-being – possibly by taking a stance against these exogenous influences?

The Western paradigm decidedly takes an Archimedean point,[25] one that is localized in the *future*. This is not the future understood as a wide range of potentialities entailed within today's realities (some of which are probably still unknown). It is a future specified by institutionalized science through the analysis and combination of new technological possibilities and expected market tendencies. Hence the interventionist nature of Western agricultural sciences. For *sui generis* reasons they know the future better than others and – since this same future denies today's reality – they must intervene and 'reset' agrarian realities (some would even claim they have the moral obligation to do so). The Chinese view this differently. They focus on agriculture as it is and by doing so they will repeatedly encounter an extended set of possible factors that may or may not be of relevance: these could be the natural endowments, demographic patterns, history, the main institutional arrangements (especially the main land–labour institutions), town–countryside relations, culinary traditions, collective memory, diversity in labour objects, possibilities to experiment, the presence of novelties and the possibility to build upon them, work routines, traditions, or market constellations, etc. One never knows beforehand what cocktail of these many 'ingredients' will be relevant and important. One knows even less about how they operate. But then, the joy is trying to identify and understand them, in order to build upon them.

One consequence of western agricultural science adopting this exclusive Archimedean point is that it completely reverses the normal relationships between past, present and future.[26] It also means that agricultural development stifles the agriculture that is practiced today. Since there is only one possible future, development becomes a unidirectional process and the policies and technologies designed to move agriculture towards this future broach no deviation. They become a straightjacket[27] within which any deviation is seen as a mistake, doomed to end in failure, and placing unnecessary obstacles on the path of development. Such deviations should be abandoned and shunned. As a result, the Western paradigm's view of the processes of development is largely *selective*. There are 'good' practices and there are 'bad' ones; there are 'modern' farmers and there are 'traditional' ones or 'laggards'. By contrast, alternative paradigms allow for and encourage variety and differentiated development trajectories. They do not result in coercive policies (as the Western view does), but typically result in multiple spaces for manoeuvre that allow for different and often contrasting courses of action. Instead of being coercive, alternative agricultural and rural policies tend to be *enabling*.

This helps explain why the Western paradigm frequently results in the formulation and implementation of mega-projects (Scott, 1998) and the Chinese paradigm in 'small works' (Donaldson, 2011).[28] Mega-projects are 'episodes of [large-scale] state-initiated social engineering' (Scott, 1998:4). They arise from a combination of four elements: the administrative ordering of nature and society, a highly modernist ideology, an authoritative state and a prostrate civil society that lacks the capacity to resist. Small works, on the other hand, build on the networks of 'small-scale, family based enterprises and petty traders' (Donaldson, 2011:165) that are strengthened by well-targeted state policies. Mega-projects usually look very impressive. They promise, and sometimes realize, what seem like amazing increases in, say, production – but they are only able to do this in limited and well-controlled spaces. In many instances mega-projects can increase production by an impressive 100 per cent.[29] Yet this may only possible on, say, 5 per cent of the available agricultural area. On the remaining 95 per cent production remains as it was, though now at a greater competitive disadvantage to the areas that see the increases. Small works, by contrast, tend to include all areas but may realize only small increments in production, say 5 per cent. At the end of the day though, the result is the same (a 5 per cent increase in total production). If small works achieve growth levels that exceed 5 per cent, then they are outperforming the mega-projects.

Interestingly, one can find examples of both of these radically different approaches (and their differentiated impacts) within China itself. In an empirically rich and theoretically well-articulated book, Donaldson (2011) compares two provinces in the south-west of China: Yunnan and Guizhou. From geo-physical, socio-cultural and institutional points of view these two provinces are very similar. There is, though, one overriding difference. Over recent decades Yunnan has experienced significant economic growth without any reduction in poverty levels. In fact 'economic expansion in Yunnan left a higher proportion of its people poor' (Donaldson, 2011:6). Diametrically opposed to this is the experience of Guizhou province where economic growth has been slow (well below the Chinese average) but 'the rate of severe poverty declined from 41.9% in 1990 to below 20% by the end of the decade' (Donaldson, 2011:4). To put it differently: in Guizhou the number of very poor people was reduced by 2.4 million, while in Yunnan the ranks of the very poor *grew* by 2.2 million. Both examples are at odds with the thesis that economic growth generates poverty reduction. They are, what statisticians term, *outliers*. So how can this enigma be explained?

Donaldson examined the development strategies of the two provinces and convincingly demonstrates that this apparent mystery can be explained by differences in economic strategy. Yunnan's strategy is dominated by large-scale urban development, organized through mega-projects. Guizhou, by contrast, is building and promoting small-scale economic opportunities with particular emphasis on the rural. This strengthens the regional economy as opposed to leading to an increase in export orientation (which is what has happened in Yunnan). Yunnan is outward-oriented, while Guizhou is typically inward-looking and seeking opportunities for endogenous growth.

Donaldson discusses, illustrates and analyzes these basic differences by looking at four fields of activity. Road construction is one of these. Yunnan developed a highway system that linked the province with the capitals of three other provinces and with the three countries that border the province (Vietnam, Laos and Myanmar). These highways successfully stimulated exports from the province but did little to reduce poverty. Guizhou, on the other hand, concentrated on constructing a dense web of rural roads that made market places accessible for farmers living and working in remote areas. Much of the road construction occurred through the *Yigong-Daizhen* programme, a food-for-work scheme. Thus, this particular type of infrastructure creation had both direct and indirect effects on poverty reduction. Donaldson also looked at migration. Guizhou actively assisted labour migrants to move, for longer or shorter periods, out of the province towards the south-eastern coastal towns where comparatively high wages could be earned. This not only triggered a large flow of remittances – on their return the migrant workers also invested their savings into the local economy. Yunnan did not have a policy for migration and had far less migrant workers: 'In 2005 Guizhou had 0.58 migrants per household, dwarfing the percentage in Yunnan of 0.19, a gap that only increased in the next year.' (Donaldson, 2011:98).

A third field of economic activity is tourism. Here again there is a major and strategic difference. Yunnan focussed on constructing large hotels for elite international travellers, while Guizhou stimulated *nongjiale* (literally 'happy farmer's houses'), a decentralised form of agro-tourism that only requires small investments but opens up opportunities for many rural residents to supplement their incomes.

Finally, Donaldson looked at another area where the contrasts are marked: coal mining. Yunnan opted for large and well-equipped coal mines. Guizhou, by contrast, continued with smaller scale mines of the TVE type – especially in the poorer counties. And although the latter came with a price – the dangerous mines caused many tragedies – their economic impact was again significantly higher.

Taken together, the contrasting patterning of these four areas of activity clearly influenced the structure and distribution of economic growth. They patterned economic growth in different ways – the result of differently designed and implemented development policies. This brings the questions of *how*, *where*, *what for* and *by whom* firmly back into the discourse on development, growth and poverty reduction. What *kind* of roads? Accessible to *whom*? *Who* builds them? *For what* will they be used? And so on. Considerable synergies can be created by carefully linking development activities to local opportunities. Migration from Guizhou allowed for savings and small investments. The dense web of newly constructed roadways – that built or improved the linkages between villages, townships and counties – helped to stimulate local markets allowing for surplus production to be sold. At the same time this web supports the decentralized flows of tourists enjoying *nongjiale*. In this way, 'small works' do work – and they work very well.

In short, Yunnan and Guizhou each adopted a 'distinct approach to the rural economy' (Donaldson, 2011:151). In general terms we can say that Guizhou embodies the Chinese paradigm while Yunnan is more reflective of the Western paradigm. Donaldson's comparison highlights the strikingly different impacts of the two paradigms. The mega-projects of Yunnan tended to be exclusive and the numbers of poor people grew. Guizhou's approach, on the other hand, included the people, areas and activities that were neglected in Yunnan. The comparison also shows that not everything that happens in China is a result of following a Chinese paradigm. There is paradigmatic struggle in China as much as there is elsewhere. Some political factions, institutions, expert-systems, interest groups and provinces favour an approach that reflects the Western paradigm (and are often directly inspired by it). Others criticize this approach and opt for an alternative that follows the Chinese paradigm. This mirrors the situation in the West where there are also competing paradigms that inspire and inform contrasting development trajectories, farming styles, town–countryside relations and institutional arrangements. Some features of the Chinese paradigm can also be seen in grassroots activities in Europe that promote multifunctionality and endogenous rural development. Similar features can be found in the struggles of the Brazilian *Movimento dos Sem Terra* (MST – the Movement of Landless People). Internationally, the agroecological movement is also, in a way, a carrier of something akin to the Chinese paradigm. The big difference, though, between the East and the West, is in the relative weight given to these two contrasting paradigms.

Methodological aspects

The Western paradigm regards agriculture primarily as a techno-economic activity. All other dimensions are put between parentheses or deemed to be irrelevant. In the Chinese paradigm agriculture is multidimensional. There are, of course, technical and economic dimensions as well but these are definitely not dominant. There is no a priori hierarchy of dimensions. In addition, there is not just one way to structure the economy – as argued by Webber (2012) (see Chapter 6 of this book) and by Donaldson (2011) (see above) – and not just one assumedly superior technology. Instead there are different ways to pattern the economy, just as there are different techniques and technologies that can be employed.

The grammar of the Chinese paradigm is the subjunctive. It is about what *might be* (Kessel, 1990). It is about different possible uses of available resources and different ways in which farming (i.e. resource use) can unfold over time. In consequence, the Chinese paradigm has no difficulty in dealing with heterogeneity (without any a priori established hierarchy). By contrast, the Western paradigm is about how things *are to be*. 'This *is* the best way of farming; this is how it *must* be implemented'. The grammar used is the indicative, which often slips into the imperative. This relates both to the paradigm's coercive character, as well as to the Archimedean point of reference.

In the paradigm that underlies the historical and contemporary evolution of Chinese agriculture, flows and processes are central. Often there is no predefined vision of where exactly these flows and processes will lead: the activity of moving often makes the flow ('*caminando se hace el camino*', as the Spanish say). Chapter 11 recounts the dramatic events that occurred at the end of the 1970s in Anhui. Nobody knew what the outcome would be of following a new course, but through a multitude of actions, at different levels and by different actors, Chinese agriculture as we know it today was actively constructed.

This is in stark contrast to how the Western paradigm operates. It doesn't hold with flows, instead it focusses on how things are to be.[30] And in so doing it comes to define how things *necessarily* will be. An optimal situation is constructed, located somewhere in the future, and development is (necessarily) deemed to follow a linear route that leads towards this a priori specified optimum. Flows (and the knowledge of these flows and the capability to facilitate them) are replaced by imposition (and the associated capability of planning).[31]

The different characteristics of the two paradigms also translate into building blocks that differ – sometimes slightly, sometimes considerably. One of the substantial differences lies in the nature of the relations that are assumed to tie the worlds of farming and food together. The Western paradigm conceptualises relations between farms and markets, farmers and the state, consumers and the food industry, fields and cattle, etc. as *linkages* that hierarchically order the world. These linkages impose the dominance of higher levels of aggregation over the lower levels: the world market, for instance, governs national and regional markets. The in-between linkages are part of a somewhat mechanical whole in which activities and practices in micro-situations (say a farm) are determined by forces that operate in the macro-context in which these situations are embedded.[32] By contrast, in alternative paradigms, the same interrelations are viewed as *interfaces* rather than as *linkages*. Interfaces are points of contact where negotiations and translations occur and where communication can go in two directions (Long, 2015). There is no one-sided determination here: grey zones are actively created and particular rules might be interpreted differently, giving rise to varying and contrasting courses of action. Coalitions (that tie actors from both sides together) can have an important influence on these interfaces. The existence of these interfaces implies that economic life and agricultural development can, indeed, be patterned in different ways.

Similar differences occur when conceptualizing agricultural and food markets. In the Western paradigm they are basically understood as *a market system*, i.e. as an abstract and self-regulating constellation that fundamentally is (or at least should be) independent from the actors and agencies that operate in it. This is in stark contrast with the notion of *market places* (see Chapter 10), which conceptually allows for the simultaneous presence of more, and mutually different, markets, each of which is shaped by the actors and agencies that operate in it, by the products and services being exchanged and by the underlying socio-material infrastructure (see e.g. Hebinck *et al*., 2015).

In short: the Chinese paradigm not only accepts heterogeneity as a feature of the world that is being studied and represented – it also uses it as a major methodological principle. Theories, foresight studies, simulation models, linear programming studies, technology design or whatever cannot be grounded on simplified notions such as *the* agricultural exploitation, *the* intake of concentrates, *the* cost levels, *the* labour input, etc. They need to depart from differently moulded realities (that are equally also probably *perceived* in different ways).

Issues of representation

The two paradigms discussed so far both represent the world of agriculture, farmers, food and the countryside in specific, but sharply contrasting, ways. Some of these differences are summarized in Table 12.1 and briefly discussed below.

A first series of representational differences is located at farm level. The Western paradigm represents faming as the conversion of inputs (which are seen as having a 'passive' nature) into outputs. This conversion is governed by the available technology – and the newer the technology the better. In the Chinese

Table 12.1 Keywords that characterize different representations

	Western Paradigm	*Chinese Paradigm*
Farm level	The farm is an economic unit.	Farming is a multidimensional and highly diversified activity.
Markets	Focus on world market which is the dominant ordering principle.	Involvement in multiple markets. Differential dynamics.
Development process	Unilinear. Convergence at the global level.	Unfolding into multiple sub-processes. Divergence at global level.
Agricultural policy	Focus on competitiveness. Selective. Coercive.	Multiple objectives. Balancing and searching for synergy. Allowing for experimentation.
Governance model	Central direction.	Self-regulation.
Wider society	Negative externalities.	Positive externalities.

situation the view is that agriculture occurs through *co-production*: the ongoing interaction and mutual transformation of man and living nature. In and through this process resources are developed (see Chapter 4) and new institutional arrangements are constructed (see Chapter 10 on new, nested markets and Chapter 11 on the emergence of the Household Responsibility System). Such developments can be progressive, but they can also be regressive (see Chapter 1 on environmental decay). Another key difference at this level is between the farm being represented as primarily an economic unit that necessarily follows the logic of the market (the Western paradigm) or as a complex, multidimensional constellation (the Chinese paradigm).

At the level of the markets, the Western paradigm holds the *export* market (i.e. the world market) as the primary principle guiding the development of both farms and the sector as a whole.[33] Through the notion of competitiveness, the 'beacon' of the export market is translated into the ongoing need to reduce the costs per unit of end product, to enlarge the scale of production, to further increase specialization etc. The script is already written and wherever it is presented, whether in Europe, the USA, Canada or China, it shows breathtaking resemblances. By contrast, national food provisioning, the regional economy and the labour market provide very different points of reference that produce development programmes which differ significantly from those generated by the Western paradigm.

In China policy is guided by a number of objectives, which may be summarised thus:

- providing the national food market with sufficient and high quality food;
- making sure that incomes in farming are sufficient in order to strengthen the rural economy and contribute to the internal market needed for national industry;
- contributing significantly to the creation of productive employment (also to having a fall-back position for periods of industrial recession); and
- making the countryside into an attractive place with a good quality of life.

Meeting these objectives simultaneously requires *balancing* and *the construction of synergies*. Both of these features are at the heart of *sannong*. This is a significant difference from the imposition and compartmentalization[34] that characterizes policies emerging from the Western paradigm.

Another thematic difference at the level of agricultural and rural policies is that the paradigm which can be derived from Chinese agriculture stresses the notion of *unfolding* when it comes to development. In the Western paradigm *imposition* is the modus operandi.

With the focus on the global market and with policy being imposed it is self-evident that state-controlled resources almost inevitably flow to the 'growth pole' of the largest farms.[35] In the Chinese paradigm the emphasis is on achieving a far wider and more equitable distribution (from which it follows that development is an inclusive, rather than a selective, process). In turn this clearly relates to differences in what is assumed to be the main driving force of agricultural and rural development: technological change (in the Western paradigm) or the indomitable emancipatory drive of the peasantry (in the Chinese paradigm).

The Chinese paradigm stresses, to a considerable degree, the mechanism of self-regulation. This may sound ironic (in the West we are accustomed to perceiving China as a highly centralized state), but in reality market places, the complex interfaces between the peasantry and the multi-layered and dual state, the multiple ways in which rules and regulations are interpreted, *guanxi*, village committees and the proverbial stubbornness of Chinese peasants create considerable space for manoeuvre within which multifaceted forms of self-regulation are discernible – especially at village level. This is in stark contrast, not only to the centralized prescription and control faced by, say, European farmers, but also with the role that policy is assumed and required to play within the Western paradigm.[36]

Finally, there is a stark contrast in how the two paradigms represent relations with society at large. In the Chinese paradigm they are conceptualized as a complex set of positive externalities (at least, this is what it strives to achieve) whereas the Western paradigm mostly deals in negative externalities.

If we are to derive some guiding principles which can be applied elsewhere from the paradigm that underlies China's agriculture and its long, historical evolution, a number of simple and clear programmatic points emerge. These are listed below.

1) Protect, enrich and enlarge ecological capital, making it accessible (preferably through common pool arrangements)[37] to the population that is economically active in agriculture and recognize it as the basis of food production.
2) Recognize farm labour as central in the process of converting ecological capital into food.
3) Improve both the quality of labour and the quality of life of those who are actively involved in farming and make it easy and attractive for young people to find a way into farming.
4) Make the 'art of farming' (the body of knowledge that specifies the sustainable use of the available ecological capital), which has been developed and shared by those working in agriculture, into the main guideline for agricultural development.
5) Maintain, strengthen and mainstream the family farm as the central land–labour institution (and apply agrarian reforms wherever needed).
6) Make sure that all the actors involved in agricultural production (especially women, young people and minority groups) have the same rights and are treated equally.
7) Reorganize agricultural R&D in order to support the above-mentioned objectives.
8) Reorganize existing market systems into a widespread and decentralized network of market places that are nested in the mutual recognition and understanding of food producers and consumers.
9) Ensure that agricultural R&D and the reorganization of marketing contribute to the emancipation of the peasantry.

10) Structure the process of production as peasant agriculture: with low levels of usage of (and expenditure on) external inputs, little reliance on external capital and a high value added (or labour income).

Seven decades ago Karl Polanyi (1957) argued that 'leaving the fate of man and the land to the market, would be tantamount to annihilating them' (slightly paraphrased). Without any doubt it is now time to actively restore the centrality of both man and the land in the production of food. China's agriculture proves that this is both a valid and a viable path forward to achieving food and nutritional security and food sovereignty.

On capital and paradigms: an addendum

Farming is intimately connected to, and grounded upon, three different fields or pillars. These are the local ecosystems from which natural resources are drawn, the people involved in the labour process (who embody key social resources such as practical knowledge, cultural repertoire, patterns of co-operation, etc.) and society as a whole (which is provided with food and delivers, in turn, the goods and services needed for farming). Historically, the reproduction of farming depends on these three fields – and on each one individually as well as collectively. We therefore refer to them here as 'reproductive fields'.

A central feature of the Chinese paradigm[38] is that farming is primarily oriented to *securing its own reproduction* (and thus its own autonomy and continuity). Natural resources are reproduced (and at the same time further developed) through farming. The same applies to those working in agriculture: their lives, prospects, knowledge, etc. are reproduced through farming (the self-provisioning of food being a strategic component), just as the main land–labour institutions are reproduced in and through the practice of farming. The relations with society at large are also reproduced – through e.g. the delivery of enough food of good quality which generates incomes, trust and recognition.[39] If there were no consumers interested in buying the food and other products produced by peasants, peasant agriculture would quickly come to an end.

In the Chinese paradigm, farming is first and foremost oriented to the reproduction of its own roots, its own foundations. This is evidently reflected in the millenary tradition of China's agriculture. It could only exist, and continues to exist, because it has mastered the reproduction of delicate ecosystems, the evolution of precarious man–land relations and the survival of its peasantries through many periods of harsh repression and exploitation. Without such mastery, engrained in manifold institutional arrangements, China's agriculture would have vanished a long time ago.

Reciprocity is decisive in this reproduction. The activities of Chinese peasants are a form of *giving* to the resources upon which agriculture is grounded. They are reproduced through the goal oriented and knowledgeable actions of the peasantry (and others). And it is only through this *giving* that there can be *taking*. Only after having *given* the ecosystems (more specifically, the natural resources), the social groups involved and society at large what they need, can there be *taking*. Attention and care are given to the productive fields and this subsequently allows for prosperous production. As a Chinese saying goes: 'if man works hard the land will not be lazy.' (Arkush, 1984). The more that is given, the more that can be taken: the unfolding of farming occurs through the enlarged and improved reproduction of its resources. The agricultural economy is thus typically *circular*. What is given allows for subsequent, but always proportionate, taking.

It is precisely at this point where we find the main difference between the Chinese paradigm of farming and the Western paradigm. In the practice of western agriculture farming increasingly draws in a one-directional way from the three reproductive fields. It is constantly draining them and does not reconstitute them. Thus these decisive reproductive fields are being emptied – and losing their relevance for the future of farming. This implies an economy of exploitation, if not exhaustion. Yet the Western paradigm does not perceive or represent this as being damaging or regressive. On the contrary, *it sees this as opening up new spaces for science, technology and markets*. Once exhausted (or destroyed), the natural resources are replaced by artificial growth factors; those who once did the work are made redundant and replaced by new technologies; and the relation with wider society is increasingly mediated by food industries and retail chains that construct images and symbols which suggest quality and safety. Thus Western agriculture is made to devour its own foundations and replace them with new patterns of dependency *that subordinate farming to financial, industrial and commercial capital.* Ecosystems, peasantries and the synergy with society at large exit the stage and are replaced by capital, technology and marketing. Instead of reproducing its foundations, Western agriculture consumes them and then becomes dependent on external elements and relations.

The Western paradigm not only reflects this basically extractive process (the impacts of which still cannot be assessed fully) – it actively drives it. This is because the image of how-agriculture-ought-to-be lies at the very core of the Western paradigm. This is a *disruptive* image that is in strong contrast with agriculture-as-it-is, which is central to the Chinese paradigm. In the Western paradigm agricultural development is thought of as an ongoing process of modernization. This is grounded in (and creates) a *permanent disequilibrium*. Agriculture-as-it-ought-to-be (derived from the future, a 'modern' approach) systematically differs from agriculture-as-it-is. The latter is at odds with new technological possibilities and economic opportunities which are the very features that drive the former. Agriculture-as-it-ought-to-be makes new technological possibilities and market opportunities come true. And this is not a one-off operation: it is ongoing. The reorganization of farming into what-it-ought-to-be is a constant feature; it views the representation of agriculture-as-it-is as inadequate, insufficient, outdated and unable to meet new demands. Agriculture needs to be remade, to jump – time and again – into a future defined by the interests of science and dominant economic forces.

Undeniably this disruptive image has triggered considerable dynamism, but at a significant price. Once the preferred future is superimposed on the 'inadequate' present, the structural disequilibrium translates into an array of societal and natural domains – and the consequences of this have often been disastrous. Agricultural modernization, which is supposed to span the disequilibrium, has more often than not generated stress, low incomes, high risks, monotonous labour routines, solitude and the permanent threat of exclusion for those who work in agriculture. Equally, it has brought nature into disarray leading to natural resources being contaminated, biodiversity threatened and scenic landscapes destroyed. And society as a whole is, in turn, confronted with often fatal food scandals and food scares and significant decreases in the quality of life. What was assumed to be replaceable has actually suffered considerable, albeit highly varied, processes of degradation.

The dialectics of paradigm and practices coincide with the logic of capital accumulation. Agriculture is not just a source of the generation of wealth that is centralized elsewhere. It is more: it is a space where *constantly growing flows* of wealth are generated in order to feed processes of accumulation located elsewhere. This explains why modernization is not a one-time operation but a permanent process. Agriculture is permanently 'lagging behind' the needs of capital – therefore it always needs to be 'modernized'. Capital's appetite is insatiable – the more it drains the countryside the hungrier it becomes.

Under the Western paradigm agricultural development is a disruptive process – a chain of disruptions. Existing practices and their materializations need to be continuously replaced, and this process requires the introduction of ever more artificial growth factors, technologies, energy and refabricated food products. Thus capital eats its way deeper and deeper into agriculture, while simultaneously subordinating it.

The ongoing organization of disruption differs radically from ongoing and multi-level balancing. The first exemplifies both the paradigm and dominant practices in the West,[40] while the latter is central to and characteristic of the Chinese paradigm.[41] For historical *and* contemporary reasons it cannot be otherwise. There is balancing at the macro-level – as synthesized in *sannong* ('the three ruralities') under which agricultural development is balanced with food sovereignty *(nongye)*, which is deeply rooted in China's geo-political need to remain independent –especially with regard to food. Agricultural development is also balanced by promoting relative prosperity among the peasantry (*nongmin*). The need for reasonable incomes is deeply rooted in the potential explosiveness of the Chinese peasantries.[42] If they are exploited for too long and too harshly – and especially if rural–urban inequalities become too manifest – this might well threaten the continuity of the regime.[43] Finally, there is *nongcun*, the need for an attractive countryside. This is increasingly recognized as a material and undeniable need. An attractive countryside is necessary to counterbalance the rapid and seemingly uncontrollable process of urbanization – it is equally needed so that peasants can produce food in the coming decades.

The three ruralities represent a range of countervailing powers with which agricultural development needs to be balanced. If this doesn't happen then major frictions, inequalities and contradictions will emerge. The need for balancing is materially built into Chinese history and is probably even stronger in today's constellation. The need to ground agricultural development in agriculture-as-it-is (i.e. to take agriculture-as-it-is with its inbuilt balances as the point of departure for further development) is reflected in the Chinese paradigm that puts agriculture-as-it-is centre stage (as opposed to agriculture-as-it-ought-to-be, which is central to the Western paradigm). The main consequences are disruption in the latter case and balancing in the former.

The balancing at the macro-level is accompanied by an active and many-sided balancing at the micro-level. This involves a search for adequate balances between farming and labour migration, finding the right balance between different generations and assessing appropriate equilibria between labour investments and the foundations needed for the next generation. There is also a balancing between 'getting rich in the countryside' and the reinventing of peasant farming, in the interrelations between men and women and those between producers and consumers (especially when it comes to the construction of new nested markets). Then there is a search for new overall balances between town and countryside: many young people are actively looking for it. Together this multi-level and multi-actor search for adequate balances and the construction of associated equilibria make China's agriculture as a whole more resilient.

There are many countervailing powers in Western societies as well. But at the paradigmatic level these are brushed aside. Food sovereignty is deemed to be a non-issue. What is decisive is food security, which only refers to the possibility of acquiring sufficient food – through imports if needs be. Securing adequate incomes for farming families is another non-issue. Incomes are to be derived from the markets and if they are inadequate, farmers should adapt better to the market, or they should quit. The same applies for the attractiveness of the countryside. If there is a market for it, the necessary 'attractions' will be produced, and possibly by those best equipped to do so. Farmers' unions increasingly identify with the prospects of modernization and, consequently, align themselves with the view that such issues are non-issues. At the EU policy level, the domains of agriculture and the countryside have been *separated* into two independent programmes: pillar one that addresses agriculture and pillar one that seeks to secure the attractiveness of the countryside. This again denies the need for *balancing* – the two goals are regarded as different compartments. Nor are nature organizations a real countervailing power: they have their own 'reserves' but their influence elsewhere is limited. Consumer organizations are held at arm's length by the reiterated statements that food has never been as safe as it is today and that it is impossible to achieve 100 per cent safety. At the micro-level balancing has also been made superfluous, partly through redefining professional roles in terms of procedures and protocols instead of responsibilities.

With the potential countervailing powers demobilized, the Western paradigm becomes the main vehicle for actively subordinating agriculture to the needs and logic of capital. It opens the gates for the continued submission of farming to capital and the associated restructuring of farming according to the logic and needs of capital. It does so, first, by consistently proposing the introduction of new and more artificial growth factors (symbolized today by genetically modified organisms).

This reduces the reproduction of agriculture through farming itself and opens up large spaces for capital. Second, it does so through the development of new technologies (currently centring on automation and the application of global positioning systems) that make labour and *savoir faire* obsolete. This again opens considerable spaces for capital. Third comes the processing of agricultural products in food industries and their subsequent distribution through large retail organizations. The reproduction of farming through on-farm processing and direct delivery is broken and new spaces for capital are created. Finally comes the *financialization* of farming. By reducing farming to a mere financial operation that critically depends on credit, new space for (financial) capital is opened. The existence of high risks (associated with volatile markets), and the use of derivatives (swaps, futures and so on) to reduce these risks, further enlarges this space. The disruptive images at the heart of the modernization paradigm (agriculture *not yet* fully developed, etc.), ensure that these processes are continuously widened and deepened.

The Chinese paradigm is not simply the opposite of the Western one. It certainly is not a mechanism to exclude capital. Rather, it is a vehicle to discipline capital, to delineate its radius and to condition its role. Just as it presumes balances, the Chinese paradigm also acts like a balance. It is a balancing force since it assesses the equilibria required to develop an agriculture that is in line with societal priorities. The Chinese paradigm *allows* capital a function but also *restricts* it. Capital *can* play a role so long as it contributes to the overall balances that secure societal objectives such as *nongye, nongmin* and *nongcun.* And if capital is unable or unwilling to do so the Chinese paradigm provides alternatives, and thus avoids an overall dependency on capital.

While the Western paradigm leaves 'the fate of man and the land' to capital – in an increasingly unconditioned way – the Chinese paradigm helps to protect them. That might sound like a modest contribution – and maybe it is. But as modest as it may be, it is – as we have tried to show throughout this book – far from irrelevant.

Notes

1 This article is partly based on Ye, 2015.
2 This, for instance, is reflected in the terms much used in Latin America during the last four decades of the twentieth century. Two factions could be distinguished: the *campesinistas* and the *descampesinistas*. The two factions engaged in bitter polemics. A similar debate was waged previously in revolutionary Russia with Chayanov and Lenin being the main exponents (See Bernstein, 2009 for an excellent summary). One intriguing aspect of these debates is that the peasantry is considered to be a 'static bloc' that either is, or is not, there. Shanin's proposition (1971) that the peasantry should be understood as a process, as a flow through time, seems far more elegant. Such a flow might well be intrinsically contradictory and historically contingent: it might entail both ongoing de-peasantization and re-peasantization. The ratio between the two can vary and change over time and from place to place.
3 These are, respectively, the Marxist narrative on rural history and the neo-classical or liberal one.
4 The modernization of agriculture is defined in many different ways, although two common elements can be identified: the inflow of capital and the outflow of labour.

The inflow of capital (new capital goods, new technologies) allows for production to be reorganized and increased and farms to be enlarged. This reduces the need for labour input so labour can – and must – flow from the countryside to the urban economies. In Marxist versions the inflow of capital is associated with a (possible) penetration of capital–labour relations (replacing peasant farms by capitalist farms) and/or with a subordination of farming to external capital groups located for example in food processing and trading. Financialization is another mechanism that brings subordination. When looking specifically at the process of production, modernization is understood as embodying the increased importance of artificial growth factors and a reduction in the relevance of ecological capital. In sociological theories of the 1950s, 1960s and 1970s, which opposed modern and traditional culture, modernization was understood to involve farmers becoming open to exogenous renewal (accepting new technologies, credit, etc.).

5 Modernization always assumes an outflow (or expulsion) of labour from agriculture.

6 Contrasting patterns are outlined and discussed in HLPE, 2013; see especially Figure 12.

7 For example creating schools in the countryside and improving rural education opportunities (instead of obliging young children to go the cities to attend classes).

8 This does not exclude the possibility that it was needed from *particular* points of view and for, or by, *particular* interests. Much was written, in the 1960s and 1970s, about the previous period of decolonization and the subsequent modernization of farming. Decolonization implied that national capital groups (involved in and partly dependent on these colonies) needed new sources to feed the process of capital accumulation. Agriculture emerged as a new, interesting and promising field – provided that it was 'modernized'. Thus, the agricultural sector became an 'internal colony'.

9 Modernization involved an enormous expenditure of public funds (Priebe, 1985) in order to reach goals that were only minimally related to societal needs.

10 Here modernization is typically about accelerating time: it shifts future benefits to the present, just as it takes future constellations as a guide for actions in the present.

11 This is exactly what is currently happening with the major Green Revolution crops.

12 As long ago as the early 1980s Bennett (1981) identified similar dynamics in plant breeding.

13 In this respect it is telling that one of the reasons for opting for large-scale farms (as promoted by the Mansholt Plan) was their relatively extensive nature. It was thought that by making intensive small farms redundant the overproduction of milk could be contained.

14 Many scenic landscapes were destroyed; biodiversity suffered; a huge and persistent environmental crisis was created; the quality of food came under threat; through aggressive exports many developing countries' agricultural sectors were 'outcompeted'; and many small farmers were compelled to stop farming and became very disillusioned.

15 If the modernization of agriculture is about bringing capital in, and moving labour out, then the bitter conclusion is, in the end (and all details aside), that there is *not enough* capital to move in, and the labour moved out has *nowhere to go*. Ironically, after modernization the agricultural sector frequently lacks *both* capital and labour.

16 SRI (System of Rice Intensification) and, more generally speaking, agroecology are other examples that show that things could be different.

17 This is because it primarily focuses on yield increases (and associated improvements in the cost/benefit ratios) while neglecting to ask: Where?, By whom?, How? and For what end?

18 Of course, the structure and dynamics of agriculture are not independent from class relations and issues of transformation, but we use the former as our point of entry.

19 See especially points 1, 2, 3, 5 and 6 listed above.

20 Previously, classical agronomy dominated. It basically came down to knowing what farmers (of different kinds) were doing. See e.g. Ploeg, 1987.

21 Exceptions are found in the literature that explores the structure and dynamics of the Chinese economy as a whole. See e.g. Arrighi (2007:32–4) who puts an emphasis on labour, i.e. industriousness, as a key factor. Compare this with the discussion in Chapters 2, 4 and 5 of this book.

22 Socio-economic constellations that work well entail an inbuilt coherence. There is coherence because the actors involved are coordinating with each other. Through this coordination many relations are constructed (between people, ecosystems, things, institutions, markets, etc.) and these relations make up a pattern. There is no need for people to be conscious of this pattern. It mostly simply represents 'the way things are'. However, the same pattern can be understood as a paradigm – it becomes a paradigm when it is made explicit. When deriving a paradigm from a particular practice (such as Chinese agriculture) one constructs a theory of practice. This can help us to better understand this practice and contribute to its further unfolding and development.

23 One special complication here is that this position often leads institutionalized science to align itself with dominant economic interests.

24 The *cattedratici ambulanti* ('wandering agricultural professors') of the end of the nineteenth century are important examples. They operated in Denmark, the Netherlands, Italy and elsewhere.

25 An Archimedean point is a (hypothetical) vantage point from which an observer can *objectively* view the subject of inquiry.

26 Some of this has been outlined in *The Virtual Farmer* (Ploeg, 2003; see especially Chapter 1.

27 An exemplary case is the contemporary pressure to widely introduce genetically modified organisms. They are presented as per se superior and any critique or resistance as per se stupid if not immoral.

28 This section builds on a book review in Ploeg, 2015b.

29 This is why many politicians and most scientists like this kind of approach. It also makes them look powerful.

30 Alternative paradigms are built with verbs, whereas the Western paradigm uses nouns. The difference is between the *process* of constructing and the constructed *things*.

31 In reality, of course, this moves in a different way. However rigid the imposed model might be, there will still be differential responses (many of these, ironically, triggered by the imposition of the model) that generate diverging trajectories and contrasting outcomes.

32 In other words, the Western paradigm assumes and suggests a high degree of systemness. Whether or not this is reflected in empirical reality does not matter. It is assumed and is to be created (or reaffirmed). Low degrees of systemness are seen as threatening, allowing for all kind of misfortune and disaster. Here the Western paradigm shows remarkable similarity to the earlier Marxist paradigm which viewed society as very strictly determined and governed by the capitalist mode of production and the associated process of accumulation.

33 This guiding principle evidently reflects interests and prospects that are partially alien to the primary sector.

34 This means that when two or more objectives apply, reality is segmented into different compartments, each reserved for one particular objective. The management of nature and landscape, as opposed to farming (each translated into separate zones) is exemplary of this.

35 As is well known, in Europe 80 per cent of the available budget went to 20 per cent of the largest farms.

36 See, for example, Hayami and Ruttan (1985) who focus on the need for institutional and technological changes in order to align farming with the markets. The state is de facto the central actor, 'directing' these changes. In neo-liberal versions it is international agreements (such as the Transatlantic Trade and Investment Partnership) and

organizations (such as the World Trade Organization) that take over this role which once belonged to nation states.

37 See for example Ostrom, 1990 and Polman *et al.*, 2010.

38 We are talking here about a paradigm. The practices of farming might deviate, at least partly, from this paradigm. This has been especially the case during recent decades. It is important to consider whether the frictions and contradictions between the paradigm and practices will lead to corrections in the latter.

39 Or the delivery of soldiers, as has been the case throughout much of history and across much of the world.

40 It goes without saying that there are many 'alternative' empirical practices that differ from the dominant ones. These emerge as actively constructed responses to policies and trends that embody the dominant paradigm.

41 There have, of course, also been major imbalances. The 'Great Leap Forward' was a clear example, when agriculture was heavily squeezed to allow for large-scale and urban industrialization (and repayments to the Soviets). Later on the TVE policy (decentralizing industries towards villages and townships, thus providing more opportunities for peasants for peasants to gain an income) represented a partial resetting of this balance between agriculture and industry.

42 The huge number of so-called mass incidents (70 per cent of them located in rural areas) is a continual reminder of this.

43 The need for relative well-being is also rooted in historical relations between the Communist Party and the peasantry. The Chinese Revolution was a peasant revolution and the peasantry represented the masses and gave strength to the Party. A current reason for improving the peasantry's income levels lies in the need to enlarge the internal market. Since 2004 the purchasing power of the peasant population has been much enlarged.

Bibliography

Abramovay, R. (1992), *Paradigmas do Capitalismo Agrario em Questao*. Estudios Rurais, Editoria HUCITEC, ANPOCS, Editoria da UNICAMP, Sao Paolo, Rio de Janeiro and Campinas.

Agricultural Price Survey (2013a), *Chart of Price Movements 1*. China Academic Journal Electronic Publishing House, Beijing.

Agricultural Price Survey (2013b), *Producer Price Indices of Agricultural Products 2*. China Academic Journal Electronic Publishing House, Beijing.

Agricultural Price Survey (2013c), *Prices of Agricultural Products at the Rural Market Fairs 3*. China Academic Journal Electronic Publishing House, Beijing.

Akram-Lohdi, A.H. and C. Kay (2010), Surveying the agrarian question (part 1): unearthing foundations, exploring diversity. *Journal of Peasant Studies*, 37(1):177–202.

Alkire S., R. Meinzen-Dick, A. Peterman, A.R. Quisumbing, G. Seymour and A. Vaz (2012), The *Women's Empowerment in Agriculture Index*. IFPRI Discussion Paper 01240, December 2012.

Alvin Y. So (2007), Peasant conflict and the local predatory state in the Chinese countryside. *Journal of Peasant Studies*, 34(3–4):560–81.

An, Miao (2011), The Social Processes of Agriculture in Li Village: A Sociological Research on Farmers' Agricultural Production from the Perspective of Food Security. College of Humanities and Development Studies, China Agricultural University, Beijing.

Anderson, K., J. Huang and E. Ianchovichina (2004), Will China's WTO accession worsen farm household incomes? *China Economic Review*, 15:443–56.

Anon (2015), The Industrialization of Seed. *Ethical Consumer*, 153(Mar/April):24–5.

Appadurai, A. (1986), *The Social Life of Things: Commodities in cultural perspective*. Cambridge University Press, Cambridge, UK.

Arkush, D. (1984), If man works hard the land will not be lazy: entrepreneurial values in North Chinese peasant proverbs. *Modern China*, 10(4):461–479.

Arrighi, G. (2007), *Adam Smith in Beijing: lineages of the twenty-first century*. Verso, London/New York.

Ash, R.F. (1998), *Agricultural Development in China 1949–1989: The collected papers of Kenneth R. Walker (1931–1989)*. Oxford University Press, Oxford, UK.

Asis, Maruja M.B. (2003), International Migration and Families in Asia, in R. Iredale, C. Hawksley and S. Castles (eds), *Migration in the Asia Pacific, Population, Settlement and Citizenship Issues*. Edward Elgar, Cheltenham, UK, pp. 99–117.

Asis, Maruja M.B. (2004), *When Men and Women Migrate: Comparing Gendered Migration in Asia*. United Nations Division for the Advancement of Women (DAW) consultative meeting on 'Migration and mobility and how this movement affects women', Malmö, Sweden, 2003, pp. 1–20.

Beardson, T. (2013), *Stumbling Giant: The threats to China's future*. Yale University Press, New Haven and London.

Bennett, J. (1981), *Of Time and the Enterprise, North American family farm management in a context of resources marginality*. University of Minnesota Press, Minneapolis, MN.

Benvenuti, B. (1982), De technologisch-administratieve taakomgeving (TATE) van landbouwbedrijven. *Marquetalia*, 5:111–36.

Berendsen, B., T. Dietz, H. Schulte Nordholt and R. van Veen (2013), *Asian Tigers, African Lions: comparing the development performance of Southeast Asia and Africa*. Brill, Leiden.

Bernstein, H. (2009), V.I. Lenin and A.V. Chayanov: Looking back, looking forward. *Journal of Peasant Studies*, 36(1):55–81.

Bernstein, H. (2010), *Class Dynamics of Agrarian Change*. Fernwood, Halifax and Winnipeg.

Bernstein, H. (2015), Some Reflections on Agrarian Change in China. *Journal of Agrarian Change*, 15:454–77. doi: 10.1111/joac.12116 (last accessed 3 February 2016).

Berry, S. (1985), *Fathers Work for Their Sons: Accumulation, mobility and class formation in an extended Yoruba community*. University of California Press, Berkeley, CA.

Bock, B. (2004), Fitting in and Multi-tasking: Dutch Farm Women's Strategies in Rural Entrepreneurship. *Sociologia Ruralis*, 44(3):248–57.

Booth, D. (2013), Policy and Governance in Africa's Economic Transformation: Firm Findings and Remaining Questions, in B. Berendsen, T. Dietz, H. Schulte Nordholt and R. van Veen (2013), *Asian Tigers, African Lions: comparing the development performance of Southeast Asia and Africa*. Brill, Leiden, pp. 449–516.

Borras, S., J. Franco and J.D. van der Ploeg (2013), *Land concentration, land grabbing and people's struggles in Europe*. TNI, Amsterdam.

Bramall, C. and M.E. Jones (1993), Rural Income inequality in China since 1978. *Journal of Peasant Studies*, 21(1):41–70.

Brandth, B. (2002), Gender Identity in European Family Farming: A Literature Review. *Sociologia Ruralis*, 42(3):181–200.

Brauw, A. de, Jikun Huang and S. Rozelle (2004), The sequencing of reforms in China's agricultural transition. *Economics of Transition*, 12(3):427–66.

Brauw, A. de, Jikun Huang, Linxiu Zhang and S. Rozelle (2012), *The Feminization of Agriculture with Chinese Characteristics*. IFPRI Discussion Paper 01189, June 2012, Washington.

Bray, F. (1986), *The Rice Economies: Technology and development in Asian societies*. Blackwell, Oxford, UK.

Broekhuizen, R. van, H. Oostindie, H. Dirksen, M. Klever and J.D. van der Ploeg (2013), *Bouwen aan een betere balans: een analyse van bedrijfsstijlen in de melkveehouderij*. Wageningen University/DMS, Wageningen, Netherlands.

Brown, L. (1995), *Who will feed China? Wake-up Call for a small Planet*. W.W. Norton & Co., World Watch Institute, New York.

Brox, O. (2006), *The Political Economy of Rural Development: Modernisation without centralisation?* (ed. and intro. by J. Bryden and R. Storey). Eburon, Delft, Netherlands.

Brush, S., J.C. Heath and Z. Huaman (1981), Dynamics of Andean potato agriculture. *Economic Botany*, 35(1):70–88.

Bryden, J.M., C. Bell, I. Gilliatt, E. Hawkins and N. MacKinnon (1992), *Farm Household Ajustment in Western Europe 1987–1991*. Final report on the Research Programme on Farm Structures and Pluriactivity, Volume 1 and 2, ATR/92/14, European Commission, Brussels.

Burt, R.S. (1987), Social contagion and innovation: Cohesion versus structural equivalence. *American Journal of Sociology*, 92:1287–1335.

Camus, A. (1998), *Notebooks, 1935–1951*. Perseus Books Group, New York.

Carillo B. and D.S. Goodman (2012), *China's Peasants and Workers: Changing class identities*, (CSC China Perspectives series). Edward Elgar, Cheltenham UK/ Northampton, MA.

Carter, B.L., McGoldrick, M. (2005), *The Expanded Family Life Cycle: Individual, family, and social perspectives*. Allyn and Bacon, New York.

Chang, L.T. (2008), *Factory Girls: From village to city in a changing China*. Spiegel & Grau, New York.

Chayanov, A.V. (1966), *The Theory of Peasant Economy*. Manchester University Press, Manchester, UK.

Chen, F. and J. Davis (1998), Land reform in rural China since the mid-1980s. *Land Reform*, 1998(2):123–37.

Chen, H. and C. Zhang (2013), Making capitalism in China, a review. *Journal of Peasant Studies*, 40(3):606–9.

Chen, Po-Ci, Ming-Miin Yu, Ching-Cheng Chang and Shih-Hsun Hsu (2008), Total factor productivity growth in China's agricultural sector. *China Economic Review*, 19:580–93.

Chen, X.W. (1987), The rural economy in China is transferring from supra conventional growth to conventional growth. *Economic Research Journal*, 12, Beijing (in Chinese).

Cheng, S. (1998), Nongye nüxinghua qushi yu nongcun nüxing renli ziben guanxi de shizheng yanjiu (Empirical study of the relationship between agricultural feminization trends and rural labour capital). *Zhengzhou daxüe xüebao (zhexüe shehui kexüe ban) Journal of Zhengzhou University (Philosophy and Social Sciences)* 3:83–8.

CIDA (1973), Bodennutzung und betriebsfuhrung in einer latifundio–landwirtschaft, in E. Feder (ed.) *Gewalt und Ausbeutung, Lateinamerikas Landwirtschaft*, Hofmann und Campe Verlag, Hamburg, Germany.

Clayton, A. (2015), Monsanto: Guardian of seeds or profits? *Ethical Consumer* 153 (Mar/April):29.

Copus, A.K., M. Shucksmith, T. Dax and D. Meredith (2013), Cohesion Policy for rural areas after 2013. A rationale derived from the EDORA project (European Development Opportunities in Rural Areas) – ESPON 2013 Project 2013/1/2.

Cristóvão, A., H. Oostindie and F. Perreira (1994), Practices of endogenous development in Barroso, northern Portugal, in J.D. van der Ploeg and A. Long (eds), *Born from Within*. Van Gorcum, Assen, Netherlands, pp. 38–58.

Davidova, S., A. Bailey, J. Dwyer, E. Erjavec, M. Gorton and K. Thomson (2013), Semi-Subsistence Farming – Value and Directions of Development. European Parliament, Policy Department B: Structural and Cohesion Policies, Brussels.

Davis, M. (2006), *Planet of Slums*. Verso, London.

Day, A.F. (2013), *The Peasant in Post-socialist China: History, politics, and capitalism*. Cambridge University Press, Cambridge, UK.

Deere, C.D. (2005), *The Feminization of Agriculture? Economic Restructuring in Rural Latin America*. Occasional Paper 1 – The Feminization of Agriculture? Economic Restructuring in Rural Latin America, United Nations Research Institute For Social Development, Geneva, Switzerland.

Deininger, K., D. Nizalov and S.K. Singh (2013), *Are Mega-Farms the Future of Global Agriculture? Exploring the farm size-productivity relationship for large commercial farms in Ukraine*. Policy Research Working Paper 6544, The World Bank Development Research Group, Agriculture and Rural Development Team, Washington.

Donaldson, J.A. (2011), *Small Works: Poverty and economic development in south-western China*. Cornell University Press, Ithaca and London.

Du, Runsheng (2006), *The Course of China's Rural Reform*. International Food Policy Research Institute, Washington.

Duncan, K., I. Rutledge and C. Harding (eds) (1977), *Land and Labour in Latin America: Essays on the development of agrarian capitalism in the nineteenth and twentieth centuries*. Cambridge University Press, Cambridge, UK.

Elder, G. (1974), *Children of the Great Depression: Social change in life experience*. University of Chicago Press, Chicago, IL.

Eversole, R. and M. Johnson (2014), Migrant remittances and household development: an anthropological analysis. *Development Studies Research: An Open Access Journal*, 1(1):1–15. doi: 10.1080/21665095.2014.903808 (last accessed 3 February 2016).

Eyferth, J., P. Ho and E.B. Vermeer (2003), The opening-up of China's countryside. *Journal of Peasant Studies*, 30(3–4):1–17.

Fan, S. and C. Zhu (2007), Xin nongcun jianshe beijingxia de Fujian sheng nongye nüxinghua wenti yanjiu [Agricultural feminization in Fujian against the background of building a new socialist countryside]. *Fujian nonglin daxüe xüebao Zhexue shehui kexüe ban [Journal of Fujian Agriculture and Forestry University – Philosophy and Social Sciences*], 6:28–33.

FAO (2011), *The State of Food and Agriculture: 2010–2011. Women in Agriculture: Closing the gender gap for development*. Food and Agriculture Organization of the United Nations, Rome.

Fei, Hsiao-Tung (1939), *Peasant Life in China: A field study of country life in the Yangtze Valley*. George Routledge and Sons, London.

Fei, Hsiao-Tung and Chih-I Chang (1945), *Earthbound China, A study of rural economy in Yunnan*. University of Chicago Press, Chicago, IL.

Fuller, A.M. (1975), Linkages between social and spatial systems – the case of farmer mobility in Northern Italy. *Sociologia Ruralis*, 15(1–2):119–24.

Gao, Q. (2001), Role of rural women in agricultural industrialization. *Gansu Agriculture*, 3:13–15.

Garcia Marquez, G. (1970), *One Hundred Years of Solitude*. Harper & Row, New York.

Gereffi, G., J. Humphrey and T. Sturgeon (2005), The governance of global value chains. *Review of international political economy*, 12(1):78–104.

Goodman, D. (2004), Rural Europe redux? Reflections on alternative agro-food networks and paradigm change. *Sociologia Ruralis*, 44(1):3–16.

Gudeman, S. (1978), *The Demise of a Rural Economy: From subsistence to capitalism in a Latin American village*. International Library of Anthropology, London.

Gulati, A. and S. Fan (2007), *The Dragon and the Elephant: Agricultural and rural reforms in China and India*. John Hopkins University Press, Baltimore, MD.

Gürel, B. (2014), Changing relations of production in Chinese agriculture from decollectivization to capitalism. *McGill Sociological Review*, 4:67–92.

Guthman, J. (2008), Thinking inside the neoliberal box: the micro-politics of agro-food philanthropy. *Geoforum*, 39(3):1241–53.

Haggblade, S. and P. Hazell (1989), Agricultural technology and farm-nonfarm growth linkages. *Agricultural Economics*, 3(4):345–64.

Haggblade, S., P. Hazell and P.A. Dorosh (2007), Sectoral growth linkages between agriculture and the rural non-farm economy, in S. Haggblade, P. Hazell and T. Reardon (eds), *Transforming the Rural Nonfarm Economy, Opportunities and threats in the developing world*. John Hopkins University Press, Baltimore, MD, pp. 141–82.

Halamska, M. (2004), A different end of the peasants. *Polish Sociological Review*, 3(147):205–68.

Harvey, D. (2003), The 'new' imperialism: accumulation by dispossession. *Socialist Register*, 40:63–87.

Haugen, M.S. and J. Vik (2008), Farmers as entrepreneurs: the case of farm-based tourism. *International Journal of Entrepreneurship and Small Business*, 6(3):321–36.

Haugen, M.S. and M. Villa (2006), Rural Idylls or Boring Places?, in B.B. Bock and S. Shortall (eds), *Rural Gender Relations: Issues and case-studies*, CABI, Oxfordshire, UK, pp. 181–95.

Hayama, Y. and V. W. Ruttan (1985), *Agricultural Development: An International Perspective* (revised and expanded edn). John Hopkins University Press, Baltimore, MD and London.

Hebinck, P., J.D. van der Ploeg and S. Schneider (2015), *Rural Development and the Construction of New Markets*. Routledge ISS Studies in Rural Livelihoods 12, Routledge, London and New York.

Heerink, N., R. Li, S. Feng, K. Lu and X. Bao (2010), The impact of soil and water conservation investments on agricultural development in western China, in V. Beckmann, N.H. Dung, X. Shi, M. Spoor and J. Wesseler (eds), *Economic Transition and Natural Resource Management in East and Southeast Asia*. Shaker, Aachen, Germany, pp. 67–94.

Henley, D. and J.K. van Donge (2013), Diverging Paths: Explanations and Implication, in B. Berendsen, T. Dietz, H. Schulte Nordholt and R. van Veen (eds), *Asian Tigers, African Lions: comparing the development performance of Southeast Asia and Africa*. Brill, Leiden, Netherlands, pp. 27–50.

Hessels, L.K. and H. van Lent (2008), Re-thinking new knowledge production: a literature review and a research agenda. *Research Policy*, 37:740–60.

Hinton, W. (1990), *The Privatisation of China, the Great Reversal*. Earthscan, London.

HLPE [High Level Panel of Experts] (2013), *Investing in Smallholder Agriculture for Food Security*. Committee on World Food Security (CFS), FAO, Rome.

Hobsbawm, E. (1994), *The Age of Extremes: The short twentieth century, 1914–1991*. Michael Joseph, London.

Hofman, I. and P. Ho (2012), China's 'Developmental Outsourcing': A critical examination of Chinese global 'land grabs' discourse. *Journal of Peasant Studies*, 39(1):1–48.

Holloway, J. (2010), *Crack Capitalism*. Pluto Press, Sidmouth, UK.

Hoste, R., N. Bondt and P. T. M. Ingenbleek (eds) (2004), *Visie op de varkenskolom, Wetenschapswinkel Wageningen UR Rapport 207*. Wetenschapswinkel WUR, Wageningen.

Howell, J. (1998), Prospects for village self-governance in China. *Journal of Peasant Studies*, 25(3):86–111.

Hu, Zhanping (2014), *Socio-Economic Drivers of Agricultural Production in a Transition Economy: a Case-Study of Hu Village, Sichuan Province*, China. PhD thesis. Plymouth University, Plymouth, UK.

Huang, P. (2010a), Small farms and vertical integration in China's new era: Dragon-head enterprises or cooperative organizations? *China Rural Studies*, 8:11–30 (in Chinese).

Huang, P. (2010b), *China's Hidden Agricultural Revolution*. Law Press, Beijing.

Huang, P. (2011), The modern Chinese family: in light of economic and legal history. *Modern China* 37(5), 459–97.

Huang, W. (2008), Role of rural women's human resource development. *China Science and Technology Forum* 5:132–5.

Hugo, G.J. (2000), Migration and Women's Empowerment, in H.B. Presser and G. Sen (eds), *Women's Empowerment and Demographic Processes: Moving beyond Cairo*. Oxford University Press, New York, pp. 287–317.

IAASTD [International Assessment of Agricultural Knowledge, Science and Technology for Development] (2008), *Global Summary for Decision Makers*. www.agassessment.org/docs/Global_SDM_050508_FINAL.pdf (last accessed 2 February 2016).

Iacoponi, L. (2007), Impresa e azienda agraria: vecchi e nuovi paradigmi. *Rivista di Economia Agraria*, Anno LXII, 3:333–40.

IFAD [International Fund for Agricultural Development] (2013), *Smallholders, food security and the environment*. IFAD, Rome.

ISMEA (2012), *La Competitività dell'Agroalimentare Italiano, Checkup 2012*. ISMEA, Rome.

Janvry, A. de and E. Sadoulet (1993), Market, state, and civil organizations in Latin America beyond the debt crisis: the context for rural development. *World Development*, 21(4):659–74.

Janvry, A. de., E. Sadoulet and N. Zhu (2005), *The Role of Non-Farm Incomes in Reducing Rural Poverty and Inequality in China*. Working Paper. www.are.berkeley.edu/sadoulet (last accessed 2 February 2016).

Jollivet, M. (2001), *Pour une science sociale a travers champs: Paysannerie, ruralité, capitalisme* (France XXe siècle). Editions Arguments, Paris.

Judd, Ellen R. (2007), No change for thirty years: the renewed question of women's land rights in rural China. *Development and Change*, 38(4):691–712.

Kabeer, Naila (2005), Gender equality and women's empowerment: a critical analysis of the third Millennium Development Goal 1. *Gender and Development*, 13(1):13–24.

Karshenas, M. (1996), Dynamic economies and the critique of urban bias. *Journal of Peasant Studies*, 24(1–2):60–102.

Kearney, M. (1996), *Reconceptualizing the Peasantry: Anthropology in a global perspective*. Westview Press, Boulder, CO.

Kelkar, G. (2009), *The Feminisation of Agriculture in Asia: Implication for women's agency and* productivity. United Nations Development Fund for Women, New Delhi.

Kessel, J. van (1990), Herwaarderen om te herleven: produktieritueel en technologisch betoog bij de Andesvolken. *Derde Wereld*, 90(1–2):77–98.

Kimhi, A. and R. Bollman (1999), Family farm dynamics in Canada and Israel: The case of farm exits. *Agricultural Economics* 21:69–79.

King, F.H. (1911), *Farmers of Forty Centuries: Organic farming in China, Korea, and Japan*. Mrs F.H. King, Madison, Wisconsin (reprinted in 2011 by Benediction Classics, Oxford).

Kung, J. K. and Y. Lin (2007), The decline of township-and-village enterprises in China's economic transition. *World Development* 35:569–84.

Kuns, B., O. Visser and A. Wästfelt (2014), Swedish black earth: The trials and tribulations of the publicly listed Western farming model in Russia and Ukraine. Conference paper presented at the *ISS/EURAL* conference *Food, Farmland and Forests in Transition: The Eurasian Countryside 25 Years After*. F.I. Rainer Institute, Bucharest, 7 November 2014.

Lastarria-Cornhiel, S. (2006), *Feminization of Agriculture: Trends and driving forces*. Background Paper for the World Development Report 2008. Rimisp-Latin American Center for Rural Development.

Lee, C-K. (2007), *Against the Law: Labor protest in China s rustbelt and sunbelt*. University of California Press, Berkeley, CA.

LEI (2008), *Agricultural Economic Report 2008 of the Netherlands*. P. Berkhout and C. van Bruchem (eds). Den Haag, LEI, Rapport 2008-030.

Lerman, Z. and D.J. Sedik (2007), Productivity and efficiency of corporate and individual farms in Ukraine. Selected paper prepared for presentation at the *American Agricultural Economics Association Annual Meeting*, Portland, OR, 29 July–1August 2007.

Leung, P.P. and A.Y. So (2012), The making and re-making of the working class in South China, in B. Carillo and D.S. Goodman (eds), *China's Peasants and Workers: Changing class identities*. CSC China Perspectives series. Edward Elgar, Cheltenham, UK and Northampton, MA, pp. 62–78.

Li, T. (2010), Peasants and globalization: political economy, rural transformation and the agrarian question. *Canadian Journal of Development Studies*, 29(3–4):519–21.

Li, W. (2009), Zhongguo nongcun funü jingji diwei shizheng yanjiu: yi jiangsusheng he gansusheng weili (Research on the economic status of rural women in China: Case of Jiangsu and Gansu Provinces). *Nongye jingji wenti (Issues in Agricultural Economy)* 4:26–32.

Li, Xiande (2003), Rethinking the peasant burden: evidence from a Chinese village. *Journal of Peasant Studies*, 30(3–4):45–74.

Li, Xiaoyun, Q. Gubo, T. Lixia, Z. Lixia, J. Leshan, G. Zhanfeng and W. Jin (2012), *Agricultural Development in China and Africa: A Comparative Analysis*. Routledge/Taylor and Francis Group, London and New York.

Lin, Zhibin and Zhang Lixin (2006) Gender, Land and Local Heterogeneity. *Journal of Contemporary China*, 15(49):637–50.

Lipton, M. (1977), *Why Poor People Stay Poor: Urban bias in world development*. Temple Smith, London.

Little, D. (1989), *Understanding Peasant China: Case studies in the philosophy of social science*. Yale University Press, New Haven and London.

Long, H. (2014), Land consolidation: An indispensable way of spatial restructuring in rural China. *Journal of Geographical Sciences*, 24(2):211–25.

Long, H., J. Zou, J. Pykett and Y. Li (2011), Analysis of rural transformation development in China since the turn of the new millennium. *Applied Geography*, 31(3):1094–1105.

Long, H., Y. Li, Y. Liu, M. Woods and J. Zou (2012), Accelerated restructuring in rural China fueled by 'increasing vs. decreasing balance' land-use policy for dealing with hollowed villages. *Land Use Policy*, 29(1): 11–22.

Long, N. (2015), Activities, Actants and Actors: Theoretical Perspectives on Development Practice and Practitioners, in P. Milone, F. Ventura and Ye Jingzhong (eds), *Constructing a New Framework for Rural Development*. Research in Rural Sociology and Development Series (vol. 22), Emerald, UK, pp.31–58.

Long, N. and B.R. Roberts (1984), *Miners, Peasants, and Entrepreneurs: Regional development in the central highlands of Peru*. Cambridge University Press, Cambridge, UK.

Long, N. and J. D. van der Ploeg (1994), Heterogeneity, actor and structure: towards a reconstitution of the concept of structure, in D. Booth (ed.), *Rethinking Social Development: Theory, research and practice*. Longman Group, Harlow, UK, pp. 62–90.

Losch, B. (2013), Structural change and employment in Africa: situation and development issues. Paper for *Rural Development Forum in Africa*, Cotonou, 2–4 May 2013.

Lou, B., Z. Zheng, R. Connelly and K.D. Roberts (2004), The migration experiences of young women from four counties in Sichuan and Anhui, in A.M. Gaetano and T. Jacka (eds), *On the Move: Women and rural to urban migration in contemporary China*. Columbia University Press, New York, pp. 207–42.

Lu, M., and X.J. Dai (1987), An analysis of peasant households' economic behaviour at present. *Economic Research Journal*, Beijing 17 (in Chinese).
Lu, Xiaobo (1997), The politics of peasant burden in reform China. *Journal of Peasant Studies*, 25(1):113–38.
Lu, Zhigang and S. Song (2007), Rural–urban migration and wage determination: the case of Tianjin, China, in B.M. Fleisher, H. Li and S. Song (eds), *Market Development in China, Spillovers, Growth and Inequality*. Advances in Chinese Economic Studies. Edward Elgar, Cheltenham, UK, pp. 209–21.
Ma, Z. (1999), Temporary migration and regional development in China. *Environment and Planning A*, 31:783–802.
McKinley, T. and K. Griffin (1993), The distribution of land in rural China. *Journal of Peasant Studies*, 21(1):71–84.
Mamdani, M. (1987), Peasants and Democracy in Africa. *New Left Review*, 1(156):37–49.
Martin, L.R. (1977), *A Survey of Agricultural Economics Literature, Vol 1. Traditional fields of agricultural economics, 1940s to 1970s*. University of Minnesota Press, Minneapolis, MN.
Mazoyer, M. and L. Roudart (2006). *A History of World Agriculture: From the Neolithic Age to the current crisis* (trans. J.H. Membrez). Earthscan, London.
Mendras, H. (1976), *Societes paysannes: elements pour une theorie de la paysannerie*. Colin, Paris.
Meng, Xiangdan (2014), *Feminization of Agricultural Production in Rural China. A sociological analysis*. PhD thesis. WASS, Wageningen University, Netherlands.
Ministry [Ministry of Agriculture of the People's Republic of China] (2010), *China Agricultural Development Report, 2010*. China Agricultural Press, Beijing.
Mohapatra, S., S. Rozelle and R. Goodhue (2006), The rise of self-employment in rural China: development or distress? *World Development*, 35(1):163–81. doi:10.1016/j.worlddev.2006.09.007 (last accessed 3 February 2016).
Mons W., and K. Le (2006), 'Gangster capitalism' and peasant protest in China: The last twenty years. *Journal of Peasant Studies*, 33(1):1–33.
Mosher, A.T. (1966), *Getting Agriculture Moving: essentials for development and modernization*. Praeger, New York, pp. 209–27.
Mu, R. and D. van de Walle (2009), *Left behind to farm? Women's labor re-allocation in rural China*. Policy Research Working paper 5107, The World Bank, Washington.
Myrdal, J. (1965), *Report From a Chinese Village*. Vintage Books, New York.
National Bureau of Statistics (2010), *China Yearbook 2010*. Department of Rural Surveys, China Statistics Press, Beijing.
Netting, R. McC (1993), *Smallholders, Householders: Farm families and the ecology of intensive, sustainable agriculture*. Stanford University Press, Stanford, CA.
Newby, H. (1980), *Green and pleasant land? Social change in rural England*. Pelican Books, London.
Nguyen, M.T.N. and C. Locke (2014), Rural–urban migration in Vietnam and China: gendered householding, production of space and the state. *Journal of Peasant Studies*, 41(5):855–76.
Niu, R. (2006), Management and development of industrialized agriculture in China: Observations and comments. *Issues in Agricultural Economy*, 3:8–15 (in Chinese).
OECD (2010), *OECD Territorial Reviews: Guangdong, China 2010*. OECD. www.oecd-ilibrary.org/urban-rural-and-regional-development/oecd-territorial-reviews-guangdong-china-2010_9789264090088-en (last accessed 2 February 2016).

Oostindie, H., R. Van Broekhuizen and J.D. van der Ploeg (2013), *Buffercapaciteit: Bedrijfsstijlen in de melkveehouderij, volatiele markten en kengetallen*. WUR/ALFA, Wageningen, Netherlands.

Osti, G. (1991), *Gli innovatori della periferia, la figura sociale dell'innovatore nell' agricoltura di montagna*. Reverdito Edizioni, Torino, Italy.

Ostrom, E. (1990), *Governing the Commons: The evolution of institutions for collective action*. Cambridge University Press, Cambridge, UK.

Pasternak, B. (1972), *Kinship and Community in Two Chinese Villages*. Stanford University Press, Stanford, CA.

Pearse, A. (1975), *The Latin American Peasant*. Frank Cass, London.

Peng, Mai. (2007), The impact of 'empty nest village' phenomenon on new countryside construction. *Academic Journal of Zhongzhou* 3:125–7.

Peterman, A., J. Behrman and A. Quisimbing (2010), *A Review of Empirical Evidence on Gender Differences in Non-land Agricultural Inputs, Technology, and Services in Developing Countries*. International Food Policy Research Institute (IFPRI).

Plas, L. van der and M. Fonte (eds) (1994), *Rural Gender Studies in Europe*. Van Gorcum, Assen, Netherlands.

Ploeg, J.D. van der (1974), *De Gestolen Toekomst; imperialisme, landhervorming en boerenstrijd in Peru*. Uytbuyt, Wageningen, Netherlands.

Ploeg, J.D. van der (1987), *De Verwetenschappelijking van de Landbouwbeoefening*. Mededelingen van de vakgroepen voor sociologie 21, Landbouwuniversiteit, Wageningen, Netherlands.

Ploeg, J. D. van der (1990), *Labor, Markets and Agricultural Production*. Westview Press, Boulder, CO.

Ploeg, J.D. van der (1993), On potatoes and metaphors, in M. Hobart (ed.), *An Anthropological Critique of Development: The growth of ignorance*. Routledge, London and New York, pp. 209–27.

Ploeg, J. D. van der (2003), *The Virtual Farmer: Past, present and future of the Dutch peasantry*. Royal van Gorcum, Assen, Netherlands.

Ploeg, J.D. van der (2006), *El Futuro Robado: Tierra, agua y lucha campesina*. Instituto de Estudios Peruanos (IEP), Lima.

Ploeg, J. D. van der (2008), *The New Peasantries: Struggles for autonomy and sustainability in an era of empire and globalization*. Earthscan, London.

Ploeg, J.D. van der (2013), *Peasants and the Art of Farming: A Chayanovian manifesto*. Fernwood Publishing, Halifax and Winnipeg.

Ploeg, J.D. van der (2014), Peasant-driven agricultural growth and food sovereignty. *Journal of Peasant Studies*, 41(6):999–1030.

Ploeg, J.D. van der (2015a), Newly emerging, nested markets: a theoretical introduction, in P. Hebinck, J.D. van der Ploeg and S. Schneider (eds), *Rural Development and the Construction of New Markets*. Routledge, London and New York, pp. 16–40.

Ploeg, J.D. van der (2015b), Book Review. Small works: poverty and economic development in southwestern China by John A. Donaldson. *Journal of Peasant Studies*, 42(2), pp. 449–52.

Ploeg, J.D. van der and Ye Jingzhong (2010), Multiple job holding in rural villages and the Chinese road to development. *Journal of Peasant Studies*, 37(3):513–30.

Ploeg, J.D. van der, Ye Jingzhong and S. Schneider (2012), Rural development through the construction of new, nested, markets: comparative perspectives from China, Brazil and the European Union. *Journal of Peasant Studies*, 39(1):133–73.

Ploeg, J.D. van der, Ye Jingzhong, Wu Huifang and Wang Chunyu (2014), Peasant-managed agricultural growth in China: mechanisms of labour-driven intensification. *International Journal of the Sociology of Agriculture and Food* 21(1):155–71.

Polanyi, K. (1957), *The Great Transformation, The political and Economic Origins of Our Time*. Beacon Press, Boston, MA.

Polman, N., K. Poppe, J.W. van der Schans and J.D. van der Ploeg (2010), Nested markets with common pool resources in multifunctional agriculture. *Rivista di Economia Agraria*, Anno LXV, No. 2:295–318.

Poulantzas, N. (1975), *Classes in Contemporary Capitalism*. Fernback, London.

Priebe, H. (1985), *Die Subventionierte Unvernunft, Landwirtschaft and Naturhaushalt*. Siedler, Berlin.

Qu, F., A. Kuyvenhoven, X. Shi and N. Heerink (2009), *Sustainable Natural Resource Use in Rural China: Trends and policies* (working paper). Wageningen University, Wageningen, Netherlands.

Quisumbing, A. (1994), *Improving Women's Agricultural Productivity as Farmers and Workers*. Education and Social Policy Department, World Bank ESP Discussion Paper Series No. 37, July.

Rampini, F. (2005), *Il secolo cinese. Storie di uomini, città e denaro dalla fabbrica del mondo*. Mondadori, Milano.

Ray Liaw, H. (2008), Women's land rights in rural China: transforming existing laws into a source of property rights. *Pacific Rim Law & Policy Journal*, 17(1):237–64.

Renting, H., W.A.H. Rossing, J.C.J. Groot, J.D. van der Ploeg, C. Laurent, D. Perraud, D.J. Stobbelaar and M.K. Ittersum (2009), Exploring multifunctional agriculture: a review of conceptual approachers and prospects for an integrative transitional framework. *Journal of Environmental Management* 90:112–23.

Rete Rurale Nazionale (2012), *Young people's perception of rural areas*. A European survey carried out in 8 EU member states. Rete Rurale Nazionale, Rome.

Rip, A. and R. Kemp (1998), Technological change, in S. Rayner and E.L. Malone (eds), *Human Choice and Climatic Change*, Vol. 2. Batelle Press, Columbus, OH, pp. 327–99.

Rip, A. and J.W. Schot (2001), Identifying loci for influencing the dynamics of technological development, in J. Williamson and P. Sorensen (eds), *Social Shaping of Technology*. Edward Elgar, London, pp. 155–72.

Rooij, S. J. G. de (1992), *Werk van de Tweede Soort, boerinnen in de melkveehouderij*. Van Gorcum, Assen, Netherlands.

Rooij, S. J. G. de (1994), Work of the second order, in L. Van der Plas and M. Fonte (eds), *Rural Gender Studies in Europe*.Van Gorcum, Assen, Netherlands, pp. 69–79.

Rooij, S. J. G. de, E. Brouwer and R. Van Broekhuizen (1995), *Agrarische Vrouwen en Bedrijfsontwikkeling*. LUW/WLTO, Wageningen, Netherlands.

Rooij, S. de, F. Ventura, P. Milone and J.D. van der Ploeg (2014), Sustaining Food Production through Multifunctionality: The Dynamics of Large Farms in Italy. *Sociologia Ruralis*, 54(3):303–20.

Rozelle, S, J., E. Taylor and A. de Brauw (1999a), Migration, remittances and productivity in China. *The American Economic Review*, 89(2):287–91.

Rozelle, S., Guo Li, Minggao Shen, A. Hughart and J. Giles (1999b), Leaving China's farms: survey results of new paths and remaining hurdles to rural migration. *China Quarterly*, 158:367–93.

Saccomandi, V. (1998), *Agricultural Market Economics: A neo-institutional analysis of exchange, circulation and distribution of agricultural products*. Royal Van Gorcum, Assen, Netherlands.

Safilios-Rothschild, C. (2006), CAP Regulations and Farm Household Relations, in B.B. Bock and S. Shortall (eds), *Rural Gender Relations: Issues and case-studies*. CABI, Oxfordshire, pp. 260–78.

Saith, A. (1995), From collectives to markets: restructured agriculture–industry linkages in rural China: some micro-level evidence. *Journal of Peasant Studies*, 22(1):201–60.

Saito, K.A., H. Mekonnen and D. Spurling (1994), *Raising the Productivity of Women Farmers in Sub-Saharan Africa*. World Bank Publications, Washington.

Sandro, G. di (1981), Per una nuova política di gestione della stalla sociale. *Rivista di Política Agraria*, 28(3):33–46.

Scherer, F. (1975), *The Economics of Multiplant Operation*. Harvard University Press, Cambridge, MA.

Schneider, M. (2013), *Dragonhead enterprises and the state of agribusiness in reform era China*. Paper presented at a workshop on 'Agrarian Development in China: Legacies and Prospects', The Arrighi Center for Global Studies & The East Asian Studies Program, Johns Hopkins University, Baltimore, MD, 1 November 2013.

Schneider, M. (2014), What then, is a Chinese peasant? Nongmin discourses and agro-industrialization in contemporary China. *Agriculture and Human Values*, 32(2):331–46. DOI 10.1007/s10460–014–9559–6 (last accessed 3 February 2016).

Schoneveld, G.C. (2014), The geographic and sectoral patterns of large-scale farmland investments in sub-Saharan Africa. *Food Policy* 48:34–50.

Schultz, T.W. (1964), *Transforming Traditional Agriculture*. Yale University Press, New Haven, CT.

Schumilas, T. (2014), Alternative Food Networks with Chinese Characteristics. PhD thesis. University of Waterloo, Ontario, Canada.

Schumpeter, J.A. (1951), *Essays: On entrepreneurs, innovations, business cycles, and the evolution of capitalism*. Addison-Wesley, Cambridge, MA.

Schutter, O. de (2013), The agrarian transition and the 'feminization' of agriculture. *Food Sovereignty: A Critical Dialogue*, conference paper No. 37, international conference 14–15 September 2013, Yale University.

Scott, J.C. (1998), *Seeing Like a State: How certain schemes to improve the human condition have failed*. Yale University Press, New Haven and London.

Servolin, C. (1989), *L'Agriculture Moderne*. Seuil, Paris.

Shanin, T. (1971), Peasantry: Delineation of a sociological concept and a field of study. *European Journal of Sociology*, 12(2):289–300.

Si, Zhenzhong (2014), *Alternative Food Networks and Rural Development Initiatives in China: Characterization, Contestations and Interactions*. PhD thesis. University of Waterloo, Ontario, Canada.

Si, Zhengzhong, T. Schumilas and S. Scott (2014), Characterizing alternative food networks in China. *Agriculture and Human Values*, 31(3). doi 10.1007/s10460–014–9530–6 (last accessed 3 February 2016).

Sitko, N.J. and T.S. Jayne (2014), Structural transformation or elite land capture? The growth of 'emergent' farmers in Zambia. *Food Policy*, 48:194–202.

So, A.Y. (2007), Peasant conflict and the local predatory state in the Chinese countryside. *Journal of Peasant Studies*, 34(3–4):560–81.

Stuiver, M., J.D. van der Ploeg and C. Leeuwis (2003), The VEL and VANLA cooperatives as field laboratories. Wageningen Journal of Life Sciences, 51(1–2):27–40.

Talbot, H. and M. Shucksmith (2011), European Development Opportunities for Rural Areas (EDORA). Applied Research 2013/1/2, Final Report, Parts A, B and C. ESPON and UHI Millennium Institute.

Tamang, S., K.P. Paudel and K.K. Shrestha (2014), Feminization of agriculture and its implications for food security in rural Nepal. *Journal of Forest and Livelihood* 12(1):20–32.

Taylor, J.E., S. Rozelle and A. de Brauw (2003), Migration and incomes in source communities: A new economics of migration perspective from China. *Economic Development and Cultural Change*, 52(1):75–101.

Thiesenhuisen, W.C. (1995), *Broken promises: Agrarian Reform and the Latin American Campesino*. Westview Press, Boulder, CO.

Thomas, F., P-F. Vaquié, V. Lucas and P. Gasselin (2015), *Coopération agricole de production: renouvellement des modalités de coopération de proximité entre agriculteurs*. Paper presented at the Colloque SFER SA 2015, 12–13 February 2015, Rennes, France.

Tittonell, P. (2013), *Farming Systems Ecology*. Inaugural lecture upon taking up the position of Chair in Farming Systems Ecology at Wageningen University on 16 May 2013, Wageningen University, Wageningen, Netherlands.

Toledo, V.M. (1990), The ecological rationality of peasant production, in M. Altieri and S. Hecht (eds), *Agroecology and Small Farm Development*. CRC Press, Ann Arbor, MI.

Toledo, V.M. (1995), *Campesinidad, agroindustrialidad, sostenibilidad: los fundamentos ecológicos e históricos del desarrollo rural*. Cuadernos de Trabajo 3, Grupo Interamericano oara el Desarrollo Sostenible de la Agricultura y los Recursios Naturales, México.

UNDP [United Nations Development Programme] (2003), *China's Accession to WTO: Challenges for women in the agricultural and industrial sector*. Collaborative Research Report by UNDP, UNIFEM, All China's Women Federation, National Development and Reform Commission and Center for Chinese Agricultural Policy, Beijing.

USDA [United States Department of Agriculture] (1998), *A time to act*. National Commission on Small Farms, USDA, Washington.

Ventura, F. (1994), Women in Italian Agriculture: New Roles, New Problems, in Leendert van der Plas and Maria Fonte (eds), *Rural Gender Studies in Europe*. Van Gorcum, Assen, Netherlands, pp. 80–90.

Vera Delgado, J. (2011), *The ethno-politics of water security: contestations of ethnicity and gender in strategies to control water in the Andes of Peru*. PhD thesis. Wageningen University, Wageningen, Netherlands.

Vermeer, E.B. (2003), Determinants of income from wages in rural Wuxi and Baoding: a survey of 22 villages. *Journal of Peasant Studies*, 30(3–4):93–120.

Via Campesina (2007), *Declaration of Nyéléni*. Forum for Food Sovereignty, Senegal.

Vincent, J. (1977), Agrarian society as organized flow: processes of development, past and present. *Peasant Studies*, 6(2):56–65.

Visser, J. (2010), *Down to Earth: A historical-sociological analysis of the rise and fall of 'industrial' agriculture and of the prospects for the re-rooting of agriculture from the factory to the local farmers and ecology*. PhD thesis. Wageningen University, Wageningen, Netherlands.

Visser, O., M. Spoor and N. Mamonova (2014), Is Russia the emerging global 'breadbasket'? Re-cultivation, agroholdings and grain production. *Europe-Asia Studies*, 66(10):1589–1610.

Walker, K. L. M. (2006), 'Gangster capitalism' and peasant protest in China: the last twenty years. *Journal of Peasant Studies* 33(1): 1–33.

Wang, L.X. (2009), Theory and Reality of Agrarian Capitalist: agricultural development in Punjab of India during Green Revolution. *Social Sciences in China*, 5:189–203 (in Chinese).

Wang, W.W. and C.C. Fan (2006), Success or failure: selectivity and reasons of return migration in Sichuan and Anhui, China. *Environment and Planning A*, 38:939–58.

Wang, Yunxian (2013), Women's land rights in China: an unpromising future, in Govind Kelkar and Maithreyi Krishnaraj (eds), *Women, Land and Power in Asia*. Routledge, India, pp. 103–21.

Warriner, D. (1969), *Land Reform in Principle and Practice*. Oxford University Press, Oxford.

Webber, M. (2012), *Making Capitalism in Rural China*. Edward Elgar, Cheltenham, UK and Northampton, MA.

Weersink, A., C. Nicholson and J. Weerhewa (1998), Multiple job holdings among dairy farm families in New York and Ontario. *Agricultural Economics*, 18:127–43.

Wen, Tiejun (2004), Jiegou xianjiang lu [The Deconstruction of Modernity], in Jiegou xiandaihua – Wen Tiejun yanjiang lu [*The Deconstruction of Modernization – talks by Wen Tiejun*], Guangdong renmin chubanshe, Guangzhou, pp. 8–22.

Winders, B. (2009), The vanishing free market: the formation and spread of the British and US food regimes. *Journal of Agrarian Change*, 9(3):315–44.

Wiskerke, J.S.C. and J.D. van der Ploeg (eds) (2004), *Seeds of Transition. Essays on Novelty Production, Niches and Regimes in Agriculture*. Royal van Gorcum, Assen, Netherlands.

Wittman, H., A.A. Desmarais and N. Wiebe (2010), *Food Sovereignty: Reconnecting food, nature & community*. Pambazuka, Oxford.

White, B. (2011), *Who will own the countryside? Dispossession, rural youth and the future of farming*. Valedictory address delivered on 13 October 2011 on the occasion of the 59th Dies Natalis of the International Institute of Social Studies, The Hague.

White, L. (1998), *Unstately power*. M.E. Sharpe, Armonk, New York.

Woods, M. (2005), *Rural Geography: Processes, Responses and Experiences in Rural Restructuring*. Sage, London.

Woods, M. (2011), *Rural*. Routledge, London/New York.

Wortman, S. (1974), *The Rural Development System of the People's Republic of China: Impressions and Questions*. Paper, November 1974, Rockefeller Foundation, Washington.

Wu, Huifang and Xiandan Meng (1996a), The direct impact of relocation of farm labour on Chinese grain production. *China Economic Review*, 7 (2):105–22.

Wu, Huifang and Xiandan Meng (1996b), Do Chinese farmers reinvest in grain production? *China Economic Review*, 7 (2):123–34.

Wu, Huifang and Jingzhong Ye (2010), Zhangfu waichu wugong dui nongcun liushou funü de xinli yingxiang fenxi [Analysis of the psychological impact of husbands' migration on the women left at home in rural China]. *Zhejiang Daxüe Xüebao* [*Journal of Zhejiang University – Humanities and Social Sciences*], 1:13–14.

Wu, Huifang and Jingzhong Ye (2014), Hollow lives: Women left behind in rural China. *Journal of Agrarian Change*. doi: 10.1111/joac.12089 (last accessed 3 February 2016).

Wu, Huifang, Baoyin Ding and Jingzhong Ye (2015), The construction of new nested markets and rural development in China, in P. Hebinck, J.D. van der Ploeg and S. Schneider (eds), *Rural Development and the Construction of New Markets*. Routledge, London and New York, pp. 99–114.

Xu, C. (2009a) Huiyin guanxi manyidu: Liushou funü yu feiliushou funü de bijiao yanjiu [Marital satisfaction: A comparative study of left-behind and non-left-behind women]. *Funü Yanjiu Luncong* [*Collection of Women's Studies*], 5:25–32.

Xu, C. (2009b), Nongcun liushou funu yanjiu: Huigu yu qianzhan [Studies on left-behind women in rural areas: Review and prospects]. *Renkou Yu Fazhan* [*Population and Development*], 15:55–6.

Xu, Yingfeng (1999), Agricultural productivity in China. *China Economic Review*, 10:108–21.

Yang, C. K. (1959), *A Chinese Village in Early Communist Transition*. Technology Press, MIT/Harvard University Press, Cambridge, MA.

Yang, M. C. (1945), *A Chinese Village: Taitou, Shantung Province*. Columbia University Press, New York.

Yang, S. and Lui Heyuan (2009), On farmer subject vacancy and transfer of rural labor in 'empty nest' villages. *Research of Agricultural Modernization* 3:325–28.

Ye, Jingzhong (2000), Farmers, rural communities and technology: the conflict between cognition and reality. *Science and Technology Review*, 10:6–10.

Ye, Jingzhong (2002), Processes of Enlightenment: Farm Initiatives in Rural Development in China. PhD thesis. Wageningen University, Wageningen, Netherlands.

Ye, Jingzhong (2011), Left-behind children: the social price of China's economic boom. *Journal of Peasant Studies*, 38(3):613–50.

Ye, Jingzhong (2015), Land transfer and the pursuit of agricultural modernization in China. *Journal of Agrarian Change*, 14(3):314–17.

Ye, Jingzhong and He Congzhi (2008), *Lonely Sunsets: The elderly left behind in rural China*. Social Sciences Academic Press, Beijing.

Ye, Jingzhong and Pan Lu (2008), *Differentiated Childhoods: Children left behind in rural China*. Social Sciences Academic Press, Beijing.

Ye, Jingzhong and Wu Huifang (2008), *Dancing Solo: Women left behind in rural China*. Social Sciences Academic Press, Beijing.

Ye, Jingzhong and Pan Lu (2011), Differentiated childhoods: impacts of rural labor migration on left-behind children in China. *Journal of Peasant Studies*, 38(2):355–77.

Ye, Jingzhong and Pan Lu (2015), *Concepts and Realities of Family Farming in Asia and the Pacific*. Paper prepared for the FAO Conference on Family Farming, October 2014, Rome.

Ye, Jingzhong, Y. Wang and N. Long (2009), Farmer initiatives and livelihood diversification: from the collective to a market economy in rural China. *Journal of Agrarian Change*, 9(2):175–203.

Ye, Jingzhong, Rao Jing and Wu Huifang (2010), Crossing the river by feeling the stones: rural development in China. *Rivista di Economia Agraria*, Anno LXV, No. 2:261–94.

Ye Jingzhong, J.D. van der Ploeg and S. Schneider (2015), Rural development: actors and practices, in P. Milone, F. Ventura and Ye Jingzhong (eds), *Constructing a New Framework for Rural Development*. Research in Rural Sociology and Development Series (vol. 22), Emerald, UK, pp.17–30.

Ye, Jingzhong, Cunyu Wang, Huifang Wu, Congzhi He and J. Liu (2013), Internal migration and left-behind population in China. *Journal of Peasant Studies*, 40(6):1119–46.

Ye, Yumin and R. LeGates (2013), *Coordinating Urban and Rural Development in China: Learning from Chengdu*. Edward Elgar, Cheltenham, UK and Northampton, MA.

Yep, R. (2002), *Maintaining Stability in Rural China: Challenges and responses*. Center for Northeastern Asian Policy Studies, The Brookings Institution, Washington.

Yuan, J. W. (2010), *The Household Responsibility System and Social Change in Rural Guizhou, China: Applying a cohort approach*. PhD Thesis. Wageningen University, Netherlands.

Yuan, L. (2006), Nongcunu funü sushi yu nongcun shehui de quanmian xiaokang [Rural women's quality and well-to-do society]. *Zhonghua Nüzi Xüeyuan Xüebao* [*Journal of China Women's University*], 2:6–9.

Zhang, J., L. Zhang, S. Rozelle and S. Boucher (2006), Self-employment with Chinese characteristics: the forgotten engine of rural China's growth. *Contemporary Economic Policy*, 24(3):446–58. doi:10.1093/cep/byj034 (last accessed 3 February 2016).

Zhang, K., J.A. Dearing, T.P. Dawson, Xuhui Dong, Xiangdong Yang and Weiguo Zhang (2015), Poverty alleviation strategies in eastern China lead to critical ecological dynamics. *Science of the Total Environment*, 506–7:164–81.

Zhang, L. X., S. Rozelle, C. F. Liu, S. Olivia, A. De Brauw and Q. Li (2006), *Feminization of Agriculture in China: Debunking the myth and measuring the consequences of women's participation in agriculture*. Background paper for the World Development Report (WDR).

Zhang, Q.F. and J.A. Donaldson (2010), From peasants to farmers: peasant differentiation, labor regimes, and land-rights institutions in China's agrarian transition. *Politics & Society*, 38:458–89.

Zhao, Yong and J.D. van der Ploeg (2009), Telling data: an analysis of the note book of a Chinese farmer. *Journal of China Agricultural University*, 3:89–103.

Zhao, Yong and J.D. van der Ploeg (2014), Telling data: the accountancy record of a Chinese farmer. *NJAS* [*Wageningen Journal of Life Sciences*] 68:21–8.

Zhang, Q. F. and J.A. Donaldson (2013), China's agrarian reform and the privatization of land: a contrarian view. *Journal of Contemporary China*, 22(80):255–72.

Zhou, Q. and C. Song (2008), On the trend of feminization and aging in agricultural sector and countermeasures for risk-eliminating. *Journal of Shijiazhuang University of Economics*, 2:68–72.

Zhu, L. and Jiang, Z.Y. (1993), From brigade to village community: the land tenure system and rural development in China. *Cambridge Journal of Economics*, 17:441–61.

Index